Physical Diagnosis

Sixteenth Edition

Combining the Tradition of
Two Institutions of Service in Medicine

The Bulfinch Building
Massachusetts General Hospital, 1821

The Milton S. Hershey Medical Center, 1981

Physical Diagnosis

An Introduction to Clinical Medicine

Sixteenth Edition

John W. Burnside, M.D.

Professor of Medicine
Chief, Division of Internal Medicine,
The Milton S. Hershey Medical Center of
the Pennsylvania State University

WILLIAMS & WILKINS
Baltimore/London

Copyright ©, 1981
The Williams & Wilkins Company
428 E. Preston Street
Baltimore, Md. 21202, U.S.A.

Made in the United States of America

11th edition, 1934
12th edition, 1938
13th edition, 1942
14th edition, 1958
15th edition, 1974

Library of Congress Cataloging in Publication Data

Burnside, John W
    Physical diagnosis.

    A revision of Physical diagnosis by F. D. Adams.
    Includes index.
    1. Physical diagnosis. I. Adams, Frank Dennette, 1892- Physical diagnosis. II.
Title. [DNLM: 1. Diagnosis. WB200 B967p]
RC76.A3 1981   616.07′54   80-16941
ISBN 0-683-01137-5

Composed and printed at the
Waverly Press, Inc.
Mt. Royal and Guilford Aves.
Baltimore, Md. 21202, U.S.A.

To
Wayne Daniel Burnside
My Father and Friend

PREFACE TO THE SIXTEENTH EDITION

The fifteenth edition of this text, the first for which I was solely responsible, was a considerable departure from its predecessors in content but not in intent. Although it was no longer desirable to attempt to cover all of clinical medicine, the need to provide the young physician with a framework upon which to base a rational approach to physical diagnosis remained. That edition provided a method by which the student might lead him or herself to the understanding of a disease process even if the student had never before encountered that process.

The fifteenth edition met with a gracious acceptance by coursemasters and young physicians. With their endorsement then, this edition was prepared to improve on the basic concept. In so doing, very heavy reliance was placed on the comments which so many readers were kind enough to submit. Strong points have been emphasized, corrections made, and some new material added.

Emphasis on illustrations has been preserved. The artwork has been greatly improved through the efforts of Diane Abeloff, whose artistic talent is coupled with the remarkable ability to have her drawings relay technique as well as fact.

Two of my colleagues, Mark Widome and Thomas Johnson, have contributed much needed material to this edition. Dr. Johnson is a master at the physician-patient interaction and interpersonal skill. His chapter provides valuable assistance for the interviewing process. Dr. Widome has a way with children which he lucidly displays in his chapter on the examination of the child.

Most of the chapters now contain a specific practice session. I hope that these exercises will be helpful "walk through" sessions to help establish a rhythm to the examination.

I am again grateful for the assistance of Toni M. Tracy and all of the professional staff at The Williams & Wilkins Company and the Waverly Press.

PREFACE TO THE FIFTEENTH EDITION

This is less the fifteenth edition of a textbook than it is the fifteenth editon of a concept. In 1900, Dr. Richard C. Cabot wrote the first edition of a text on physical diagnosis. Thirty-eight years later he invited his young colleague, Dr. F. Dennette Adams, to collaborate on the preparation of the twelfth edition. Subsequent editions were the product of Dr. Adams' sole labor of love. Content and form changed completely but the basic precept of presenting tools of diagnosis to young physicians remained.

Dr. Adams, whose diligence and patient sensitivity is surpassed only by his enormous experience, has now passed the baton to me. The charge is the same and the latitude provided an expression of confidence of which I hope to be worthy.

Dr. Adams enormously expanded on the clinical material as a reflection of the explosion of medical information. The exponential growth of that body of information prohibits a similar expansion of this textbook. How, then, does one provide the where-with-all of physical diagnosis without attempting to encompass all of clinical medicine?

The solution, it appears to me, is to provide a framework of technique and a pathophysiologic thought process which might allow the student to define a disease even if he has never encountered it or read of it before. Most texts of physical diagnosis have been either "how to" books or have been disease oriented, i.e. given disease X, the findings are Y and Z. The former lacks stimulation and the latter lacks reality.

A patient presents first with symptoms; therefore a consideration of the history for each area or organ system seems appropriate. Basic methods of inspection, palpation, percussion and auscultation then elicit signs. Every sign has a pathophysiologic reason for its presence. Understand the reason and a pattern must surely emerge, pointing the physician toward the nature of the illness.

At the very least, the student should arrive at the index point for his text and journal search. Hopefully, he will come to the process itself—not the eponym but the aberrations responsible. It matters little that the process has been fully described and recorded. Conversely, deducing this on his own perpetuates the thrill of discovery and retards the insinuation of preconceptions on data collection.

This edition, then, is old and new. Old with the idea of providing the tools; new with de-emphasis of rote and more concern with the thought process. This old-new tack is perhaps reflected by the two institutions from which it arises. The Massachusetts General Hospital, steeped in tradition, forwards excellence as the only acceptable tenet of training. The Milton Hershey Medical Center of the Pennsylvania State University accepts that tenet with a challenge to time-burnished concepts by the question "What now is the best way?"

How do I thank those who have helped me directly or indirectly with the preparation of this book? Obviously, a roll call will offend those inadvertently overlooked and a listing will be of little note to the reader. If my teachers recognize parts of themselves enclosed within, I hope they will be pleased. The professionals of The Williams & Wilkins Company have extended the opportunity to write this book and their advice and counsel during its preparation befitted the most acknowledged author—all extended to an untried novice. Their only charge, one I accept willingly—sans tache.

John W. Burnside, M.D.
Hershey, Pennsylvania

CONTENTS

chapter 1

INTRODUCTION

The purpose of this book is to introduce to you the techniques of physical diagnosis, to explain the findings of common or particularly illustrative disease states and to help you construct a framework within which to grow as a clinician. The techniques of physical diagnosis are not hard to learn. The information you glean from those techniques depends on perseverance, repetition, and reproducibility of methods. Cursory examinations yield cursory data. Single observations rob you of the parameter of time. Sloppy methods are best left undone. The time expended is great at the onset but an economy as you become more proficient.

Many illustrations of common or meaningful physical findings are included to help you put the techniques into context. Some illustrations are presented because the physical findings are, if not diagnostic, highly suggestive of a given disease. Some are shown to evoke your interest because we cannot explain the findings—the state of the science does not yet allow it. Many of the photographs are of common findings. You will soon see them in living patients, and the pictures are to help you recognize them. Finally, the pictures are fun. Leaf through and study some of the photogaphs. Think about what you are seeing and then read the captions. The latter are purposely lengthy so that they will be informative without the text.

Physical diagnostic technique is the framework upon which to build your clinical career. A firm stucture encourages expansion. Embellishments come with experience and your individuality. Without a foundation, cantilevering into a subspecialty is hazardous. The framework, however, should not become constrictive or a labor of compulsion. If the findings fail to fit what is expected, look for another explanation. It is startling how many physicians stop thinking when faced with discrepancies between what is found and what is expected. Their framework is too rigid to allow challenge and new construction.

This book is written for students in medicine. This is a very tall order for any author since all good physicians are life-long students. More specifically, this book is written for students in medical school and house officers recently graduated. This makes the challenge no less impressive. There is no similar 5- or 6-year span during which you will learn as much or during which your labors will be as intense. A conscious effort has been made to allow for selective reading according to your needs and experience. While certainly less than encyclopedic, I hope that this will remain a useful reference book both for your own continuing education and for the time when you are privileged to teach your younger colleagues.

Many medical schools are departing from the traditional separation of preclinical instruction and patient-oriented activities. While it is not my purpose to comment on these changes, they do blend in another variable. Here, too, allowance is made for selective reading. The only suppositions are that you are eager to learn, sensitive to discomfort in others, and mature in your relations with people. Basic pathophysiology is included not to offend the wise but to help the novice understand the logic of the physical findings.

Medicine makes a number of demands. Observations, data recording, and communication keynote a scientific approach. We will be concerned primarily with observations—how to make them and their significance. Most observations are obtained through our special senses occasionally aided by devices to heighten our sensitivity, e.g., the stethoscope, ophthalmoscope, etc. Interpreting significance depends on groupings and associations of various findings all explained by the pathophysiology of the illness.

Data recording involves the medical record. This is the log and protocol of the physician. This vital document has largely been ignored or at best mistreated by your predecessors. Doctors faced with many uncontrolled variables resort to similarly random methods of data keeping. If we are to make progress in understanding the patients with complex diseases or patients with multiple disease interfaces, a reproducible way of recording observations is mandatory. We will focus more on this shortly.

Communication involves diction, using the right word or words to express what we mean. Throughout this text an attempt is made to provide you with specific meanings. Clarity is the goal in communication with colleagues and patients.

Regrettably, eponyms are common in medicine. They substitute familiarity for understanding. We all take solace in knowing that someone before us has made a similar observation. For some reason, however, attaching a name to a disease or physical finding retards inquisition. Try always to define diseases to the level of understanding. It is humbling but stimulating. For instance, hereditary hemorrhagic telangiectasia is a far more telling phrase than the syndrome of Osler-Weber-Rendu. Subsequent chapters will include some eponyms. I apologize in advance—it is a service to communication and history.

You and Patient

You have opted to be a public servant. People will seek you out for a service which you can provide. At the onset, the patient may have little choice but to find you at the bedside. Later, he will have a choice. The lay public have been accused of being inept judges of quality of care. This may be true, but what they do know is whom they can talk to, who listens, and who makes them feel better just by his presence. Ignorance of, or lack of attention to, the matter of the doctor-patient relationship will prevent the brightest doctor from helping a single patient.

Interested Physician

People are interesting. Your career choice indicates that you are interested in human disease. Try always to remember that your patients are people with illnesses and not "interesting cases."

Most students of medicine pass through three stages in their acquaintance with a given disease. Initially, they are excited by seeing a disease for the first time. There is an exhilaration of discovery. Later comes the petulance of "just another case." Finally, and most productive, is the recognition of the subtleties and nuances which declare that no two patients are quite alike.

Interest in the individuality of patients will prevent you from offending them. The latest in fashions may assert your individuality but may not help you communicate with the octogenarian. Your patient expects moderation from you and sensitivity to what frightens or annoys him. Avoid the latter.

Interest in a patient involves more than concern about his illness. A good doctor-patient relationship occurs when the physician attends to the patient's feelings about his life and illness. Basic emotions of anger, fear, pleasure, joy, helplessness, and rage are seldom far from the surface. These feelings are worth ferreting out: they color the disease and give it more meaning than the black and white of a text. The doctor whose concerns are for more than the disease soon has the patient as an ally.

Privileged Communication

The patient expects, and the law demands, that the dialogue is confidential. The medical record is a privileged document and is accorded legal status. Information should be shared among those professionals responsible for the care of the patient. When shared with other professionals, the patient's identity deserves protection. Failure to do this is the genesis of many hospital coffee shop tragedies. Candor results when your patient understands the respect you have for what he tells you.

Patient Controls

To a large extent, the patient controls his care. At the outset, he controls the interview by how much he divulges. Later, he may prohibit what he considers assaults on his person and, finally, he may decide not to take medicines or undergo surgery. You must function within these restraints and limitations.

The patient's control of the interview depends on many factors. You can influence these by an attitude of acceptance and understanding. When the patient knows that his story is confidential, is free from moral judgement, and is heard sympathetically, his reticence will usually dissolve. Patients frequently look on illness as payment for moral transgressions. In their attempts to understand, "Why me?" they will look to their past, their families, their business, etc. Levity about their illness is an affront. Belittlement of their attempts to explain their disease is intolerable. Gentle attempts to provide the patient with insight into the nature of the illness are rewarded by a patient who is less fearful, perhaps less selfdeprecating, and usually more cooperative.

The patient may prohibit parts of the physical examination. Success, or lack of it, in obtaining a history affects the ease with which you can do the physical examination. Physical contact is frightening to some patients, especially strange contacts by strange people. A reasonable explanation of what you are going to do before you do it will facilitate things. Much of the reluctance is resolved by leading the patient to look upon the examination as a joint effort.

Strong emotions in the doctor are quickly sensed by the patient and negatively interpreted. If you become angry and demanding, your patient will withdraw or respond in kind. When a patient cries, look on it as a reaction to be accepted. When he is angry, look for the genesis. When he is demanding, find out what he really wants. Emotions are not impediments to understanding illness; they are part of the illness.

In spite of all efforts, an occasional patient will withhold historical information, prohibit a complete physical examination, and thwart therapeutic efforts. When this occurs, it is a signal of other problems, usually psychiatric, which should be recognized.

Your Controls

Most of the determinants of the doctor-patient relationship are yours. Most are obvious and undisputed. Some require comment, and a few are hard to learn.

Patients usually expect that your questions and actions will be directed toward taking care of them. An occasional patient will remind you how much control that requires. There should be no question who is conducting the interview.

During the interview, control the questions and answers. Difficulty arises with the verbose patient or with the opposite, the silent stare. Is the verbosity a smoke screen or merely difficulty in sticking to the point? Is the silence a manifestation of anger, fear, or inattention? You have the right to expect answers to your questions. Prodding and interruptions may be necessary. It is sometimes even necessary to show authority. You might be reticent to exert these controls as a student not primarily involved in the patient's care. Reticence, however, only makes this difficult skill harder to learn.

Another privilege which is yours is access to any and all information about the patient. All of the patient's past history should be available to you. The patient may be unable to judge relevance. Once again, however, when the patient understands how you will use such information, he generally speaks freely. Here, too, your status as a student must not be an impediment.

Moral Judgements in Medicine

I allude to this topic primarily to condemn it. There is no illness, behavior pattern, or life style which deserves comment on its rightness or wrongness. Although any of these might be personally abhorrent to you, strive to separate these feelings from dealings with patients. Questions of morality interfere with sound medical judgement.

This is not to say interference is to be avoided. On the contrary, we wish to stop the alcoholic from drinking, to stop the self-destructive patient from suicide, and to stop the antisocial person from harming others. Generally, our goal is to correct the abnormal. It is the attitude of "You made your bed now lie in it," which is to be condemned. The only valid question we ask is, "Are the conditions contributing to disease and will a change contribute to his well being?"

Honesty in Medicine

Honesty is a clear virtue, but we are all sometimes sorely tempted to make exceptions. Distortions of the truth sometimes seem appropriate under the guise of "what's best for the patient." Rarely is this true.

If we are honest about honesty, the real temptation is to protect ourselves. Who likes to tell a patient that he is only a student doctor? Who likes to admit that he doesn't know what is wrong with a patient? Who likes to tell a patient he has a fatal disease? Examine the motives for the distortion of the truth and then be honest.

History

The characterization of disease occurs through its symptoms and signs. *Symptoms* are the patient's complaints, or his recognition of something abnormal. *Signs* are findings elicited by physical examination. The medical history, then, is the elucidation of the patient's symptoms.

Done with care and direction, history taking provides the greatest source of information for the time expended. Although we are primarily concerned with the present illness, we need to know data about the milieu of the disease. Hence, we ask about prior departures from health, about the life style, and about the family history. In addition, we recognize that the patient may not be aware that what he experiences is abnormal. That is why we perform a "systems review," a series of system-oriented questions designed to uncover abnormalities.

Setting the Stage

Much of what was alluded to in the section You and the Patient now comes to bear. An understanding, cooperative, and motivated patient gives a far more cogent history.

Physical comforts require attention. A quiet, well lighted room espouses confidence. Attention to privacy helps the patient appreciate your concern that his words will be taken seriously and confidentially. Physical discomfort taxes the patient. Usually such discomfort is obvious and an expression of concern tells the patient that you are aware of his plight. Discomfort secondary to the illness may be hard to bear. It is even worse when discomfort is added by you. A distended bladder, for instance, caused by a long interview might be reflected in terse responses to your questions.

Present Illness

The *present illness* is the reason for the hospitalization or office visit. Foremost on the patient's mind, the present illness is readily explored. The history of the present illness may establish the diagnosis; it almost always

points to the major area of pathology or the major mechanism of disease. Dissection of the present illness proceeds from a broad base and progressively homes in on and defines the symptoms.

The *chief complaint* stated by the patient is the broad base from which to start. Most chief complaints are characterized by *pain, dysfunction, a change from the steady state*, or *an observation made by the patient.*

Pain, dysfunction, and observations usually point to an identifiable area. Pain in the chest obviously calls attention to the chest and its associated structures. Difficulty urinating or blood in the stools as examples of dysfunction and observation likewise have anatomic connotations. Changes from a steady state do not localize well. Fever, malaise, weight loss, or emotional lability may defy anatomic localization. Instead, a blood-borne malady or abnormality of an ubiquitous tissue, septicemia, or occult neoplasms may be at fault.

Encourage the patient to be as specific as possible about his chief complaint. Ask him to relate the quality and quantity of the symptom to a common reference. Is the pain like a toothache or stabbing? Is the bloody sputum a teaspoonful or a cupful?

If a gross anatomic localization is possible, logic dictates that we pinpoint the organ or organ system involved. The relation of the symptom to the structure and function of the organs in question is next in order. Thus, abdominal pain which is crampy suggests disease of a hollow peristaltic viscus. Chest pain aggravated by deep breathing might indicate irritation of a serous membrane.

What aggravates or alleviates the symptom? Here, as well, the structure and function of organs or organ systems point the way. Midepigastric pain relieved by milk or antacids suggests acid peptic disease. Chest pressure aggravated by exercise might be angina pectoris.

Time relationships are very helpful. The duration of the present illness separates the acute from the chronic. Symptoms persisting for years are unlikely to be caused by malignancy or infection. Changes in the symptoms as a function of time help to establish the progression of a disease. Have the symptoms been peristent, recurring, getting worse or better? If periodic, have the cyclic lengths been changing?

Patients frequently volunteer associated symptoms or events. These must always be taken seriously no matter how bizarre or nonphysiologic they sound. This not only expresses your respect for the patient as an observer but may be fruitful in uncovering occult disease. Muscular weakness associated with heat intolerance and weight change, all of which follow an emotional upset, sounds unrelated, yet this is a common history of thyrotoxicosis.

The patient cannot be expected to link all possible relationships. Here, your freedom to follow various lines of questioning is valuable. Avoid developing a rote form which is so restrictive that you cannot follow leads in the conversation. This naturally presupposes that you listen and think simultaneously.

In summary, then, the *present illness* is developed from the chief complaint by asking: What is the problem? Where is the problem? What is it like and how severe is it? How does it relate to implicated organ stucture and function? What are the temporal relationships? What makes it better or worse? What are the associated manifestations? A few of these points are demonstrated in the following example:

A young woman was admitted to the hospital because of chest pain. She related that the pain began suddenly 1 week earlier and was getting progressively worse. She described it as sharp pain located anteriorly, and she further volunteered that it was relieved by sitting up and leaning forward. On further questioning, a history of fever and malaise was elicited. The examiner, now considering inflammatory diseases of the mediastinum, especially the pericardium, found that the patient had complaints referable to other serous membranes. Pleurisy had been diagnosed 2 years earlier, and a mild nondeforming arthritis prompted the patient to take aspirin every day. This history led the physician to collate the findings of a facial rash and pericardial friction rub into a presumptive diagnosis of systemic lupus erythematosus.

We will return to the specific applications of these principles in subsequent chapters.

Past History

The *past history* is a survey of all previous illnesses and contacts with physicians. Specific emphasis may be indicated by information obtained in the present illness.

Review the past history chronologically.

Prenatal, parturient, and postnatal information is relevant to investigation of hereditary or congenital illness. Thus, a heavy birth weight might indicate parental diabetes for which the patient is at risk. Birth trauma might explain retardation or limb paresis. Breech birth can be a clue to congenital anomalies.

Childhood diseases are so common as to serve as an early tolerance test to illness. In both this and the review of infant health, it is wise to remember that the patient knows only what he was told by his parents or guardians. By reducing the questions to simple terms, significant data may be revealed. The patient with valvular heart disease frequently cannot recall having had acute rheumatic fever. He might, however, remember spending weeks in bed as a child because of "growing pains." When reviewing common childhood diseases, remember to inquire about sequelae to the illness.

The sequelae of earlier illnesses are important in another sense. How did your patient tolerate pain, discomfort, dependency, or weakness? What impressions did other doctors make on him? This is important for proper interpretation of the present illness. To be sure, this is interpretive information but valuable. A complaint registered by the stoic may be quite different in significance from the same complaint brought to you by the doctor shopper.

A few items not viewed as illness by the patient should be reviewed with the past history. Immunizations, reactions to tuberculin tests, military record, allergies (especially reactions to drugs), date and place of the last chest x-ray, and routine physical examinations all round out the survey.

Review of Systems

A review of systems represents an effort to ferret out symptoms which the patient has not recognized as such, symptoms he has forgotten or thought unimportant. Nocturia, for instance, may not be volunteered but when specifically asked for the patient may say with surprise, "Well, yes, but it's probably just the amount of water I drink." Yet with further questioning, congestive heart failure, benign prostatic hypertrophy, diabetes mellitus, or renal failure could become the more likely explanation.

Questioning revolves about the stucture

and function of the system being reviewed similar to the technique in eliciting the present illness. Listing questions for you to ask without considering the implications of the answers leads us nowhere. Therefore, to help make the review of systems more meaningful, the subsequent chapters will begin with considerations of the medical history related to the system to be examined.

Family History

The family history is the past medical history of blood relatives. The same points relevant to the patient's past history are relevant to the family history. Obviously, our concern is for data relevant to the patient at hand, both in terms of the patient's present illness and in terms of risk factors.

All diseases can be reviewed as having some genetic background. Consider, for example, the young patient with fever, nausea, and left lower abdominal pain. Her doctors were perplexed until the mother volunteered that she herself had had a similar illness and that her doctors found a left-sided appendix. More obvious is the patient whose parents both had diabetes mellitus. Such a patient has a 90%-plus probability of developing the disease. Single diseases and clusters of related diseases often appear in families. Hypertension, early coronary artery disease, or manic-depressive psychosis may each thread through an individual family. Similarly, different but related diseases—rheumatoid arthritis, lupus erythematosus, aplastic anemia, thymomas, and myasthenia gravis—may appear in various combinations in the same family.

The family history is also important because of the physical proximity implied. Infectious disease, then, may also cluster. Tuberculosis, streptococcal infections, and organisms or toxins carried in the family meal are examples.

Whenever possible, construct a pedigree to record the family health. Many surprising associations and fruitful research have come from this simple, inexpensive exercise (Fig. 1.1).

Social History

If we are interested in the contributions of the gene, we need to know also the effects of the environment—not only the character of the milieu but how the patient reacts to it.

The educational background, financial status, employment record, marital status, place in the family unit, the daily routine all reflect on the patient and his environment. Much of this information comes out in "small talk" throughout your contact with the patient. We separate it out for emphasis. Some of this is sensitive information to the patient and may require several interviews for full evaluation.

In the record system to be discussed later, the contents of the social history are accorded a prominent place in the beginning of the record under the title of *patient profile.* The patient profile contains the information obtained in the social history and some interpretation by the examiner about the significance of the life style.

Medical Records

The legal importance of the medical record has already been alluded to. Far more significant is the importance of the record in patient care and medical education.

The medical record, in theory, serves two functions. First, it acts as a means of communication. The explosion of sophistication in patient care has brought with it an array of experts, all of whom bring their special talents to bear on the patient's problems. More and more, we practice committee medicine. The profusion of people involved in patient care makes it less possible to talk together on a one-to-one basis. The record, then, becomes the common vehicle used by all for communication. We need to be certain that the record says precisely what we mean. Questions need to be clearly stated, statements well supported, and logic easily followed (Fig. 1.2).

The second major function of the medical record is to store data. Just as many people participate in a patient's care, data accumulate from many sources. Recorded information is worthless unless it is applied to the problems at hand. Attaching data to problems, assigning relevance, makes the exercise more than just storage. Now we have the start of a retrieval system. A trip through most record rooms is a sad journey. The answers to many perplexing questions in medicine lie locked in thousands of records. Most of the answers will remain there because to find them out requires hours and hours of diligent search.

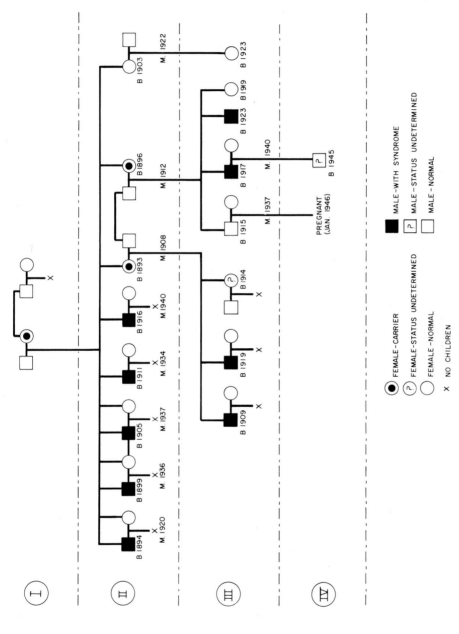

Fig. 1.1. Pedigree. This is a sample pedigree to show how such a recording can be graphically represented. If you study this, you will soon recognize that it is a typical pedigree of an X-linked recessive trait.

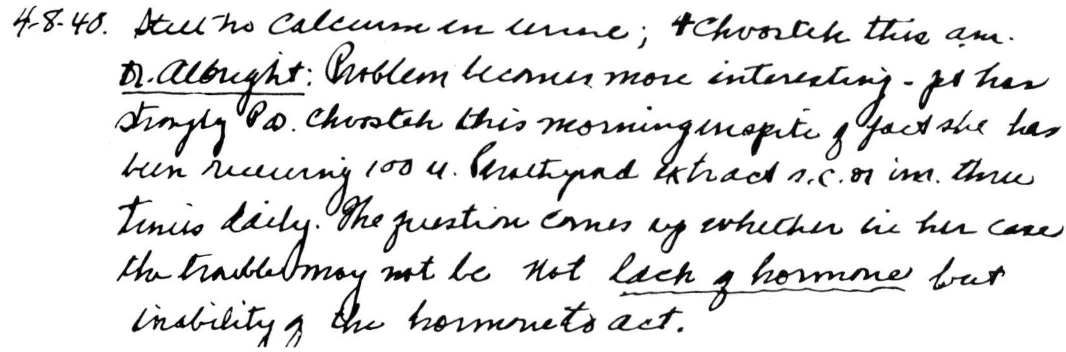

4-8-40. Still no Calcium in urine; + Chvostek this am.

Dr. Albright: Problem becomes more interesting - pt has strongly P.O. Chvostek this morning inspite of fact she has been receiving 100 u. Parathyroid extract s.c. or im. three times daily. The question comes up whether in her case the trouble may not be Not lack of hormone but inability of the hormone to act.

Fig. 1.2.   An original clinical observation made by Dr. Fuller Albright and recorded in a patient's record. This is the first clinical recording of the supposition of end organ unresponsiveness to a hormone. In this case, the hormone in question was parathyroid hormone. Since then, this mechanism has been demonstrated for several endocrine humoral substances. (Courtesy Endocrine Unit, Massachusetts General Hospital.)

Problem-Oriented Medical Record

What format will allow us to make the record perform both of its functions? Dr. Lawrence Weed has developed such a system called the *problem-oriented medical record.* The appeal of his system lies in its logic, reproducibility, and ease of use. There are four components to the problem-oriented medical record: the data base, the problem list, plans for each problem, and progress notes. Since your course involves medical records and medical recording, it is appropriate to discuss the elements of the problem-oriented system.

Data Base. The data base consists of all the information available about a patient at a given time. Initially, it includes the history and physical examination. As new information emerges, it becomes part of the data base. Although the data may be arranged chronologically in the history, and by systems in the physical examination, they are still crude data until interpreted and sorted out in terms of problems.

Problem List. The problem list is a dynamic table of contents. Problems are identified from the data base. The term *problem* is preferred to diagnosis. A problem might be a diagnosis but it might also be a symptom, abnormal laboratory test, risk factor, or social circumstance. The problems are listed only to the limits of the data. This demands honesty and pays in pointing the direction for further investigation of the problem. Any problem must be supported by the data base. A problem may be as sophisticated as glucose 6-phosphate dehydrogenase deficiency or as simple as fatigue.

In this context, a problem is anything which is abnormal, requires management, or contributes to the illness. The problems are numbered consecutively, and the numbers are used throughout the record as a method of indexing them. The problems are also listed separately on a master problem list prominently placed in the record for reference and review.

As the data base expands through further physical examinations, laboratory tests, more history, or the passage of time, the problem list will change. Changes are recorded on the master problem list and dated. For instance:

#1) Chest pain (9/12), (9/15) myocardial infarction.

The reader then knows that the patient had chest pain on 9/12 which proved to be a myocardial infarction on 9/15. Looking in the record, he will find the substantiating data in the progress note of 9/15.

Occasionally, several problems become recognized as different manifestations of the same disease. For instance:

Problem
#1) Rheumatic heart disease with mitral stenosis (1971)
#2) Fever (9/12/73) → (9/15/73), subacute bacterial endocarditis
#3) Splenomegaly (9/12/73) → (9/15/73) #2
#4) Anemia (9/12/73) → (9/15/73) #2
#5) Hematuria (9/12/73) → (9/15/73) #2

This patient had mitral stenosis diagnosed in 1971 and presented on 9/12/73 with fever, anemia, splenomegaly and hematuria. Blood cultures confirmed subacute bacterial endocarditis on 9/15 and the problem list was changed appropriately.

The problem list allows for the designation

of *resolved* problems as well. These are included for completeness and because they might indicate an important association.

	Active	Resolved
#1)	Jaundice (9/12/73)	Cholecystectomy for cholethiasis (7/18/73)

The fact that this jaundiced patient had a previous cholecystectomy would be an important clue to the proper approach to his jaundice.

Plans. Now that we have identified the problem, we can plan a rational approach to its solution(s). The plan for each problem has three parts: diagnostic, therapeutic, and patient education.

Diagnostic considerations include all efforts to obtain further definition of the problem. The object is to understand the problem at its most basic level—then have the problem list closely approximate the actual pathology. The plan is the place to consider differential diagnosis or "rule outs" (R/O).

Problem
   #2) Chest pain
Plans
   Diagnostic
      R/O myocardial infarction with serial electrocardiograms and cardiac enzymes.
      R/O dissection of the aorta—follow pressures in both arms, chest fluoroscopy, cardiac consultation re: angiography.
      R/O acute pulmonary embolism with arterial blood gases.

Note in the example above that the problem is listed as chest pain until sufficient data are available to be more specific. If we were to list R/O myocardial infarction as the problem, it would imply that once the diagnosis is excluded the problem no longer exists when, in fact, the patient might still have chest pain.

Therapeutic. The therapeutic part of the plan includes all efforts to interfere with the natural history of the problem. Thus, not only drugs or surgery is listed but also diets, physical therapy, aid from social, religious, or governmental agencies, and so on.

   #2) Chest pain, *continued*
Plans, continued
      Will treat as a myocardial infarction for safety while collecting data. Coronary precautions, monitoring of cardiac rhythm, vital signs, central venous pressure, intake, and output.
      Prophylaxis with antiarrhythmics, begin anticoagulation with Coumadin and treat pain with morphine sulfate. Will have social service find support for family while patient in hospital.

Patient education has long been ignored, if not in fact, at least in designation in the record. It is virtually axiomatic that the patient who understands his illness will play a more active role in the solution of the problem. The plan for diabetes mellitus would include teaching the nature of the disease, insulin administration, urine testing, diet instructions, and perhaps genetic counseling. It is also important to designate these efforts in the record so that similar information is enforced by all members of the care team. Does, for instance, the patient know what his disease is, and if does not is there a specific reason for keeping it from him?

Progress Notes. Progress notes are the diary of the problem list: not a random free flow of disjointed thoughts or observations but a structured reproducible way of recording data.

Progress notes are dated. Problems are titled by number and name. Under this, the data are organized as follows—subjective, objective, assessment, and plan (S.O.A.P.). Subjective information is what the patient relates about the problem. Objective data are factual findings—physical, laboratory, x-ray, etc. The assessment is a statement of your interpretation of the problem, and the plan is an update of how you wish to proceed. Not every problem needs listing with every progress note—only those problems about which you have something to say.

9/15/73
   #2) Chest pain
      S. No further pain, slept well through the night, but still feels fatigued.
      O. B.P.[1] 110/70, P 110 and regular (NSR or monitor) Central venous pressure 10 cm. Pulses equal. Few basilar rales in both lungs. Soft S3 gallop sound. EKG shows evolving anterior myocardial infarction. Enzymes confirm myocardial necrosis.
      A. Clearly, a myocardial infarction—so far no complications.

-----

[1] B.P., blood presssure; P., pulse; NSR, normal sinus rhythm; and EKG, electrocardiogram.

P. Change problem list #2—Myocardial infarction. Will obtain a glucose tolerance test and lipid profile in near future. For now, continue coronary precautions: monitoring, antiarrhythmics, and anticoagulation. Patient told that he had a heart attack—will discuss the implications with him at a later date.

Some of the advantages of the problem-oriented system should be obvious. In addition to communicating better and storing data in a more meaningful fashion, it allows a rapid record review, avoids overinterpretation of the data, displays the logic of the physician, and prevents duplication of effort.

chapter 2

INTERPERSONAL SKILL IN PHYSICAL DIAGNOSIS

Thomas M. Johnson, Ph.D.*

Physical diagnosis is a process in which the physician uses a variety of tools and techniques to gather data in order to determine the nature of a patient's illness and to initiate appropriate therapy. As a medical student, you will struggle to master a set of technical diagnostic skills such as auscultation of the heart, performing a sigmoidoscopic examination, or ordering from among a myriad of sophisticated laboratory tests. You may be inclined to spend proportionately less effort learning and practicing interpersonal skills, in part, because it is often assumed that communicating with patients "comes naturally" or that effective interpersonal skill is of secondary importance in diagnosis and treatment.

In fact, the best clinicians recognize that 80% of most diagnoses can be derived solely by talking with the patient, that half of all patients value "physician interest in the patient" more than "technical skill," and that many patients who feel that care was delivered in an unacceptable manner will fail to follow treatment recommendations. When patients are dissatisfied and physicians are frustrated, both are likely victims of a communication breakdown, and the ability of even the most able physician to help patients is compromised. Although we have traditionally taught medical students the skills to look for failure of liver function or cardiac output, the same cannot be said for failure of communication.

This chapter is intended to enhance your understanding of medical care as a social process and to *provide you with a "tool kit" of interpersonal techniques to increase your clinical effectiveness*. This is not simply a lesson in "social sensitivity." The improvement of interpersonal skills is a many-sided challenge which must involve learning at both intellectual and behavioral levels. The most effective and sought after clinicians are most adept at interpersonal skills. They are able to elicit patient feelings about illness and treatment, to create an atmosphere in which both patient and physician perspectives can be addressed, to gather and convey information efficiently, and to maintain a positive professional self-image at a time when patient problems are commonly psychologic or sociologic rather than bacteriologic in nature.

Changing Physician-Patient Relationship

In our society, there is increasing emphasis on, and many changes taking place in, the physician-patient relationship. Patients are less willing to unquestioningly accept the authority of the physician, and there is a diminution of what used to be implicit trust in physician expertise and ethics. Both of these changes can be related to a "demythologizing" of the physician role, in part through media exposure. In both the popular literature and on television, physicians are no longer invariably portrayed as infallible, omniscient, and ethical. Patients are also better educated in health matters, with nutrition and health classes a common feature of school curricula. The consumer rights movement and an increasingly litigation-conscious public also contribute to the present state of physician-patient relations.

Within medicine, the rapid increase in medical knowledge has mandated specialization at the expense of primary care and, when coupled with increased mobility of the patient population, has tended to mitigate

* Assistant Professor, Department of Psychiatry, University of South Florida, College of Medicine, Tampa, Fla. 33612

against long-term physician-patient relationships. Specialization also tends to make physicians dependent upon, and primarily responsible to, referring physicians rather than patients. In addition, "fee-for-service" medicine encourages spending less time with each patient at a time when chronic diseases, which demand more care over longer periods of time, are beginning to predominate.

Interaction with Patients

There are inherent conflicts in the interaction between physicians and patients. Traditionally, patients were expected to reveal their chief complaints and then become the passive recipients of physician ministrations. This active/passive orientation still exists when acutely ill or comatose patients cannot actively interact, but it also exists because most of the power in the physician-patient relationship belongs to the physician. In almost all cases, physicians have greater mastery of medical knowledge, greater familiarity with the treatment environment, more confidence in their role (physicians practice "being a doctor" much more than people practice "being a patient"), the exclusive sanction to treat illness, and the ultimate power to refuse patient requests. Ill patients are also functioning at a less than optimal level.

This asymmetry of power in the physician-patient relationship tends to diminish any sense of control in the patient. Your patients who are less seriously ill will have perspectives on their illness and treatment. The challenge to you will be to create a climate in which they can freely communicate them to you. Unlike the seriously incapacitated patient who is unable to assume any control in the relationship, the majority of patients will respond negatively to assuming a passive role. These patients will react to a controlling physician and to their illness with anxiety, frustration, anger, and depression. If you fail to negotiate treatment plans which address patient needs for some measure of control, your patients will likely be dissatisfied, noncompliant, or they may change physicians.

Your goal for most interactions with patients should be to encourage their involvement in treatment. A therapeutic relationship based on mutual cooperation and shared control will only be achieved if the *process* of interaction with patients is as important as *content*. The guidelines for working with patients presented in this chapter reflect that priority.

If you follow these guidelines your most successful interviews will have several disconcerting characteristics: (1) 90% of the talking will have been done by the patient; (2) much of the conversation will be about topics which seem "nonmedical," such as the social and psychological context of the illness; (3) there will be relatively little discussion of symptoms; and (4) the interview will seem to be too long and inefficient. These characteristics will be disconcerting because you will feel that you have too little control over the interchange and that you are not an active enough participant.

In reality, you will be actively involved in structuring the interview and you will be eliciting greater amounts of more valuable information in the process. Most interactions involve seven tasks: (1) preparing for the interview; (2) meeting and offering introductions; (3) soliciting initial information and patient expectations for the visit; (4) elaboration of initial information through interviewing and physical examination; (5) problem solving with the patient; (6) contracting with the patient for treatment; and (7) reviewing and terminating the interview.

Preparing to Meet Patient

Preparing for the interview involves several tasks. *Before meeting the patient you should thoroughly review available records.* It is bad form to be obviously reading a patient's records for the first time in front of the patient. The patient will view such activity as signifying greater interest in the chart and its problems than with the patient's problems. *You should also consult intermediary sources of information*, particularly the nurse who may have already met and talked with the patient or the patient's family. Because patients are usually anxious, and anxious people tend to be hypervigilant, patients often sit behind the closed door straining to hear clues to what will soon transpire. Any discussion of the patient's case must be conducted quietly and away from the examining or hospital room door. It is all too easy for you to not observe this caveat and unintentionally offend or frighten a patient, or breach confidentiality. Finally, *you should reflect on your own feelings and initial expectations.* This would include expectations of the amount of time you can spend and the scope of the interaction (for example, detailed medical history and physical exam versus prescription refill).

Meeting Patient

The task of meeting the patient, seemingly a formality, is a time to set the stage for the remaining interaction. The first minute of any interaction between two people has been shown to be a strong predictor of what transpires thereafter. When meeting the patient, you will be: (1) establishing an emotional tone; (2) signaling your interest; (3) arranging for physical comfort of the patient; and (4) initiating data gathering information. *When first meeting the patient you should appear calm, smiling, and unhurried.* The rhythm of clinical activity tends to result in tremendous time pressure, and you will feel the urge to communicate this to the patient by acting accordingly. Unfortunately, patients confronted by an obviously hurried physician usually feel guilty at taking time and will not volunteer information readily. Other patients will resent your approach and subconsciously "force" you to spend more time by a number of communication-blocking techniques. *Always, introduce yourself and any other person with you,* but give only essential information such as your name and position. If you elaborate about yourself, you signal to the patient that their primary concern should be you, rather than vice versa.

To promote and convey respect, *patients should be dressed and seated when you first meet them.* Meeting a gowned, supine patient should not occur in the office; it exacerbates the patient's sense of powerlessness, increases anxiety, and conveys no feelings of respect from you.

Great attention should be given to nonverbal aspects of communication which include facial expressions, postures, gestures, spatial distributions, and autonomic signals such as rapid breathing or perspiring. Verbal and nonverbal messages occur simultaneously but nonverbal cues are important because they are more powerful and "honest" indicators of actual feelings. Although nonverbal cues are usually transmitted and received unconsciously, *you should consciously utilize nonverbal behaviors to facilitate interaction with the patient. You should maintain eye contact and adopt a relaxed yet alert facial expression.*

Your eyes should always be at or below the level of the patient's eyes and you should position yourself just outside the patient's "personal space."

In our culture, all people have a "personal space" which extends elliptically 3 feet in front and 1 foot behind the body. Unexpected intrusion into a personal space is anxiety provoking. *You should attempt to move gradually into a patient's personal space,* a handshake being a ritualized gesture which nonthreateningly prepares the patient for the violation of personal space which will occur later in the physical examination. *The environment of the room, itself, should be made comfortable for the patient* by rearranging chairs, lighting, or curtains. It is better not to position yourself between the patient and the door, because patients in that position often feel "trapped." If appropriate, reassurance of confidentiality should be given. *You should begin to observe and evaluate the patient's nonverbal behaviors* for signs of anxiety, pain, etc. *If there are time limitations on the visit you should express them to the patient at this point:*

"Mrs. Jones, before we start I should tell you that we will have to be finished in a half hour, if it is at all possible."

"I am afraid that we can only spend fifteen minutes today…Do you think we can meet your needs in that much time?"

Time limitations are a fact of life to which patients can readily adapt and, in terms of giving the patient some control, a short but definite time limitation is preferable to unspecified or ambiguous expectations.

Soliciting Information

A third task in interviewing a patient includes soliciting initial information and defining objectives with the patient.

Questioning Techniques

Questioning of patients is a skill crucial to data gathering which must be honed through careful practice. Skill in questioning will increase your ability to control the interview and simultaneously establish rapport. Questions can be phrased in a number of ways, from "open" to "closed." Open questions can never be answered with a simple "yes" or "no" and are utilized to learn about attitudes and fears as well as facts, to ascertain a level of patient understanding, to increase patient motivation and rapport, or to help structure and guide the patient's thoughts. Open-ended questions also sustain a high level of patient involvement. Closed questions, on the other hand, demand specific facts and limit patient expression. In asking about a chief complaint of "headache" you might ask any of the following four questions, arranged on a continuum from open to closed:

"What can you tell me about this headache?"
"How would you describe the pain?"
"Where is the pain located?"
"Is the pain only on one side of your head?"

Similarly, questions about home environment can be asked in either an open or closed format:

"What is your living situation like?"
"Who lives with you in your house?"
"How many people live with you?"
"Are you married?"

There are several general rules for using open and closed questions. *You should always ask open-ended questions first, and move to increasingly closed questions only if you need additional specific information.* Asking an initial closed question may yield specific information about a headache, but in so doing you may never learn that the patient's main concern is that his headache keeps him awake at night or that he fears he may have a malignancy. A questioning technique which works very well involves "chaining" open questions and empathic responses. This involves responding to a statement made by a patient with an empathic response followed immediately by an open-ended question. For example:

Patient: "Anyway, I've got several problems but I can't stop worrying about the headaches!"
Physician: "It sounds like the headache bothers you the most [emphatic response]…What more can you tell me about it [open question]?"
Patient: "Sometimes I think I'm going out of my mind. It's a throbbing kind of pain."
Physician: "Those can be very painful. [emphatic response] What can you tell me about the location [open question]?"

Another rule to remember is that questions asked in response to patients' feelings should be short (single word or phrase), whereas responses to facts can be more lengthy. Long interjections will tend to inhibit the expression of feelings.

Questions can also be asked in "direct" or "indirect" fashion:

Direct: "What do you do when the pain starts?"
Indirect: "I'm curious about what you do for the pain."

Direct questions can always be written with a question mark at the end and sound like an interrogation when used repeatedly. Indirect questions can be phrased as statements of interest in the problem and should be used as much as possible because they convey your interest in the patient. They are also less accusatory in tone. Indirect questions are also excellent for nonthreateningly confronting discrepancies in the patient's presentation. For example:

"I get the feeling from looking at you that this pain bothers you more than your admitting…"
"I was wondering why you haven't mentioned being worried about the biopsy…"
"I thought I heard you say earlier that your breathing bothers you…"

In general, indirect questions should always be used to introduce a topic, while direct questions should be reserved for eliciting the last bits of information about a particular topic.

It is common for interviewers to use entirely too many closed, direct questions, often as a product of insecurity. Open, indirect questions tend to allow the patient much greater freedom in choosing a response. When asking open questions, you will feel less able to predict what the response will be. Closed, direct questions are often justified with the argument that closed questions are more time-efficient. While it is true that use of closed questions may shorten the immediate interaction, in the long term this is clearly not the case. Even in the short term, there is support for the contention that closed questions actually can take more time. There is a common problem in medical interviewing which involves making assumptions, zeroing in too quickly with specific closed questions, following blind alleys in hypothesis testing, and ultimately spending valuable time collecting irrelevant information.

When questioning a patient you should avoid multiple questions:

"Do you want to keep your baby or do you want to terminate the pregnancy?"

"How did you sleep last night...Was the pain better?"

These multiple questions assume that there are no other alternatives or assume relationships which do not necessarily exist. They limit the patients ability to express true feelings by including acceptable responses in the question. *You should also avoid "flooding" or "bombarding" the patient with a long series of closed questions, asking leading questions which are accusatory, and using the word "why" which is confrontative:*

"Do you have fevers, joint pain, chills, or swelling?"

"Are you sure you took your medicine?"

"Why did you decide to get pregnant?"

When asking questions in sensitive areas, such as when taking a sexual history, always ask permission of the patient first. For example:

"In order for me to really understand your problem, may I ask some questions of a personal nature which some people find embarrassing...I hope you can understand how important this information is and that you can be as comfortable as possible discussing this subject with me."

Many physicians, themselves embarrassed by sensitive topics, tend to spring such questions on the patient unexpectedly. Again, this unpredictable strategy deprives the patient of any control in the interview and does not convey a respect for the patient.

Establishing Chief Complaint

The initial information, referred to as a "chief complaint," is a statement in the patients own words of his or her problem(s). *You should facilitate elicitation of the chief complaint by asking simple questions:*

"What brings you to see me today?"

"What seems to be the problem?"

If you requested the visit *you should state why you asked the patient to come.* If the patient is a repeat visit, *you should establish continuity by referring to previous visits:*

"Last time you were in you were having a great deal of trouble with your skin...How has it been?"

In the beginning, the patient should be allowed to fully express his or her feelings. Patients usually have rehearsed what they will say and must be given the opportunity to express this without interruption. At this point, the "direction" of conversation should be almost exclusively from the patient to the physician. *You should not ask for much clarification, instead using affirmative or encouraging leads like nodding and saying "O.K." or "Go on, I'd like to hear more."* A good technique is to ask for a "restatement." For example:

Patient: "The pain is worse at night."
Physician: "At night?"
Patient: "Sometimes I get so nervous that I think I am going to lose control!"
Physician: "Lose control?...Can you tell me about that feeling?"

The restatement technique allows you to bring focus to lengthly narrations, reinforce a specific direction of thought, to reinforce to the patient that you really are listening, and to check on your understanding of the patient's complaint.

Remember, your primary activity at this point will be listening. *It is important that you listen actively,* giving attention to content, tone, expressions, gestures, and what is inferred but not verbalized. *If what the patient says is not supported by nonverbal cues, there is likely to be special significance to the subject being discussed, and you should plan to discuss it later.*

When the patient has completed the initial statement, *you should summarize what you have heard.* To communicate that you give credence to the patient, *you should refer to the patient's complaints in his or her own words.* In addition, *you should express your objectives for the remainder of the visit and elicit approval from the patient for a specific course of interaction.*

Natural History of Chief Complaint

The next task of interaction with patients is to ask questions which are designed to further elaborate on the patient's initial "re-hearsed statement." The purpose of some questioning will be to sharpen the focus of the patient's chief complaint, to understand the patient's perceptions of his or her problem, and to obtain information needed to test and reduce multiple hypotheses leading to a clinical diagnosis. For example, if the patients chief complaint is "headache" you will want data on location, type, frequency, intensity, and duration of pain, precipitating or ameliorating factors, and concomitant symptoms.

To fully define a patient's problem *you must ask about the "natural history" of symptoms presented by the patient.* It is important to remember that many patients present with physical symptoms because our society expects them to, but most patient's medical needs extend beyond establishing a diagnosis based on physical symptoms. Unless you consciously broaden the basis for your clinical hypothesis testing to include social and psychological factors, you will likely find yourself working at cross-purposes with the patient. *You should explore the relationship of the patient's symptoms and changes in the patient's style of carrying out activities of daily living* (occupational, educational, interpersonal, sexual, leisure). Almost all disorders develop or become more severe in response to events in a patient's life, and it is impairment of ability to carry out accustomed social roles rather than a worsening of symptoms which will cause many patients to visit you.

Understanding the natural history of a symptom implies a strong commitment to diagnosing and treating *illness* rather than *disease.* Diseases are named abnormalities in the structure and function of body organs and systems, whereas illnesses are changes in feeling states and social functions of patients. Different patients with the same disease may suffer different degrees of illness (pain, distress, dysfunction), and illness may occur in the absence of disease (50% of visits to physicians are for complaints with no ascertainable biologic base). *You should ask the following questions to elicit the natural history of an illness:*

"How have these headaches affected you in your life?"

"How does your family feel about your headaches?"

"What has been different lately to cause you to come here?"

"What things in your life seem to be related to your symptoms?"

"I wonder if you have noticed any pattern to your symptoms?"

"How do your symptoms alter your normal behavior?"

"When you have your symptoms, how do others react to you?"

Some problems in communication between patients and physicians occur because of the *specialized language* of the physician. It is often assumed that if the physician uses medical terminology most patients will not understand. In fact, studies have found that patients' fund of medical knowledge is higher than that attributed to them by physicians. When the physician uses lay terms or euphemisms, the patient may interpret the informality as an assumption of ignorance. The semantic challenge in interviewing, then, is to find an appropriate balance between professional and informal terminology. *Given a choice between professional and informal phrasing, use professional terminology first.* If the patient's verbal and nonverbal response indicates lack of comprehension, employ progressively more informal phrasing.

Some patients have difficulty *remembering past medical history.* Memory for events is related to time lapse, importance of the event, and degree of the emotional threat. Patients tend to remember more recent events but, after only 1 year, under-reporting of hospitalizations is about 30%. Illnesses which did not restrict activity or which were short tend to be forgotten, and emotionally disturbing events are remembered less well.

There are several questioning techniques which will facilitate the elicitation of historical data from patients. *You should ask for data in chronological order,* attempting to cover all health problems in a given time period rather than concentrating on each health problem and asking the patient to trace a number of chronologies. *It is helpful to ask about related life events* ("marker" events such as graduation, employment, marriage, birth of children, family crisis, or deaths) which the patient is more likely to remember *and to use these events to construct a data baseline.* Not only will this help the patient to remember but it will help you to understand the interplay of life style and illness. *You should always ask, for example:*

"What was happening in your life at the time of that problem?"

You may also "chain" events together in your questioning. Insead of asking "What happened next?", ask:

"You had asthma as a child and your parents moved to the city...How was your health after you moved?"

"After you moved your asthma improved and you graduated from high school...What about other health problems at that time?"

Although you should resist asking "Have you ever had..." followed by a long list of diseases or symptoms, questions which ask for recognition rather than recall are usually better for stimulating memory, and can be utilized appropriately:

"I was curious about any childhood diseases you might have had...like mumps, measles...?"

In our culture, we tend to become anxious if there are *periods of silence* of more than 3 seconds in a conversation. When a patient stops talking you may feel anxiety, rejection, or failure. You will likely react by "filling the silence" with your own voice, or by ignoring the silence altogether. Both of these alternatives block your ability to deal effectively with the patient. When confronted by a patient who suddenly has difficulty talking: (1) *allow pauses of up to 30 seconds, and* (2) *use periods of silence to analyze nonverbal messages.* (3) *If the patient appears unwilling to talk because of anxiety or anger you should call attention to the silence:*

"You seem to be finding it difficult to express your thoughts...Perhaps it's difficult to put into words."

"I'm wondering if your silence has anything to do with what I have said or done?"

(4) *If the patient appears unable to continue due to confusion about facts or your expectations, use a summary and reflection statement:*

"Up to now, you have told me about your reaction to your seizures and it is clear to me you've come to terms with the disorder [summary]...But I'm also curious about how your family and friends have reacted to you [reflection]?"

If the patient has finished his conversational "turn," you must learn to be comfortable with a small silence during which you formulate your next question. *Do not try to listen to a patient response and formulate a subsequent question simultaneously.* Finally, *before asking any question, you should anticipate the possible responses by the patient.*

If a patient is being overly talkative, to the point that you may be unable to get essential information, *you should control the interview by use of "transitions."* A transition consists of an interruption, a statement of clarification, and a topic change:

"Excuse me Mr. Adams, [interruption] I understand you have had this stomach pain for 2 weeks and that it burns up in your chest [clarification]. It would help me to know how your appetite or sleep habits have changed because of the pain [topic change]."

The major problem in transitioning is in poor timing. Changing the flow of conversation demands that you make a judgment about the amount of redundancy in the patient's presentation, tempered by recognition that for some patients talking is highly therapeutic.

Questioning of certain patients may be difficult. *Older patients,* for example, are more likely to have mutiple medical problems, have longer histories to recall, and have difficulty in systematic review of their medical histories. Geriatric patients also will focus on problems you consider trivial, yet ignore problems of significance to you which they consider "just part of old age." Older patients may have diminished sensory activity which will challenge your ability to communicate. Because elderly patients may drift from subject to subject, you will have to spend more time with these patients, but you should try to let them present their history while you put disassociated data into chronological order. Because older people often experience communication deprivation in their lives, they may be very talkative, but because of a diminished sense of social power they will quickly become withdrawn if they sense your disapproval.

In questioning, you may need to use a translator because of sensory deficits or lack of a common language. *Always maintain eye contact with the patient.* Do not talk to the translator: simply allow time after each utterance for the translator to talk. The translator should be positioned unobtrusively in the room.

Similarly, the *pediatric* or *adolescent* patient requires special concern. It is very important to let the talking child and adolescent know that your concern is with him or her as the patient. When the parent is in the examination room, direct your questions to the patient and allow them free opportunity to respond. Encourage parent participation in providing information but do not let them dominate the interview. It will take experience and sensitivity to know when to exclude parents from the interview and/or examination. Watch for verbal and nonverbal clues from the patient to help make that decision.

When questioning a patient *your interview may be interrupted* by an outside agent (nurse, telephone call, page). *In the case of an interruption in the interview you should:* (1) *remember what was being discussed immediately before the interruption*; (2) *excuse yourself to deal with the interruption*, and when you return (3) *recall the specific topic for the patient*:

"Let's see, you were just mentioning the time you first noticed the swelling in your feet..."

This ensures that the patient will not go back over information already discussed, and communicates to the patient your attentiveness.

There are times when you will be expected to fill out forms for the patient, as is often the case with insurance or school physical examinations. *You may have the need to take notes,* particularly when the patient is reporting detailed past medical history. Writing can be a barrier to good communication but there are techniques to minimize its negative effects. Many interviewers "hide behind" their paperwork by actually positioning the paper between themselves and the patient, thereby creating a nonverbal "barrier." You may find yourself focusing on the writing task when the patient's monologue becomes boring or the subject matter emotionally threatening. *If you have to write, it is best to tell the patient what you are writing:*

"I hope you don't mind Mr. Jones, it would help me to be able to take notes as we talk so that I can be sure not to leave out important details."

You should also position your writing or other paperwork so that it is not between you and the patient. Anxiety in patients often stems from fear of the unknown. An extremely anxious patient may be so curious about your writing that it will detract from the patient's ability to communicate clearly.

Another potential problem occurs when the patient asks questions. For example:

"You do think I'm O.K., don't you doctor?"
"Is it possible that worry or stress can cause such a headache?"
"Did you see anything wrong in my throat?"
"Do you think it might cause cancer?"

When asked a question by a patient, it is tempting to answer quickly and with authority to reassure the patient. In so doing, we reassure ourselves that we are the experts and that we have the situation well under control. Unfortunately, patients frequently ask questions in an offhanded way which belies the depth of their concern. If you rush to reassure, answering with a simple "Yes" or "No" or "Of course not," the patient may never express the concern hidden behind the question. Every patient question should first be viewed as an attempt to convey information. *A good rule to follow is to ask a question in response to a question.* For example, in response to the questions above:

"You seem to be worried that you're not O.K....."
"I'm curious about your understanding of stress and headaches..."
"Is there something about your throat that worries you?"
"I'm wondering what makes you think it might cause cancer?"

When a patient asks a question you should view it as "opening a door" for you. Do not "slam the door in the patient's face" with a quick response which discourages further exploration.

Dealing with Emotions

For some clinicians, one of the most threatening aspects of interacting with patients is dealing with emotions. Yet emotions are an integral feature of most illnesses and human interchanges. Patients respond to their illness and their medical treatment with emotions like anxiety, depression, resentment, and anger. Patients will also deal with their conflicts about illness and treatment by using the defense mechanisms of repression, denial, projection, rationalization, displacement, and intellectualization. Emotions and patient defenses are not impediments to understanding illness: they are a part of illness with which you must deal.

You will be tempted to ignore obvious nonverbal indications of emotions and to rationalize this action by saying that involvement in a patient's emotions will take too much time or will reduce your ability to be "objective." In fact, such objectivity is a myth. We constantly form subjective impressions of people and react to them, often without conscious consideration of the process. Although our perceptions of, and reactions to, emotions usually are not conscious, it is possible to rationally monitor our reactions and to employ techniques to assist in dealing therapeutically with emotions.

The challenge to you will be to find an appropriate balance between emotional involvement and emotional distance. Emotional distance by a physician is usually seen by the patient as coldness or disinterest, and care received will be perceived of as inadequate regardless of treatment outcome. On the other hand, strong emotions in a physician are usually negatively interpreted by the patient and physicians who have too much emotional involvement will soon find themselves anxious or fatigued.

One of the reasons you may feel threatened by emotions in patients is a fear that emotions are relatively unpredictable. In fact, clinicians who are sensitive to emotions learn very quickly to predict, anticipate, and deal effectively with patient emotion. Generally speaking, all patients who suffer illness experience a "grief reaction." They grieve because they have suffered a loss of self-esteem. Their relationship with others, as well as with their own body has been altered. During an illness, patients can be predicted to displace their anger, express frustration, project their fears, rationalize their guilt feelings, deny emotions completely, and deal with anxiety in a number of ways.

As with physical symptoms, the major tasks in dealing with emotional symptoms in patients are recognition, diagnosis, and therapeutic intervention. *Your first task is to decide whether the patient's emotions are a character trait* (a style of overall functioning) *or whether they represent an adaptation to the stress of illness and treatment.* Most emotions with which you will be expected to deal are of the latter type. They are clear responses to changes in self concept, they ebb and flow

with the degree of stress, and they usually operate automatically or unconsciously. Emotions are maladaptive in the patient-practitioner relationship if they impede communication. You probably recognize emotions intuitively by picking up verbal and nonverbal cues from the patient. In recognition of emotions, remember that the head and face are usually the nonverbal signalling centers for the *type* of affect, while the rest of the body communicates the *intensity* of emotions.

Anger in patients can be a reflection of basic personality, but it is more likely a reaction to temporary situations unrelated to you, a reaction to illness, or voicing of a legitimate complaint related to treatment. *You must deal with patient anger immediately or the communication process will be disrupted.* There is a simple formula for dealing with anger in patients: (1) *ask about the patient's apparent anger*; (2) *allow time for the patient to verbalize their anger*; and (3) *agree with everything the patient says.* Using this technique you will defuse anger quickly. *In asking about anger, do not accuse the patient.* Say:

"I'm feeling that you are a little angry right now...What seems to be bothering you?"

rather than

"Why are you angry?"

If open questioning does not elicit a response, you should offer some hypotheses to explain the anger:

"You seem a little angry right now...Have you had to wait for me long?"

You should then allow the patient time to verbalize their feelings without argument. Attempts to alter emotions by confrontation usually intensify rather than diminish them. Even though you may be the scapegoat for patient anger directed at someone else, *you should not seek to defend yourself or argue with the patient.* Wait until later to gently correct gross misperceptions by the patient.

You will also see *depression in patients* and need to *distinguish between depression that is a transient reaction to illness and that which is endogenous in nature.* Depression in patients is accompanied by characteristic postures, speech patterns, and body movements. In their complaints, depressed patients can be self-accusatory and frequently complain of feelings of apathy or unhappiness, fatigue, gastrointestinal or appetite disturbance, sleep disturbance, lack of interest in sexual activity, conviction of helplessness, and escapist or suicidal thoughts. Diagnosis of depression can be facilitated by asking about feelings when awakening (with depression patients usually feel worse in the morning and better later), sleeping (depressed people report difficulty), dreaming (depressed patients usually do not dream), appetite, and general feelings about the future. *To distinguish depression which is a reaction to stress from endogenous depression, you can ask questions about the onset and duration of symptoms.* It is also useful to ask:

"How does your mood and appearance today differ from the way you normally are?"

"If I asked people who know you well if you are acting like yourself today, what would they say?"

Depression secondary to illness is characteristic in patients who feel guilt over anger and frustration, and who have repressed their normal emotional responses to illness. Depression is a manifestation of feelings of disappointment, futility, and anger resulting either from illness or inappropriate medical care. *In working with depressed patients it will be necessary to be reassuring and to be able to spend time.* Medication should be used with caution because the patient may interpret it as your way of avoiding him and depression may be exacerbated. *Since depressed patients usually feel helpless, you should try to orchestrate support from the family.* If depression is a manifestation of guilt (for example: because of illness, a primary wage earner believes that he is letting his family down by missing work) you may have to give your approval for their actions, thereby absolving them of their guilt. *One of the best approaches to the depressed patient is to give a full explanation of the relationship between illness, stress and depression. You should try to avoid treating the patient as weak and helpless and encourage the patient to engage in activities which increase sense of self-worth. Patients who report self-destructive thoughts or escapist thoughts (such as "needing to go away for a long vacation" or wanting to "get away from it all") should be evaluated for suicide risk.*

Anxiety is the Most Common Emotion Seen in Patients. It usually occurs as a reponse to the perceived threat or danger of illness and treatment and has distinctive behavioral and autonomic manifestations. We recognize the anxious patient by characteristic cues, but most patients in our culture try to repress signs of anxiety. In addition to signs such as perspiration, rapid pulse, shifting gaze, inappropriate laughter, or excessive questioning, the anxious patient may present with complaints of chest pain, dizziness, heart trouble, or respiratory distress.

As discussed previously, feelings of helplessness or powerlessness on the part of patients in medical situations make anxiety quite common. Many of the suggestions offered elsewhere in this chapter will serve to allay anxiety. Specific techniques can be employed when working with the anxious patient, however. As with the angry patient, *it is helpful to comment on the patient's apparent anxiety.* It may help to reassure the patient that their anxiety is justified given the situation. *Talking about anxiety does not make the patient more anxious.* Specific reassurance, giving the patient an opportunity to express fears, and physical contact with the patient often reduce anxiety. Additionally, *one of the best ways to allay anxiety is through patient education.*

Occasionally, information you are trying to elicit from the patient will arouse strong emotional responses which prevent the patient from responding adequately. In these situations, where the stimulus to emotional expression is the topic being discussed, you will need to employ techniques to increase the patient's "distance" from the emotion. This involves shifting the patient's attention toward events that are in the present rather than in the past, having them discuss their fantasies about the situation, asking them to review the event very rapidly, and requesting that the patient focus on positive aspects of the situation. At other times, patients will not exhibit enough emotion. To facilitate expression of emotions, you should ask the patient to focus on events in the past which are real, remember them in great detail, and to focus on negative or unpleasant features.

Contracting with Patient

Usually, after you have progressed this far with a patient, you will have a few ideas of the reasons for the visit and the patient's expectations. You may be surprised however, to find that confusion and misunderstanding have already arisen. For that reason, *you need* to consciously negotiate a contract with the patient regarding his or her expectations of you. Negotiating is something that has not traditionally been an explicit task in working with patients. Physician control in the traditional relationship mitigated against soliciting the patient's definition of the problem, anticipation of treatment outcome, or expectation of physician performance. We now recognize that patients who are drawn into the decision-making process have a greater interest in outcome, feel more in control, and suffer less anxiety in treatment. You should ask questions to elicit patient perspectives on illness attribution, treatment outcome, and expectations of you:

"I am curious...What do you think is causing your symptoms?"

"What do you think your symptoms mean?"

"What would you see as the best possible result of treatment?"

"What is your goal for this visit?"

"How can I best help you?"

As a student or young physician, you may be reticent to ask such questions because they seem to suggest to the patient that you do not really know what the problem is or how to deal with it. This suggestion of incompetence or lack of control will likely cause you some discomfort.

Establishing a contract with the patient does not imply that the patient will tell you how to practice medicine. Patients today like this approach because it reduces their sense of helplessness by encouraging active participation in treatment. Good clinicians like the negotiated approach because it increases their information base, encourages patient commitment to treatment, increases ability to recognize and deal with patient misconceptions or unrealistic expectations, and reduces patient dissatisfaction.

Physical Examination

A major part of data gathering for diagnosis is the physical examination. Although specific techniques for conducting physical examinations make up the bulk of this book, a few points should be made regarding your approach to the patient as you prepare for and conduct a physical examination. For most patients, the physical examination is a period of extreme vulnerability. There are

but three major rules for conducting a physical examination: (1) always ask permission to examine; (2) tell the patient what you expect to do before you do it; and (3) ensure comfort and privacy. Asking permission means not saying:

"Now I am going to do a physical exam..."

Instead, say:

"If I may, I would like to do a physical examination to check some of the problems you have told me about."

Because disrobing is more threatening than nudity, the patient should be given an opportunity to undress in privacy. You should give specific instructions, leaving no room for ambiguity. Do not say:

"If you will get undressed, I'll be back in a minute."

Instead say:

"I will leave to give you a chance to get undressed. If you would, please take off all your clothes including your underwear, and slip on this gown which opens in the back. I shall return in a minute."

This degree of explicitness is important because any ambiguity tends to raise anxiety.

During the examination, you should keep the patient informed of your intentions by asking permission before you do anything. The difference between asking and telling is subtle yet important:

Asking: "Now I would like to do a breast examination."
Telling: "Now I'm going to do a breast examination."

You need not be afraid that a patient will refuse your requests. Most will accede with either approach, but asking shows respect for the patient, acknowledges tacitly the patient's right to control access to his personal space and, because the patient actively gives approval, encourages a sense of mutual participation as opposed to helpless passivity. Asking also allows the patient to hesitate or express reticence at some time during the examination. Being able to express such reluctance is a sign of a good physician-patient relationship. A simple acknowledgment of the patients reluctance by you will almost always hold sway:

"I know a rectal exam is uncomfortable, and I know you don't really want one, but it is the only way to get some very important information."

During the physical examination you should show concern for your patient's comfort. Altering the position of the table, proper use of drapes and gowns, and a professional manner are all advisable. Asking about patient comfort and providing verbal reassurance are also appropriate.

While conducting a physical examination, experienced clinicians carry on a relaxed dialogue with the patient. Many physicians combine a "review of systems" with the examination of the various parts of the body. This strategy helps to structure the examination, and can help put the patient at ease. *One thing you must do during the examination is to contol your own nonverbal signals.* Patients who are extremely fearful about their illness will often be afraid to ask directly about their concerns. Instead, they will monitor your actions closely, reading devastating diagnoses and prognoses into your expression as you strain to listen to heart sounds.

Problem Solving with Patients

After the physical examination, you should share your diagnostic hypotheses with the patient. Many diseases will have been eliminated from consideration and it should be possible to either convey to the patient your feelings of the nature of the problem, or what additional tests need to be conducted to further eliminate hypotheses. *You should always talk* with *rather than* to *the patient, conveying the message that you are sharing information rather than lecturing.* You can monitor your approach to the patient on this dimension by attention to use of the pronoun "we" (incorporative) rather than "I" and "you" (divisive). You should not simply tell the patient that he "must get an intravenous pyelogram" or that you are "ordering a glucose tolerance test." *The information you give to your patient should be designed to help* them *make decisions regarding additional diagnostic procedures or the initiation of treatment.*

You should solicit the patient's view of the presumptive diagnosis as well as perspectives on treatment alternatives as soon as possible. For example:

"It appears probable you have a problem of hypertension which can be treated in a number of ways...I'm interested in your understanding of this problem."

"It seems to me that you probably have a problem of hypertension and I have talked with you about this problem...I'm wondering if you have other questions or some ideas about hypertension which we did not cover?"

"It is possible to treat hypertension by modifying your diet, use of certain drugs, and perhaps an exercise program...What we need is to make a decision based on what is best for you."

In this manner, you can assess the patient's fund of knowledge, fears, and treatment preferences. You can then gently correct misconceptions, allay anxiety, and agree on a treatment program which would be most acceptable, and more faithfully followed by the patient.

You should be straightforward when you present the patient with alternative diagnoses and treatment plans. Some diagnoses and treatment alternatives may be difficult to communicate because of their gravity or discomfort, but patients who are ill often misinterpret physician evasiveness or ambiguity as indicating either more optimism or pessimism than the actual clinical data warrant. Clearly, you should judge the possible effects of revealing your diagnosis and treatment recommendations to the patient and modify your approach accordingly.

In discussing your clinical impressions and recommendations for treatment you should always address the patient's chief complaint even though it may have become clear that it is not the major medical problem. You might say, for example:

"At first you indicated your problem was 'ringing in your ears,' but, as we have discussed your history, that seems likely to be a symptom of your hypertension. I just wanted to make sure that our plan to start you on antihypertension regimen, and just see if this problem disappears on its own, is agreeable to you?"

After communicating a diagnosis, you should never attempt to encourage the patient by saying:

"Well...many people have this problem and learn to live with it."

or

"I know just how you feel."

Such statements convey little in the way of concern for the individual patient or of honest empathy. Instead, you might say:

"It looks like you're feeling terrible about this problem...it *is* a difficult problem but we really have considerable experience, much of which should be of help to you in dealing with it."

This type of statement acknowledges the patient's discomfort, suggests that the patient's problems are deserving of individualized treatment, but conveys a sense of competence and precedence for treatment.

Many patients will have several problems which need attention. You may be unable to address each of them at one visit and some may not require immediate attention. *The strategy for dealing with multiple problems is to delineate them with the patient, acknowledge that not all of them can be dealt with presently, and negotiate which problem is the most deserving of immediate attention and how the others can be addressed later.*

"It seems as though we have identified three major problems which concern you: your joint pain, your concern that you will be crippled with arthritis like your sister, and the problem of not sleeping...Which of these concerns you the most?"
"You have told me about two problems: your back pain and worries about your diet. We really can't deal effectively with both today...Which problem can we deal with now, and which can we more thoroughly investigate at another appointment?"

In this manner, you do not minimize the importance of any complaint, but instead mutually prioritize them for a comprehensive treatment plan.

If the treatment plan involves a referral (to another physician or to an agency), *you should make a "personal" referral.* The two examples below illustrate this concept. Do not say:

"I think you should see a psychiatrist and attend Alcoholics Anonymous."

Instead, say:

"There is a psychiatrist for whom I have great respect, named Dr. Jim Smith. He might be someone you could find helpful, and I would be glad to discuss your problem with him should you choose to make an appointment. I also have a friend in AA who should be excellent help for you."

The former statement is likely to be construed by the patient as a personal rejection. The latter maximizes the likelihood of further treatment with another physician or agency without jeopardizing your relationship to the patient.

Review and Termination

The final task in working with the patient is review and termination. *You should explicitly review the major complaints, your clinical impressions, the patient's expectations, what you have asked of the patient* (additional tests, referral to another physician specialist, medication, etc.) *and your prognosis. After this brief review, you should ask the patient if they agree or can add to the synopsis, ask if there are any final questions, and wish the patient well.* You thus conclude what should have been a mutually beneficial patient-physician interaction.

Skills Performance Checklist

Interpersonal skills training has become an established curriculum in many medical schools. Techniques such as videotape and review, use of simulated patients, and practice with fellow students have been employed to teach interviewing. The following is a checklist of skills abstracted from the text. It is designed to be used as a guide for evaluating interpersonal performance in any type of medical interview setting.

I. Preparing for interview
    A. Review of patient records
    B. Consult other data sources (nurse, receptionist)
    C. Define personal expectations
II. Meeting the patient
    A. Greet patient by name
    B. Introduce yourself and others
    C. Express concern for patient comfort
    D. Use facilitative nonverbal cues
        1. Appropriate smile
        2. Calm, unhurried manner
        3. Maintenance of eye contact
        4. Eyes at same level as patient
        5. Position outside patient's personal space
        6. Position not blocking door
        7. Paperwork not positioned as barrier
    E. Observe patient nonverbals
        1. Fear, anxiety
        2. Pain
        3. Anger, hostility
        4. Depression
    F. Verbalize time constraints
III. Patient presentation of medical problems
    A. Question to elicit chief complaint
    B. Ask about problems from previous visits

C. Listen without interruption
  1. Pitch, stress, intonation
  2. Inference
D. Use affirmative cues
  1. Verbal encouragement to continue
  2. Head nod, eye contact approval
E. Summarize or restate problems
F. Refer to chief complaint in patient's words

IV. Expression of patient perspectives
  A. Elicit patient's perspective on cause of problem
  B. Elicit patient's expectation of treatment outcome
  C. Elicit patient's expectation of physician

V. Elaboration of chief complaint
  A. Question for distinctive features of symptoms
    1. Location on body
    2. Description
    3. Intensity
    4. Amount
    5. Frequency
    6. Setting when symptoms develop
    7. Ameliorating and exacerbating factors
    8. Concomitant symptoms
  B. Application of questioning techniques
    1. Anticipate response before questioning
    2. Use open, indirect questions to start questioning
    3. Use closed, direct questions for specific information
    4. Avoid use of multiple questions
    5. Avoid flooding with series of questions
    6. Ask permission to question in sensitive areas
  C. Use of appropriate terminology
    1. Use professional terminology
    2. Observe for patient comprehension
    3. Employ progressively informal terminology
  D. Ask for data in chronological order

VI. Physical examination
  A. Ask permission to examine
  B. Give specific instructions for patient preparation
  C. Allow patient privacy for undressing
  D. Explain task prior to each examination
  E. Express interest in patient comfort
  F. Carry on comfortable dialog
  G. Control nonverbal cues

VII. Problem solving with patient
  A. Ask questions in response to patient questions
  B. Give information—do not make decisions
  C. Solicit patient's view of diagnosis
  D. Discuss patient's opinions of treatment alternatives
  E. Refer to chief complaint in treatment plan
  F. Outline and prioritize multiple problems

VIII. Dealing with emotions
  A. Recognition of expression of emotions
  B. Comment or question about emotion
  C. Facilitate patient ventilation of feelings
    1. Do not accuse or argue with patient
    2. Tolerate silence
  D. Determine cause of emotion
  E. Reassure and support patient
    1. Give nonverbal reassurance
    2. Reassure verbally
    3. Provide patient education

IX. Review and termination
  A. Review and restate
    1. Patient complaints
    2. Clinical impressions
    3. Patient's expectations
    4. Your recommendations
    5. Prognosis
  B. Solicit patient agreement
  C. Ask for additional questions
  D. Wish the patient well

PRACTICE SESSION

You and your partner will probably be interviewing patients together with one of you doing the questioning and the other recording information. On one of these exercises, take the Skills Performance Checklist. Mark each of the categories on a scale of 1 to 5 as a critique of your partner's performance in the interview. Your partner will be accustomed to your assaults during the practice sessions of the physical examination so do not be gentle during this part.

Suggestions for Further Reading

It is beyond the scope of a chapter such as this to discuss every detail of interpersonal skill or nuance of the patient-practitioner relationship.

The following are several references which have inspired some of the content of this chapter and which offer more extensive discussion for interested students.

Benjamin, A. *The Helping Interview.* Houghton Mifflin

Co., Boston, 1974.

Bishop, F., and Froelich, R. *Medical Interviewing: A Programmed Manuel.* C. V. Mosby, St. Louis, 1969.

Bowden, L., and Burstein, A. *Psychosocial Basis of Medical Practice.* Williams & Wilkins, Baltimore, 1974.

Carkhuff, R., *et al. The Art of Helping III.* Human Resource Development Press, Amherst, Mass., 1976.

Gallagher, E. (ed) *The Doctor-Patient Relationship in the Changing Health Scene. Proceedings of an International Conference:* Joan E. Fogarty Center for Advanced Study in the Health Sciences, U. S. Government Printing Office, Washington, D.C., 1976. DHEW Publication No. (NIH 78-183).

Ivey, A. (ed) *Microcounseling: Innovation in Interviewing Training.* Charles C Thomas, Springfield, Ill., 1971.

Jaco, E. *Patients, Physicians and Illness: A Sourcebook in Behavioral Science and Health.* The Free Press, New York, 1979.

Knapp, M. *Nonverbal Communication in Human Interaction.* Holt, Rinehart & Winston, New York, 1972.

Mayerson, E. *Putting the Ill at Ease.* Harper & Row, Hagerstown, Md., 1976.

Wicks, R. *Counseling Strategies and Interviewing Techniques for the Human Services.* J. B. Lippencott Company, Philadelphia, 1977.

chapter 3

GENERAL PRINCIPLES

The physical examination, in fact, begins as soon as you greet the patient and begin the interview. You observe how he looks, what the handshake feels like, his carriage and general habitus and his manner of speech. These are the "first impression" we all make in any social contact. In the nonprofessional context, the observations are rarely pinpointed and analyzed. Now, however, you must be specific about your "first impression." This is a "noninvasive" physical examination, and it takes practice—practice you can do on the bus or on the subway.

Having completed the history, the next step is the "invasive" physical examination. Invasive is a correct term, for psychologists tell us that the physical approach of one person to another to a distance of less than 2 feet is always interpreted by the subconscious as invasion. It may be seen as intimate invasion or threatening invasion but always evokes a heightening of awareness. It bears emphasis that this phenomenon always occurs.

From the interview, you have a first approximation to the nature of the illness and perhaps its location. You may have discovered areas of special concern on the part of the patient about the physical examination. It may be necessary to proceed directly to the most probable area of pathology in the acutely ill patient. The best approach, however, is to move systematically to avoid errors of omission.

Detection of abnormalities by physical examination involves our special senses. The tools used are merely extensions of those senses. We hear certain sounds better with a stethoscope and see some things better with an ophthalmoscope or otoscope. A systematic examination is the systematic application of the special senses to each system or area. The routine includes *inspection, palpation, percussion,* and *auscultation.*

Inspection

It should not be necessary to point out that in order to inspect an area it must be visible. Unfortunately, you will see doctors try to examine the abdomen or breasts underneath clothing or bedsheets. To inspect, expose what you wish to see. Good lighting is important. Icterus may not be seen with fluorescent lighting, and cyanosis is hard to see in poor light. The angle of the light can be used to advantage. Indirect lighting produces shadows which accentuate small differences in contour.

Colors, contours, symmetry or lack of it, and events can all be seen and characterized. Is the face plethoric, the chest barrel shaped, one thigh larger than the other, the respiratory effort extreme?

A valuable tool for inspection is a ruler, preferably both a small metal or plastic rule and a long soft tape measure. Quantitate your observations wherever possible. Later we will see where such measurements are diagnostic.

Palpation

Palpation is the act of feeling with the hand or hands. Palpation further defines things we see and reveals things we cannot see. Palpation discriminates textures, dimensions, consistencies, temperature, and events. Different parts of the hand are best suited for different tasks.

Textures are best detected by the fingertips. Our ability to discriminate two points as separate points even though very close together is maximum in the fingertips. A small skin lesion is a papule if it is slightly raised and a macule if it is flat with the rest of the skin—a differentiation made by running a fingertip over it. All textures should be noted. Is the skin dry and coarse as in myxedema, are the lymph nodes single or matted, is the liver edge smooth or nodular?

It takes some practice to know how hard to palpate. Feeling someone else's body is quite a bit different from feeling your own. Generally, students tend to be too tentative in their efforts to be gentle. The best practice is to have your student partner respond verbally to your palpation and let you know if you are too rough. A good general rule is to use only the number of fingers or surface of the hand which is minimally necessary. Enlist the aid of the patient to get the most from palpation. The position of the patient is very important. You cannot palpate the abdomen well unless the patient is perfectly flat. You cannot palpate the anterior neck with the head flexed. Work for a balance between patient comfort and the position which gives you the most information.

Dimensions or contours are larger scale textures. Several fingers, the entire hand, or both hands are needed depending on the size of the part being examined. An enlarged liver beneath tense ascites may necessitate both hands to exert sufficient pressure to displace the fluid and detect the liver edge. *Ballottement* is a maneuver utilizing both hands to detect deep structures. Ballottement of the liver requires a sudden thrust of two hands against the abdominal wall. Ballottement of the kidney involves deep palpation with one hand anteriorly while the other hand thrusts forward from the flank. As in inspection, measurement is important in recording dimensions.

Consistency depends on the density of a solid object or the tension in the wall of a hollow viscus. Consistencies are felt best by the fingertips. A lymph node may be soft, rubbery, or stony hard. A cyst might be flaccid or tense depending on the pressure within.

Temperature cannot be ascertained by the hand in absolute degrees. What can be noted is change in temperature from one area to another. The backs of the fingers are best suited for this. The skin there is thin while the nerve density is still great. By running the backs of the fingers over the skin, subtle differences in temperature become apparent—warmer in the case of inflammation and cooler in the case of diminished blood flow.

Events palpated are any movements beneath the examining hand. They are usually best noted with the entire palm and fingers.

The dyskinetic anterior ventricular aneurysm manifests itself as a double impulse—two events separated slightly by time and position but felt by a single hand. The "water hammer" pulse of aortic insufficiency is a distinct event felt over the major arteries. Events are defined through discrimination of time and location, and these dimensions are the measurements of events.

Percussion

Percussion is a bit harder to define. Crudely, it is thumping with the hand or instrument on a part of the body. What we do, in effect, is to produce a noise and listen to what it sounds like—listen for the repercussion. The aim is to produce vibrations and note what happens to the sound waves.

The propagation of sound waves is determined by the density of the medium in which they are traveling. The number of interfaces of different density also affects the transmission of the sound. Sound required to traverse skin, muscle, fat, bone, fluid and air will move less well than through the same distance of a single kind of tissue. This is why we percuss the chest *between* ribs and tense the skin under the finger.

The degree to which sound propagates is called resonance. Air is very resonant, solid tissue is not. As a standard, percussion of normal lung tissue is called *resonant*, the gastric air bubble is *tympanitic*, the liver is *dull*, and the thigh is *flat*—that is, not at all resonant.

There are various ways to percuss. Most commonly, the tip of the left middle finger is laid firmly against the skin. The tip of the right middle finger delivers a quick tap to the finger on the skin. The technique, *mediate percussion*, or indirect percussion, interposes a finger, the *pleximeter*, between the area to be percussed and the finger creating the vibrations, the *plexor* (Fig. 3.1). The *immediate technique* uses only the plexor, that is, a direct blow to the area. The thumb may be used in this technique by holding the hand parallel to the skin and quickly pronating the hand (a guitarist's technique) (Fig. 3.2). Occasionally, a slap with the tips of all fingers is useful. The intensity of the note produced depends on how hard you strike. What you gain in intensity, however, you lose in definition. The intensity determines the depth of penetration.

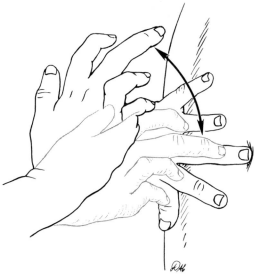

Fig. 3.1. Percussion. The middle finger of the left hand is the only point that touches the skin. The middle finger of the right hand strikes its mate with a force delivered from the wrist.

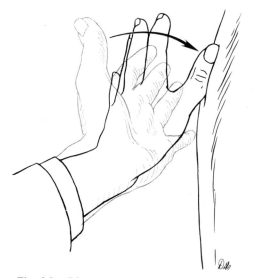

Fig. 3.2. Direct percussion. The thumb is raised approximately 4 inches from the surface to be percussed. The thumb strikes the chest and the resultant sound is noted.

Obesity requires greater intensity to penetrate to deeper structures. Approximately 7 cm is the maximum depth you can hope to discriminate.

Quantitation of the percussion note is difficult; the most helpful observations consist of change in the sound from one area to another. A change from resonant to dull is easier to appreciate than the converse. Whenever possible, percuss first a resonant area and move toward dullness to define borders. As an exercise, percuss your own right chest and move down until you hear liver dullness. Now percuss your thigh to hear flatness.

As with palpation, percussion takes some practice to get the right intensity. Here the error is generally that students hit too hard. You may tend to strike hard enough to hear the note in a noisy room. It is far better to strike softly and bend close to hear the note. With practice you will come to realize that what you feel in response to the tap is as revealing as what you hear.

Auscultation

Auscultation is listening for sounds from within the body. A few properties of sound merit discussion.

The measurements of sound are *frequency*, *intensity*, *duration*, and *quality*. *Frequency* is a measure of the number of vibrations as cycles per second. Many cycles per second, or a high frequency, produces a high-pitched sound, while few cycles per second produce low-pitched sounds. *Intensity* is a measure of the loudness of a sound as decibels (the energy equivalent is dynes/cm^2). How long the sound lasts is, of course, its *duration*. *Quality* is harder to define. Sometimes described as timbre, quality is determined by harmonics produced or the various multiples of the basic frequency. Harmonics allow us to distinguish a piano from a violin even though they are producing the same note or same number of cycles per second.

There are definite limits to our ability to hear sounds. Frequencies less than 16 cps or greater than 16,000 cps are usually inaudible. Similarly, sounds of less than 0.001 dyne/cm^2 cannot be heard, while very high intensities become painful to our ears. Further, there is a relationship between frequency and intensity and our abiliy to hear sound. The lower the frequency the greater the intensity must be for us to hear it. It is easier to hear a soft high-pitched whistle than a loud low-pitched rumble. The latter is sometimes felt better than heard.

The *stethoscope* is an instrument to aid in

auscultation. Early stethoscopes were hollow pieces of wood acting as conduits for sound from the skin to the ear. Modern stethoscopes in addition to conducting sound also collect sound and select frequencies. The head of the stethoscope applied to the skin collects the sound from beneath it. Most stethoscopes have two types of heads. The flat diaphragm applied firmly responds best to high frequency sounds and excludes low frequency ones. The bell, on the other hand, allows high frequency sounds to escape and collects low pitched sounds. To convince yourself of this, try the following exercise. Very lightly press the bell of your stethoscope against your heart. Don't worry about what sounds you are hearing yet, just concentrate on the pitch. Now gradually increase the pressure of the bell against your skin and you will hear that the higher frequency sounds get louder. What you have done is to tense the skin under the bell and converted it to the diaphragm which reacts better to high frequency sounds. The conduit of the stethoscope is a combination of rubber or plastic and metal. Leaks in the conduit will obviously waste sound. The most common leak is a poor fitting ear piece, either too large or too small. Comfort is a good guide to properly sized earpieces.

Odors

Smell is a sense occasionally useful in physical diagnosis. Halitosis may indicate poor oral and dental hygiene. A fetid breath is common in lung abscess. Fetor hepaticus is a musty odor associated with liver disease. Ketones of diabetic acidosis are quite characteristic (approximately 10% of people cannot detect the odor of ketones, perhaps a genetic trait). Ammoniacal urine might indicate infection of the urinary tract with urea-splitting bacteria. A bacteroides abscess is distinctly feculent while pseudomonas has a thick sweetness to it.

In summary, each area examined is inspected, palpated, percussed, and auscultated. Anatomically, some areas obviously preclude some techniques and lend themselves to others. Subsequent chapters will elaborate the methods most useful in the examination of specific areas or organ systems.

chapter 4

BODY AS A WHOLE

In this chapter, we will consider alterations of body habitus (Fig. 4.1). It is not enough to say that a patient looks strange. Physical diagnosis involves definition of the abnormality in quantitative terms. Certain patterns lead us to suspect specific abnormalities. In addition, certain motions and posturings are recognizable as body language—a form of nonverbal communication which may be helpful in diagnosing mental attitudes or aberrations.

Normal Body Habitus

Normal growth and development involve not only a general increase in size but a change in body proportions. If you have had occasion to study portraits by early masters, you might have noticed that the children do not look like children. Rather, they look like adults in miniature. Specifically, the arms and legs are too long for the trunk and the heads are too small. There is more, then, to the difference between adults and children than difference in height.

Determinants of Body Characteristics

What determines what we look like? Known factors include genetic determinants, hormones, connective tissue, and nutrition.

Genetics is most obviously expressed in the tall or short family. Such a characteristic, while perhaps abnormal statistically, is normal for that family. The African pygmy is genetically short, and the Masai is genetically tall. A few gross recognizable chromosomal errors produce abnormal body characteristics. Most of these involve the sex chromosomes. Less well defined at the chromosome or gene level are the inherited disorders of connective tissue.

A number of *hormones* contribute significantly to body habitus. Their effects are me-diated primarily through changes in growth and remodeling.

Growth hormone affects somatic growth (Fig. 4.2). The exact time and mode of action are not known. Thyroid hormone promotes linear growth. In addition, it affects the moulding of adult features through changes in skeletal proportions, ossification of cartilage, and maturation of epiphysis. Androgens accelerate growth and muscular development through adolescence and are responsible for secondary sex traits. Importantly, sex hormones control the maturation of the epiphysis. Epiphyses are the growth plates of long bones. Once they mature and close, longitudinal growth ceases. Too much androgen too early stunts growth, and lack of androgens allows linear growth to continue unchecked.

Nutrition, in the broadest sense, profoundly affects growth and development. Cachexia, severe childhood malnutrition (kwashiorkor), and most types of obesity are examples. The same mechanism, though less obvious, applies to the growth retardation of cyanotic congenital heart disease. Here, the nutritional deficiency is of oxygen. Likewise, any chronic debilitating illness affects nutrition. It is basically nutritional deficiency which allows us to say that a patient appears "chronically ill" or wasted. Any severe protracted illness may be operative.

Connective Tissue

Bone, fat, muscle, cartilage, collagen, elastin, and skin are the final determinants of what we look like. All of the factors discussed above involve modifications of connective tissue. Just as any factor influencing connective tissue may yield disordered growth and development, so primary abnormalities of connective tissue can produce disordered growth and development (Fig. 4.3).

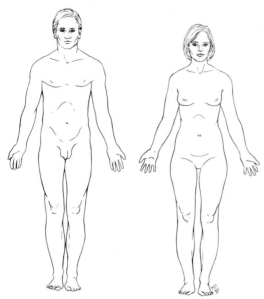

With this as background, let us examine the patient with abnormalities of the body as a whole—the tall patient, the short patient, and the asymmetric patient.

History

The evaluation of body habitus begins with clues from the history which might implicate genetic, hormonal, nutritional, or connective tissue abnormalities.

The family history might reveal that excessive height or short stature is a family trait. Endocrine or infectious diseases in the mother during pregnancy may profoundly affect the child's subsequent growth. Documentation of growth and development can be very valuable. Actual heights and weights plotted against years can be compared with charts of normals. The weight should be correlated with dietary habits. Changes in weight over time indicate the tempo of the illness. Menarche and the development of axillary, pubic, and facial hair are the indicators of puberty and denote normal sexual maturation. As mentioned previously, early or de-

Fig. 4.1. Normal body habitus. The crown-to-pubis distance equals the pubis-to-floor distance in both the man and the woman. The span of outstretched arms equals the height in both. The center of gravity of the man is through the shoulders, and through the hips of the woman.

Fig. 4.2. Acromegaly. Two photographs demonstrate the remarkable change which can occur under the influence of growth hormone. The photograph on the *right* shows the prominent nose, chin, and cheek bones as a result of this hormone effect. It is often helpful to have a patient bring photographs of previous years when suspecting such a diagnosis.

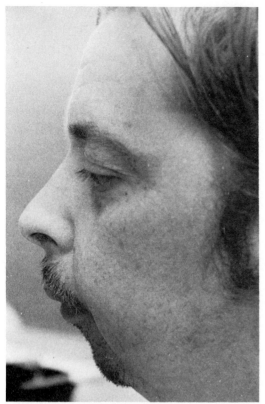

Fig. 4.3. Micrognathia. This patient suffered from juvenile rheumatoid arthritis. The involvement of the growth plates of the mandible resulted in micrognathia as an adult. This is characteristic finding of juvenile rheumatoid arthritis.

layed puberty affects growth. A history of slow development, sluggishness, anemia, and obesity in childhood might implicate thyroid deprivation as the etiology of disordered growth and maturation.

Dentition, with information about the number of teeth at 1 and 2 years and the age at which permanent teeth erupted can also serve as indicators of the timing of development.

Physical Examination

Describe first your general impression of the habitus: tall, thin, short, fat, older or younger than age, symmetrical or not, nervous, sluggish, bright or dull. The following parameters direct your attention to specifics:

1. Standing height and weight
2. Skeletal proportions
   a. Span—the distance from outstretched fingertip to fingertip
   b. Upper and lower segments—the distance from the symphysis pubis to the crown and from the symphysis to the floor
3. Circumference of the head, chest, and abdomen
4. Maturation of the features
5. Sexual development
6. Distribution and amount of fat and muscle
7. Skin color, texture, moisture, and temperature
8. Symmetry of paired structures

Height and Weight

Height should be measured standing. Estimation of the height of a patient in bed is notoriously inaccurate. Measure the weight and note the time of day and state of dress.

Skeletal Proportions

In normal adults, the span equals the height and the upper segment equals the lower segment; that is, the ratio of the segments is 1. In childhood to about age 10, the trunk is longer than the legs and the ratio is about 1.7:1. Abnormalities in the ratio may be subtle but are valuable diagnostic clues. Head, chest, and abdominal circumferences complete the specific numerical survey. Make the chest measurements in both deep inspiration and deep expiration. These two measurements will be valuable when assessing the respiratory system later.

Maturation of Features

Maturation of the features refers to the apparent age of the patient and is most easily assessed in the face. The bridge of the nose warrants scrutiny. Infantile facies characteristically show a broad flat bridge. Thyroid hormone contributes importantly to the maturation of such cartilaginous structures. Skin texture and folds similarly indicate apparent age.

Sexual Development

Sexual development is specifically related to the effects of androgens and estrogens in the female and androgens alone in the male. Specifically, note in the physical examination of the male the following androgen effects:

1. Testes (measure)
2. Penis development
3. Stippling or wrinkling of the scrotum
4. Axillary and pubic hair
5. Depth of the voice
6. Seborrhea and acne
7. Prostate development

In the female, androgens are responsible for the following developments (Fig. 4.4):

1. Labia majora
2. Clitoris
3. Axillary and pubic hair
4. Acne

Estrogens, on the other hand, contribute to the development of the following:

1. Nipples
2. Areolae
3. Breasts
4. Labia minora
5. Vulva
6. Vagina
7. Uterus
8. Ovaries

Lack of development of these structures indicates insufficient or ineffective stimulation by the respective hormones. Premature development indicates premature stimulation by the hormones. Abnormalities of timing or amount affect sexual development and, through effects on the bone epiphyses, stature.

An early release of androgens will cause premature closure of bone epiphyses and, although there is a brief growth spurt, this premature puberty results in short adult stature. This may happen in either the male or female.

Fat and Muscle Distribution

The distribution of fat and muscle means more than the simple assessment of leaness or obesity. Normal women have an apparent center of gravity through the hips and relatively more fat than muscle. Men, centered through the shoulders, predominate in muscle. Children and preadolescents show "baby fat" in the face and hands. Fat may be centripetal body fat or centrifugal limb fat.

Skin Characteristics

Skin is affected by many factors. Color is controlled by genes, vascularity, thickness, sunlight exposure, and numerous circulating pigments. Texture, moisture, and temperature are interrelated. Thyroid hormone, adrenal steroids, growth hormones, nutrition, and the environment are a few of the recognized determinants.

Symmetry

Man is basically a paired animal, and discrepancies in the size of parts are quickly noted; the atrophic limb of poliomyelitis, for instance, is all the more striking when compared with the normal arm. When such a discrepancy exists, attempt to determine if one part is abnormally small or one part abnormally large.

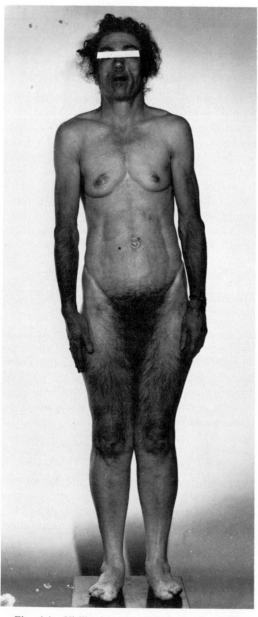

Fig. 4.4. Virilization. In addition to the marked hirsutism, note the frontal balding and the masculine habitus.

Tall Patient

A tall patient is made so by excessive longitudinal growth. This effect may be the result of an excess of stimulators of growth or the lack of factors which halt growth. Growth hormone probably stimulates growth while sex hormones halt longitudinal growth through fusion of the epiphyses. Thyroid hormone, on the other hand, probably has a permissive role in growth since its absence retards normal growth and development.

Hypogonadism

A common cause of excessively tall stature is the lack of androgens. An individual deprived of these hormones has continued longitudinal growth. Since this growth occurs predominantly in the long bones of the arms and legs, the growth is disproportionate. The span is greater than the height and the lower segment exceeds the upper segment. The appearance may be normal until past puberty.

The associated findings relate to the absence of secondary sex characteristics—infantile genitalia and absent axillary and pubic hair. These patients appear younger than their age. Fat distribution is centripetal and tends to be feminine. Facial features are juvenile. Cartilage at the base of the nose appears flat and poorly developed. The fingers are pudgy proximally and taper toward the tips. The examination should be pursued to discover, if possible, the reason for the lack of sex hormones.

The *true eunuch* has primary hypogonadism that is a primary failure of the testes. The genitalia are infantile and the testes small or absent (Fig. 4.5).

Kleinfelter's syndrome is an aberration of the sex chromosomes. Rather than the normal XX complement of the female or the XY of the male, patients with Kleinfelter's syndrome have an XXY combination. Phenotypically, they are males. A relatively common abnormality, Kleinfelter's syndrome may show few gross abnormalities. In flagrant cases, the manifestations include small testes, gynecomastia, and delayed and reduced virilization in addition to tall stature. Many observers have noted mental retardation and antisocial tendencies in some patients with Kleinfelter's syndrome (Fig. 4.6).

Secondary hypogonadism should be considered. The previous examples represent a primary defect in the testes. Remember, how-

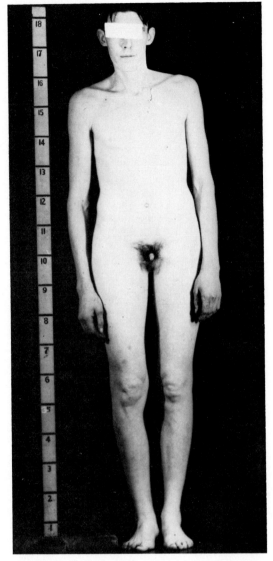

Fig. 4.5.   Eunuch. The small genitalia are apparent. Note, in addition, the relatively long arms and legs in proportion to the trunk. There is obvious lack of masculinization. (Courtesy Endocrine Unit, Massachusetts General Hospital.)

ever, that the anterior pituitary and hypothalamus stimulate the gonads. Failure at this level will also result in hypogonadism.

A selective deficiency of pituitary gonadotropins rarely occurs. When accompanied by agenesis of the olfactory lobes, anosmia results, and the constellation of hypogonadism and anosmia is called *Kallmann's syndrome.*

A rare cause of secondary hypogonadism was described by Fröhlich. In his patient, a cyst in the region of the pituitary gland re-

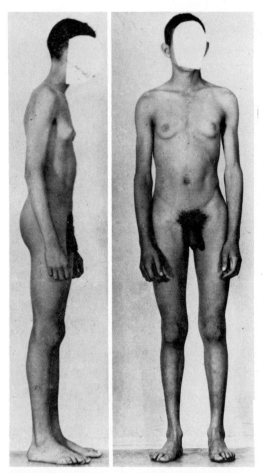

Fig. 4.6. Kleinfelter's syndrome. This man was found to have marked gynecomastia, a normal phallus, but extremely small testes. A buccal smear confirmed the presence of an additional X chromosome.

Fig. 4.7. Giant. The pituitary adenoma responsible for the extraordinary growth of this individual was present prior to puberty. Hence, the body proportions are relatively normal in spite of his enormous size. (Courtesy Endocrine Unit, Massachusetts General Hospital.)

sulted in hypogonadism and marked central obesity. Unfortunately, the label, *Fröhlich's syndrome*, has been applied to many fat boys with simple delayed puberty.

Excess Growth Hormone

An increase of growth hormone prior to epiphyseal fusion results in *giantism* (Fig. 4.7). If the responsible eosinophilic pituitary adenoma interferes with gonadotropins, the proportions will be eunuchoid in addition to tall.

Acromegaly is the term applied to the effects of increased growth hormone after epiphyseal closure. The name is descriptive—meaning enlargement of distal parts. Bone growth occurs more in flat bones. Hence, the skull enlarges, the nose is broad, the supraorbital ridges prominent, the tongue large, and the hands and feet spadelike (Fig. 4.8).

Since both giantism and acromegaly are caused by pituitary tumors, bitemporal visual loss should be looked for (see Chapter 7). Physical findings of myxedema, hypoadrenalism, and hypogonadism can occur concomitantly if destruction of the pituitary is complete. Growth hormone is diabetogenic; hence a history of polyuria, polyphagia, and polydipsia might be elicited.

Marfan's Syndrome

This disease of unknown etiology apparently represents an inherited disorder of connective tissue. An autosomal dominant disorder, Marfan's syndrome is not rare.

One of the major features is arachnodactyly, which refers to the long, spider-like fingers and toes. An asthenic, tall build is common. The joints may be hyperextensible. Skeletal abnormalities predominate in the chest and spine. Pectus excavatum or "caved in chest," pectus carinatum or "pigeon breast," mild scoliosis, and flat feet are the most common findings. Dislocated lenses of the eyes presumably result from laxity of the suspensory ligaments.

The most devastating manifestations of Marfan's syndrome involve the aorta. Degeneration of the elastic media promotes dilatation with aortic valve insufficiency or dissecting aneurysms as the sequelae.

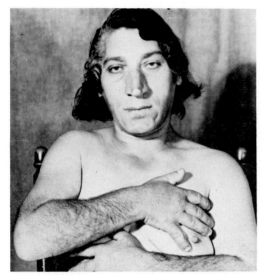

Fig. 4.8. Acromegaly. The spade-like hands, prominent nose, jaw, and cheeks are readily apparent in this woman suffering from an eosinophilic tumor of the pituitary. (Courtesy Endocrine Unit, Massachusetts General Hospital.)

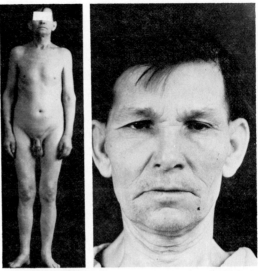

Fig. 4.9. Panhypopituitarism. The body habitus is soft. Axillary and pubic hair is absent, and the facies are best described as "young-old man." (Courtesy Endocrine Unit, Massachusetts General Hospital.)

Short Stature

Hormones, connective tissue, genes, and nutrition are central also to short stature. Here too, measurements of proportions aid in establishing the diagnosis.

Dwarfs

No firm definition of a dwarf exists. A dwarf is an abnormally undersized person. The term means very little other than that. We recognize specific kinds of dwarfs by physical hallmarks.

The *pituitary dwarf* has a short stature because of deficient growth hormone. They are normal at birth and for the first few years of life. Thereafter, growth occurs slowly. Sexual development usually does not occur. Signs of deficiencies of other pituitary hormones, however, are not common. Retardation of muscle development results in a thin "asthenic" appearance. The skin ages prematurely, but maturation of the features does not occur. This old looking skin with young looking features gives the characteristic "young-old" appearance (Fig. 4.9).

Ovarian dysgenesis, Turner's syndrome, represents a form of genetic dwarfism (Fig. 4.10). These patients possess only one chromosome—a single X. In addition to short stature, several somatic abnormalities confirm the di-

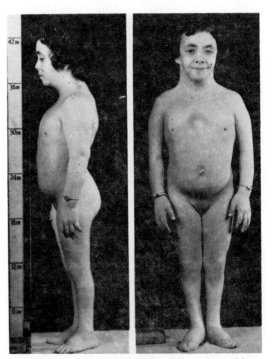

Fig. 4.10. Turner's syndrome. This patient with an XO chromosome makeup shows the typical abnormalities. The stature is short, the neck webbed, the hairline low over the neck, and sexual development minimal.

agnosis. The external genitalia are female but infantile. Pubic hair appears at puberty, but no other sexual maturation occurs. A webbed neck is characteristic. The webbing, composed of triangular folds of skin, extends from below the ears to the acromions. The hairline extends low on the neck. The chest is broad with the nipples widely separated and described as "shield-like." Abnormalities of the hand may include narrow nails with short fingers. The elbows may demonstrate a wide carrying angle called cubitus valgus. Less common findings include hypertension with or without coarctation of the aorta, edema of the hands and feet which may be unilateral, red and green color blindness, and intellectual impairment. The normal facial features of a patient with ovarian dysgenesis serve to distinguish her from a patient with pituitary dwarfism.

In *sexual precocity*, the surge of sex hormones characteristic of puberty occurs prematurely, and a growth spurt ensues. The adult result, however, is stunted growth because of premature closure of the epiphyses. The upper segment measurement is commonly slightly greater than the lower segment. Sexual precocity occurs more often in the female than in the male. The known causes vary, but most cases are idiopathic.

With respect to *hypothyroid dwarfism*, or *cretinism*, recall that thyroid hormone aids longitudinal growth and the maturation of features. Infantile deprivation of thyroid hormone stunts growth with the resultant persistence of infantile facies and upper-to-lower segment ratios. The bone epiphyses not only do not mature but become abnormal. Facial features are coarse and thickened. A protruded thick tongue, macroglossia, is seen. Physical and mental sluggishness, pale cool dry skin, puffy features, and a flat-bridged nose complete the blatant picture (Fig. 4.11). A less profound deficiency of thyroid hormone or later onset of deprivation can modify the picture so that the findings are quite subtle.

Achondroplastic or *chondrodystrophic dwarfs* are normal except for bone growth and development (Fig. 4.12). The disease is a primary disorder of connective tissue. Cartilage plates are blueprints for bone growth and bone modeling. In this disease, abnormal cartilage plates produce a recognizable altered appearance. The trunk, being less affected, is normal save for marked lumbar lordosis. This "swaybacked" posture makes the buttocks appear prominent. The arms and legs are shortened so that the upper segment is much greater than the lower segment. Joint mobility is impaired. The upper extremities resemble the forelegs of the dachshund dog, which, indeed, is the animal counterpart of this disorder. The head circumference is normal or increased with prominent frontal bossing. Sexual and mental development is normal.

Dysostosis multiplex, also called *gargoylism* or *Hurler's syndrome*, is comparatively rare. Again, the pathology is in bone cartilage, but, contrary to achondroplasia, the defect is most marked in the axial skeleton. The arms and legs are relatively long. The thorax is shortened, producing a kyphosis of the spine and pectus carinatum. The short neck makes the head appear to rest on the shoulders. The facial features are thick and grotesque. An abnormal substance, apparently responsible

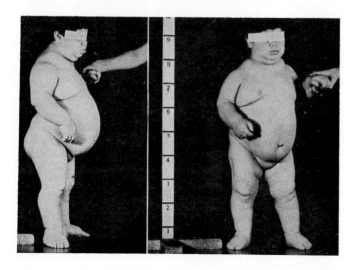

Fig. 4.11. Cretin. This patient is 24 years old. The obesity and thick coarse features are hallmarks of this disease. This patient had no detectable thyroid tissue.

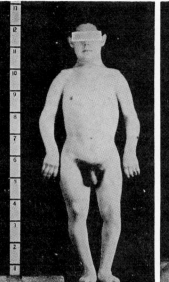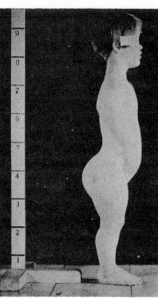

Fig. 4.12. Achondroplastic dwarfs. The patient on the left is 17 years old. Note the relatively short arms and legs for the trunk. The bow legs and cocked elbows are characteristic. On the right is a girl 8 years old with similar abnormalities. In addition, note the accentuated lumbar lordosis.

for the disease, accumulates in the reticuloendothelial structures, giving invariable enlargement of the liver and spleen.

Rickets, due to a deficiency of vitamin D, is rare. Some cases of rickets are due to resistance to the action of vitamin D and some are due to renal wasting of phosphorus. The pathologic abnormality is in the calcification of growing bone cartilage and osteoid matrix. The bone ends become soft and easily deformed. This abnormality in the adult is termed *osteomalacia*. The soft ends of ribs compressed against cartilage produce palpable lumps along the costochondral junctions—the "rachitic rosary." The lower ribs flare away from the chest producing "Harrison's groove" along the attachment of the diaphragm. The legs bow as a response to weight bearing on soft bones.

Progeria, a rare cause of short stature, presents such remarkable findings as to be easily diagnosed on physical features alone. Normal at birth, growth ceases within 3 years. Wasting ensues with thinning of the skin, loss of hair and subcutaneous tissue, and joint deformities. Early death from atherosclerosis is the rule. The patients look like tiny wizened old men. The cause is unknown.

Old age is routinely accompanied by a loss in height. The physical findings of aging deserve comment. The aging process is the most common disorder of connective tissue.

Reduced stature results from loss of bone matrix, osteoporosis, in the spine (Fig. 4.13).

The trunk shrinks more than the extremities. Kyphosis results because the loss of height of each vertebra is more marked anteriorly than posteriorly. The spinal column is then a stack of rhomboids.

Skin becomes thinned and loses its recoil. A generalized loss of elasticity is probably related to increased cross-linking of collagen fibers. This same loss of elasticity accounts for the difficulty experienced in visual accommodation—presbyopia. The pulse pressure (the difference between systolic and diastolic blood pressure) is increased because of a loss of elasticity in the major elastic arteries. These stiff pipes cannot accept systolic pressure as stretch to be returned as pressure in diastole.

Osteitis deformans, Paget's disease, is a disorder of bone which is most commonly diagnosed as an incidental finding on x-ray films. In its most blatant form, the physical findings are diagnostic (Fig. 4.14). Any bone or group of bones may be affected by this abnormality of remodeling. The axial skeleton and skull commonly are sites of involvement. The height may be reduced as the vertebrae become soft. The skull enlarges with prominent frontal bossing. The pelvis flattens, causing the legs to flare outward at the hips and inward at the feet. The bones are extremely vascular. A large dilated temporal artery, vascular hums over bone, and, rarely, high output congestive heart failure attest to this increased vascularity. Neuro-

Fig. 4.13. Osteoporosis. A normal vertebra on the left is compared with an osteoporotic vertebra on the right. The marked loss of substance is apparent. The difference in height multiplied by the number of vertebra accounts for the striking shortening which can be experienced by such patients.

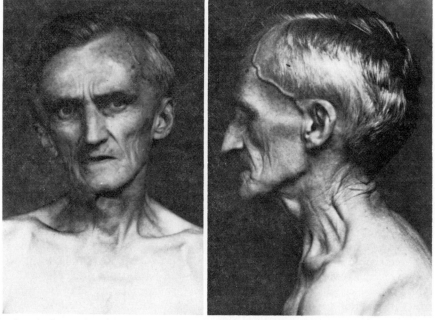

Fig. 4.14. Paget's disease. Frontal bossing presents this typical picture of Paget's disease involving the skull. On the lateral view, the enormously dilated temporal artery is obvious. This stands as mute testimony to the enormous blood supply required by this porous bone.

logic findings referable to the long spinal tracts indicate the serious complication of basilar impression. This complication results from the large soft calvaria literally sinking down onto the spine. Similarly, hearing loss and other cranial nerve deficits occur from boney encroachment on various foramina of the skull.

Hyperparathyroidism, when prolonged and severe, may result in a loss of height. The intense activity of the osteoclasts stimulated by the excess parathyroid hormone leads to loss of bone substance and a loss of structural support much as in the osteoporosis of the aged (Fig. 4.15).

Asymmetry

Abnormal fat, muscle, subcutaneous tissue, or bone may cause abnormalities of symmetry (Fig. 4.16). We are concerned not only with the comparison of the parts of a pair but also with the distribution of normal tissues.

Atrophy

When a part of the body shrinks, the term atrophy applies. An atrophic limb serves as a good example of some general principles. The loss of the vasculature or innervation of a limb will result in atrophy. If the limb is equal in length to its mate, the bone is probably uninvolved and the insult, therefore, occurred after full bone growth was attained. If the limb is shortened, the deficit occurred

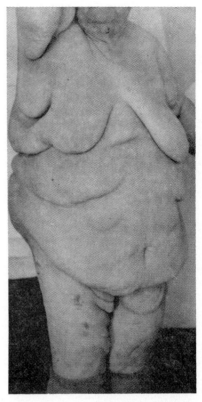

Fig. 4.16. Obesity. Fat distribution varies considerably from one patient to another. This woman was particularly marked by focal accumulations of adipose tissue. The enormous panniculus hangs like an apron.

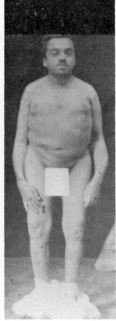

Fig. 4.15. Two photographs of Captain Martell. The good Captain Martell of the Merchant Marine is shown during his active years as a mariner and in the second photograph as he appeared after suffering the ravages of hyperparathyroidism. The profound effect on his bones caused the striking difference in the physiognomy which is apparent in these photographs. (Courtesy Endocrine Unit, Massachusetts General Hospital.)

in childhood sometime before longitudinal growth was complete. In such instances, congenital anomalies, birth trauma, and poliomyelitis are among the likely causes. Segmental atrophy of a limb may indicate segmental loss of a nerve; for instance, atrophy of the hypothenar eminence and lateral interosseous muscles indicates an ulnar nerve palsy.

Hypertrophy

Disparity in size may be the result of overgrowth of a member rather than undergrowth. Again, the elements involved are the components of the connective tissue system. There are few known causes of hypertrophy. Muscle hypertrophy is generally the result of excess use such as the hypertrophy of the arms of a paraplegic forced to use his arms to propel a wheelchair. If the vascular supply to a limb is abnormally increased as in an arteriovenous communication, hypertrophy may result.

Hemihypertrophy is an unusual but dra-

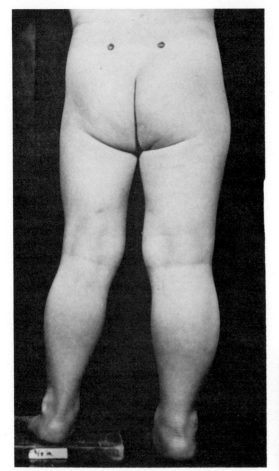

Fig. 4.17. Hemihypertrophy. The right side of this young patient is markedly larger than the left. The left foot is resting on a block which brings the sacroiliac notches into horizontal alignment.

matic example of hypertrophy leading to asymmetry (Fig. 4.17). A patient with hemihypertrophy will have enlargement of an entire side of the body and head. The growth disturbance involves all tissues and organs of that side. Subtle examples are detected by a limp and slight prominence of the involved side of the face. The importance of recognizing the syndrome is the association of various internal congenital anomalies and early neoplasms.

Polyostotic fibrous dysplasia, Albright's syndrome, is a condition of bone of unknown cause. Swelling and deformity of any involved bone, especially facial bones, gives the appearance of hypertrophy and asymmetry. The fibrous tissue which distorts these bones also makes them structurally weak, and de-

formities and asymmetry of the limbs results. Café au lait skin lesions and sexual precocity in females may be concomitant findings (Fig. 4.18).

Fat Distribution

In obesity, fat may accumulate in many different patterns. Abnormal fat distribution is found in several disease states.

Adiposis dolorosa is a primary disorder of fat tissue which accumulates in subcutaneous lumps which are painful. It is a rare disease of unknown cause.

Adrenal cortical hyperfunction, Cushing's syndrome, presents characteristic abnormalities of fat distribution. The trunk and face are obese with limb wasting. The fat in the face produces a rounded "moon face" appearance. Fat in the supraclavicular fossae and over the dorsal spine is responsible for the "buffalo hump." These findings, together with thin fragile skin, facial plethora, pigmented ab-

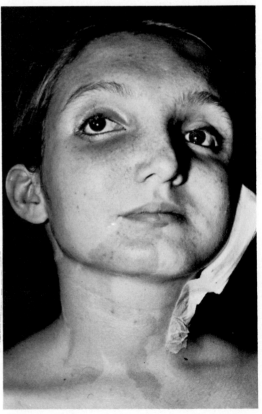

Fig. 4.18. Polyostotic fibrous dysplasia—Albright's syndrome. This young lady demonstrates the café au lait spots over her right chin and neck and the facial asymmetry of the fibrous dysplasia of the left zygomatic arch.

dominal striae, and hypertension, are almost diagnostic of the disease (Fig. 4.19).

Wasting, the loss of fat and muscle bulk, occurs whenever the caloric requirements exceed the daily consumption for a prolonged period. As mentioned previously, this may happen with any chronic debilitating illness.

The face dramatically shows the changes of wasting. Loss of tissue at the temples and cheeks makes the zygomatic arches appear prominent—the "gaunt look." Hollowing of the supraclavicular fossae and axillae occurs early. The skin falls in loose lax folds over the abdomen and upper arms. If the process includes protein depletion either through protein starvation or inability of the liver to synthesize protein, edema may be added to the picture. The edema falsely appears as fat ankles and legs. This pattern in children is called *kwashiorkor.* When the wasting has an identifiable cause, it is so designated as *cardiac cachexia* or *malignant cachexia*, for example. *Anorexia nervosa* is a peculiar disease of women, who for reasons which are unclear, stop eating (Fig. 4.20). In spite of diligent searches, no endocrine abnormality has been consistently found to explain the disease. The starvation may be profound and is occasionally fatal.

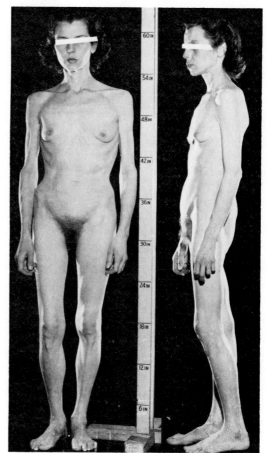

Fig. 4.20. Anorexia nervosa. The profound emaciation is apparent. Note that breast tissue and pubic hair are preserved.

Body Language

Body language is a form of nonverbal communication. We all recognize it, and we all practice it. Mannerisms, posturing, and gesturing comprise the physical aspects of body language. Vocal inflections are a form of double communication. The same statement with different inflections may take on very different meanings. Writers make liberal use of body language. Thus, we are told of an "icy stare," the "cold shoulder," the "nervous laugh," and the "sarcastic tone of voice." A few recognizable patterns merit comment since they are common and may serve as benchmarks of underlying psychodynamics.

Tension

We have all felt this particular emotion, and the physical findings of tension are common.

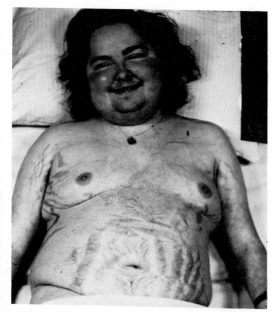

Fig. 4.19. Cushing's disease. The patient demonstrates hirsutism, moon facies, central obesity, and marked pigmented striae. (Courtesy Endocrine Unit, Massachusetts General Hospital.)

Emotional tension is frequently accompanied by physical tension. The pulse and respiratory rate may be increased. Speech is staccato, and there may be constant darting glances about the room. The patient is waiting for, looking for, and prepared for some external threat. The hands may be gripped tightly in the lap or may kneed a handkerchief or the arms of the chair. Others fidget constantly and move about. Chain smoking may be noted. Inappropriate humor or lack of humor commonly accompanies tension. This is where we see the "nervous laugh" or the lack of spontaneous smiles. If the sympathetic "fight or flight" discharge is great enough, axillary and palmar sweat pours forth. This, with skin vasoconstriction, is the genesis of the "clammy handshake."

Depression

Physical ailments are frequently depressing. Equally common, depression may be guised as physical discomforts. In either case, the recognition of depression influences therapy.

In contrast to tension, depression immobilizes the patient. Lack of spontaneity in words and gestures is very common. The patient may sit in the chair with his elbow on his knees and head down. He infrequently looks at the physician, preferring to stare at the floor.

Depression is not without visible emotion, however. Unheralded weeping, sharp outbursts of disgust, or deep sighs are common.

As the self-image is deprecated, attention to dress and personal hygiene declines. A previously meticulous individual may become sloppy and unkempt. He is saying in essence, "Why bother? It's just not worth it."

The depressed patient is difficult to satisfy and rebukes efforts to assuage his concerns. This may be frustrating to you. When you find yourself becoming angry at the patient, depression is frequently the root cause.

Seductive Patients

In the days of open wards, these patients were easy to identify at a glance. They were the most exposed, generally in filmy nightgowns or planned dishevelment of the bedclothes. The attitudes, gestures, and postures are designed to gain attention.

Seduction is not the sole province of women patients. Men may act seductive in less sexual ways. Both attempt to gain your allegiance, to gain your approval, and to have you care about them. They are frightened of being abandoned and act in any manner they feel will win your favor. Questions about your personal life and flattery are common verbal ploys.

These patients are generally immature and may be hysterical. They use sexuality or basic gratification mechanisms to hold your attention.

The proper approach to these patients involves imparting the impression that you will not abandon them while still maintaining a professional distance.

Anger and Hostility

Sometimes illness breeds anger. The "Why me?" question may mean "This isn't fair." The anger and hostility may be multidirected, and the physician is a good target. Pathologically, expressed anger and hostility are indicative of paranoia.

The demeanor is one of suspicion. As with the depressed patient, the angry patient may be relatively immobile. Unlike the depressed patient, the eyes may be staring hard at the doctor or be suspiciously directed about the examining room. Humor is absent. The angry patient requests lengthy explanations about parts of the physical examination, about your training and background, etc. The usual social amenities of greetings, handshakes, and the like are ignored or grudgingly granted.

Lest you become angry with the angry patient, remember that his anger is global and not really personal.

Supplicant

Somewhat akin to the seductive patient is the supplicant patient. This patient tells you by word and gesture that he is totally in your hands. He abandons any responsibility for his care in an attempt to abandon responsibility for his illness.

Gestures are placid and yielding. In the office, this patient typically slouches in the chair with his head and shoulders back, arms to the sides and legs open.

Patients Who Touch

A doctor's touch, the dramatized laying on of the hands, is usually gratefully expected by patients. Less expected by the doctor is the patient who touches him, yet this is quite common and carries different meanings.

The patient may mean only to emphasize his point by tapping your knee during the interview. He may wish physical assurance that you are there and you are attentive. It may be a seductive gesture, a "please don't leave me" touch. Sometimes it is to emphasize the confidential nature of the spoken word.

Note of Caution

Body language, while frequently helpful in understanding a patient, constitutes "soft data." Cultural background influences gestures and posturing. These have been exaggerated as the "emotional Italian," the "flirtatious French," the "hotheaded Irishman," and so on. Nor are any of the responses alluded to uniform in their significance. Body language is a piece of the physical examination—a piece interpreted as part of the whole.

What's in a Face?

You may be able to suggest a great deal from a careful examination of a patient's face. The drawing (Fig. 4.21) contains many clues. These are no more than hints but are worth pursuing in your history and physical.

The man is or was a heavy smoker. The cobblestone appearance of the skin of the neck and the furrow from the corner of the eye over the zygomatic arch are usually reliable indicators. They may be seen in just a weathered face from long outdoor activity.

In spite of the fact that he is a heavy smoker, he will probably not be a candidate for lung cancer. This affliction is apparently rare in men with congenital baldness. I would

Fig. 4.21. What's in a face? How many diagnostic clues can you find? See text.

appreciate hearing from anyone who has seen this combination.

He is a good candidate for coronary artery disease. A controversial sign is present in the drawing. The horizontal ear crease is seen with unusual regularity in patients with coronary artery disease although some investigators dispute its specificity.

He may have a history of peptic ulcer disease. The verticle furrows on each side of the bridge of the nose are seen frequently in men with this disorder.

Ears in excess of 12 cm in verticle dimension suggest pernicious anemia.

chapter 5

VITAL SIGNS

.

Blood Pressure

The arterial blood pressure is the lateral pressure or force exerted by the blood on the blood vessel wall. This pressure constantly changes throughout the cardiac cycle. The highest pressure occurs during the cardiac ejection and is called the *systolic pressure*. The lowest point in the cycle is termed the *diastolic pressure*. The numerical difference between systolic and diastolic pressures is the *pulse pressure*.

The several factors determining the level of the blood pressure include the cardiac output, peripheral vascular resistance, total blood volume, viscosity of the blood, and the compliance of the arterial wall. Of these, the cardiac output and the vascular resistance most profoundly affect the blood pressure.

The most accurate method of measuring blood pressure is by recording it directly from an intra-arterial needle. In clinical practice, we use the indirect method.

Blood Pressure Measuring Equipment

The sphygmomanometer consists of an inflatable bag enclosed in an unyielding cuff, an inflating bulb, a manometer from which the pressure is read, and a controlled exhaust to deflate the system.

The inflatable bag generates the pressure and the unyielding cuff ensures that the pressure is directed into the arm and not outward. The cuff must be a proper width to ensure accurate determinations. The relationship between the width of the cuff and the diameter of the limb describes a cone of pressure into the tissue. If the cone does not reach the artery, a falsely high pressure must be exerted and a falsely high reading results. If the cuff is too large the cones of pressure will overlap the artery and a falsely low pressure will be recorded (Fig. 5.1). A small cuff should be used on a small arm and a large cuff on an obese arm. A special large cuff is available for blood pressure determinations in the leg.

Manometers are of two types—the mercury and the aneroid. The *mercury gravity manometer* consists of a glass tube connected to a mercury-filled reservoir. The pressure within the cuff forces the mercury up the tube. The mercury falls with gravity and, since gravity is a constant, repeated calibration is unnecessary. The mercury gravity manometer is the most reliable. The major disadvantage is the bulk. The aneroid manometer has a metal bellows which receives the pressure from the cuff. With increasing pressure, the bellows expands, and this in turn rotates a gear that turns a needle across a calibrated dial. This type of system is easily disturbed and requires periodic calibration against a mercury gravity system (Fig. 5.2).

Taking Blood Pressure

Any artery which can be encircled proximally and palpated distally may be used. The brachial artery, because of its convenience, is most commonly used.

With the patient sitting or lying, the arm is positioned so that the brachial artery is at heart level. Abduct, externally rotate, and slightly flex the arm. Wrap the deflated cuff snugly about the arm so that the lower edge of the cuff is about 1 inch above the antecubital fossa.

Palpate the blood pressure first by feeling the pulse in the radial artery and inflating the cuff until the pulse disappears. Slowly deflate the cuff and note the point at which the pulse returns—this is the systolic pressure. The reason we palpate the pressure first is to avoid missing an *auscultatory gap*. The auscultatory gap exists in many patients and consists of a period of silence between the systolic and diastolic pressures. Were we only to auscultate the blood pressure, we might find ourselves starting in the middle of such a gap

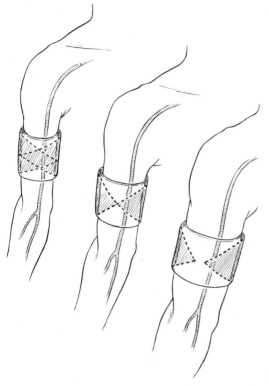

Fig. 5.1. Relationship between arm size and blood pressure. With a small arm and wide cuff, a falsely low value will be recorded because of the overlapping cones of pressure. Conversely, a fat arm with a narrow cuff gives a falsely high reading.

in Phase 3. In Phase 4, the sound is abruptly muffled, yielding a soft blowing sound. Phase 5 is the point at which sound completely disappears. This point is usually regarded as the diastolic pressure. In some patients, there is a very distinct change in sounds at the start of Phase 4 and then many millimeters of pressure before sound disappears. All three numbers are then recorded; for instance, 130/80/10 for the first, fourth, and fifth phases.

Variations in Blood Pressure

The blood pressure determination of a given patient may vary from examination to examination. The nervous patient tends to have a slightly higher blood pressure when first recorded, with lower readings when he is more relaxed. Variations from beat to beat and variations between different arteries in the same paitent may also be encountered.

If the cardiac output varies from beat to

and obtain a falsely low systolic reading. An auscultatory gap, other than being a source of observer error, has no clinical significance.

After palpating the systolic pressure, the bag is again inflated, and the stethoscope is applied lightly over the artery. There are two circumstances when no sound will be heard—when there is no flow through the vessel and when the flow is smooth and laminar. Between these two extremes, turbulence develops and vibrations are set up in the vessel wall. As the cuff is slowly deflated, these vibrations, the *Korotkoff sounds*, emerge (Fig. 5.3). The Korotkoff sounds are divided into five phases. Phase 1 begins with the first audible sound, the systolic pressure. In Phase 1, the systolic pressure is just enough to open the vessel momentarily and produce crisp tapping sounds, which increase in intensity. As the cuff pressure is further decreased, flow through the vessel increases, causing a swishing sound characteristic of Phase 2. The sounds then again become louder and crisper

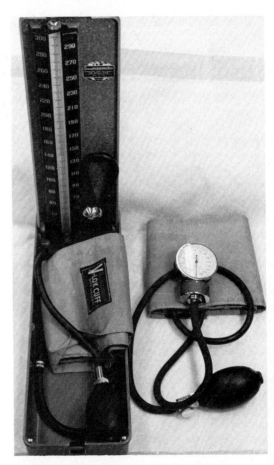

Fig. 5.2. Sphygmomanometers. On the *left* is a mercury manometer and on the *right* an aneroid manometer.

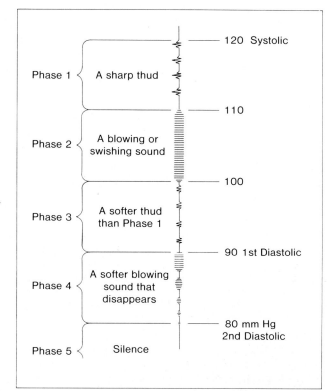

Fig. 5.3.   Korotkoff sounds.

Phase 1	A sharp thud	120 Systolic
Phase 2	A blowing or swishing sound	110
Phase 3	A softer thud than Phase 1	100
Phase 4	A softer blowing sound that disappears	90 1st Diastolic
Phase 5	Silence	80 mm Hg 2nd Diastolic

beat, so too will the blood pressure vary. When the heart beats irregularly, as in atrial fibrillation, the systolic blood pressure may span 20–30 mm Hg. During this type of arrhythmia, the ventricular filling period changes. When diastole is short, the ventricular volume is small, the subsequent beat ejects less blood, and the pressure generated is low.

When the pulse is irregular, you must deflate the cuff quite slowly and record the pressure at which the first sound is heard as the systolic pressure even though at that level the sound may be quite intermittent.

Pulsus paradoxus is a normal phenomenon of variation of the blood pressure with respiration. During deep inspiration, the systolic pressure falls about 5 mm Hg and rises again during expiration. During inspiration, the negative intrathoracic pressure and expanding lungs tend to flood the pulmonary vasculature and hold the blood in the lesser circuit, thus producing a slight fall in cardiac output. In certain diseases of the pericardium, this normal fall is greatly accentuated. When the pericardium restricts the heart either by direct scarring or the accumulation of pericardial fluid (pericardial tamponade), patho-

logic pulsus parodoxus results (Fig. 5.4). The suggested mechanism here relates to additional tensing of the pericardium as the diaphragm descends during inspiration. The interference may be so profound as to obliterate the pulse completely during deep inspiration. Normal pulsus pasadoxus is detected only by very slowly deflating the cuff and simultaneously observing the respirations. A difference of greater than 10 mm Hg between the pressure at end inspiration and that at end expiration is pathologic.

Variations in the blood pressure may occur when the patient assumes different positions. When the blood volume is low, as in severe blood loss or dehydration, the blood pressure falls when the patient stands up. Normally, when we stand up from a sitting or lying position, a reflex increase in arterial tone allows us to maintain a normal blood pressure. If that tone is already maximal because of a diminished vascular volume, standing adds the intolerable pull of gravity and the blood pressure falls—sometimes to unobtainable levels. This *orthostatic hypotension* together with a rise in the pulse rate is a good diagnostic sign of hypovolemia. When the blood pressure is low *and* the pulse rate is

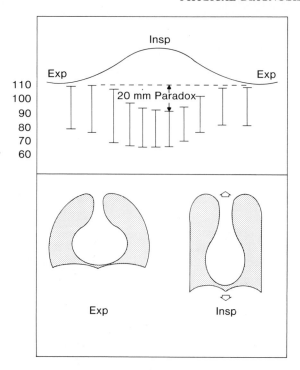

Fig. 5.4. Pulsus paradoxus, Kussmaul's sign. With a pericardial effusion, the filling of the right ventricle is impeded during inspiration and ejection from the left ventricle lessened. The blood pressure falls correspondingly. Note, too, that the pulse rate increases with inspiration. Exp, expiration; Insp, inspiration.

slow, the etiology is likely to be an increase in *vagal tone*. This increase is real in *vasovagal syncope* and relative in primary autonomic system diseases in which the sympathetic tone is pathologically decreased leaving the parasympathetic tone unopposed.

Variations in blood pressure may be detected in different arteries. A normal variation of blood pressure readings is frequently found between the arms. This variation should be no more than 5–10 mm Hg. Blood pressure readings in the legs requires a larger cuff. Wrapping the cuff about the thigh, the blood pressure is auscultated over the popliteal artery. In the leg, the sytolic pressure is normally 10–40 mm Hg higher than in the arm, but the diastolic pressure is the same. Any obstructive disease of the aorta or iliac or femoral vessels may reduce the pressure detected in the legs. The blood pressure should be recorded in the legs in every patient with hypertension—*coarctation of the aorta is a curable form of hypertension.* Similarly, obliterative atherosclerotic disease, extrinsic arterial compression, and dissection of the aorta may lead to disparity in pressure recordings between limbs.

PRACTICE SESSION

Have your partner sit comfortably in a chair with a side table or rest. Apply the cuff snuggly around the upper arm. Take the systolic pressure by inflating the cuff to a pressure of 200 mm Hg and slowly deflating it while feeling the radial pulse. Note the pressure when you can first feel the pulse.

Now, deflate the cuff all the way. Apply the stethoscope over the brachial artery after locating it by palpation. Again inflate the cuff and slowly deflate it and see if you can detect the various stages of sounds.

Now, again, inflate the cuff but only to 5 mm higher than the systolic pressure. Have your partner breath in and out slowly and deeply. *Very* slowly deflate the cuff and look for the paradox. You will first hear sounds only when your partner exhales. As the pressure comes down you will hear more and more individual beats until you hear sound throughout the respiratory cycle. The difference in pressure from the first beat to the point at which you hear each beat is the paradox and should be no more than 10 mm.

Now, check the pressure in the opposite arm. A slight difference is common.

Now, use a large cuff and place it around the thigh just above the knee. Palpate either the popliteal or dorsalis pedis pulse, inflate the cuff and deflate it until you feel the pulse return. You may also try to auscultate it in the popliteal fossa. The systolic pressure should be about 10 mm higher.

Now, check the blood pressure in the supine position, immediately after standing and again in 3 minutes and observe the changes.

Special Situations and Uses of Sphygmomanometer

Carpopedal spasm results from hypocalcemia—either real as in hypoparathyroidism or relative as in the alkalosis of hyperventilation. When hypocalcemia is suspected, Trousseau's sign may be elicited. By inflating the cuff above systolic pressure for 3 minutes, the characteristic hand contraction of carpopedal spasm develops—the so-called *main d'accoucheur* or obstetrician's hand (Fig. 5.5).

Vascular purpura results from a weakness in the walls of precapillary arterioles which allows the extravasation of red cells into the skin. The weakness is accentuated by an increase in the pressure in these arterioles. This is why the petechiae or purpura appear in the dependent portions of the body—the lower limbs and the feet. The pressure can be artificially elevated by the Rumple-Leede test. In this maneuver the blood pressure cuff is inflated to a point between systolic and diastolic pressures for 8 minutes. Venous congestion results with a backup of pressure on the arterioles. If the arterioles are weak or damaged, showers of petechiae will appear on the forearm, and the test is positive.

Blood pressure measurements in infants and children require special attention. A small cuff is required and the child should be quiet and resting. Vigorous crying can elevate blood pressure as much as 50 mm Hg. It may be necessary to apply the cuff loosely and return later if the child is crying. The sounds are usually inaudible in infants under a year of age. The *flush method* is then appropriate. Elevate the arm or leg and wrap the limb from the distal end to the proximal end with an elastic bandage to drain out the blood. With the limb elevated and wrapped, the cuff is inflated to 200 mm Hg. Next, remove the wrap and slowly deflate the cuff until a flush of color appears in the skin. The pressure at which the flush is seen is the *mean* blood pressure.

Pulse

A wealth of clinical information accrues from the examination of the arterial pulse. An Egyptian papyrus dating to the 17th century BC contains the first recorded association of the pulse with heart action. The Chinese acupuncture physicians have for centuries relied on the pulse for total diagnosis. Athough this exalts it value, there is much to be learned through careful assessment of the pulse. Many factors contribute to the formation of the pulse, and it is not always easy to dissect one factor from another. Before reviewing specific abnormalities of the pulse, we should consider what it is that makes the pulse.

Determinants of Pulse

The pulse is initiated with the opening of the aortic valve. The pressure in the aorta rises rapidly as the bolus of blood distends its walls. As the ejection slows and blood flows away into the peripheral circuit, the pressure in the aorta falls. A momentary reversal of

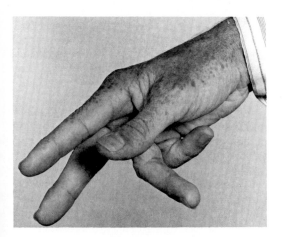

Fig. 5.5.   Carpopedal spasm. This is the position that a hand will assume with a positive Trousseau's sign in hypocalcemia.

flow closes the aortic valve and a subsequent
rebound inward of the aortic wall inscribes
the dicrotic notch. This notch is usually not
palpable.

There is a difference between the flow of
blood and the flow of the pressure wave.
Blood travels approximately 0.5 m/second
while the pressure front travels much faster
at about 5.0 m/second. Much like a freight
train starting up, the shock wave moves much
faster than the train itself. The pulse we feel
is really the pressure front rather than the
actual movement of blood. The pulse wave
arrives at the carotids 30 ms, the brachials 60
ms, the radials 80 ms, and the femorals 75 ms
after electrical activation of ventricular systole.

The contour of the pulse changes as the
pressure front moves away from the heart.
The pulse pressure increases because of an
increase in systolic pressure and, less so, because of a decrease in diastolic pressure. In
addition, the ascending limb of the pulse
contour becomes sharper and the dicrotic
notch occurs later and is more rounded.

Technique of Taking Pulse

The value of feeling the pulse is enhanced
if more than one artery is palpated. The
carotid and brachial arteries most accurately
reflect the wave form. These arteries are
closer to the aorta, and the wave form is
consequently less distorted. The radial artery
is convenient for rate determination. Other
vessels should be checked to determine the
patency of the arterial tree.

Clincians vary in their preference for the
palpating finger or fingers. The thumb is used
by some and one or more digits by others. I
find the first two or three fingers to be most
sensitive. By simultaneously palpating two or
three slightly separate points on the same
artery, analysis of the wave form is enhanced.

The pulps of the fingers apply increasing
pressure on the artery until the maximum
pulsation is felt. The muscles around the
artery should be relaxed. When feeling the
brachial artery, support the arm and flex the
elbow slightly (Fig. 5.6). The carotid artery is
best felt with the head in the neutral position,
slightly extended. *Palpate the carotid artery
with care.* Vigorous palpation of the carotid
excites the carotid sinus and reflexly slows
the heart, occasionally with disastrous results.
We will discuss the location of other arteries

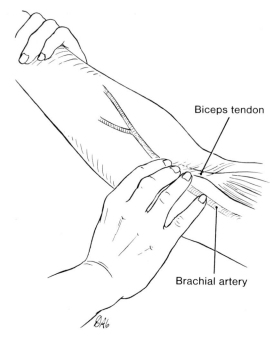

Fig. 5.6.   Palpating the brachial artery. The artery is
located just medial to the *biceps tendon.*

and the techniques of palpating them in a
later chapter.

When taking the pulse, note the *rate,
rhythm, volume, and contour.*

Rate and Rhythm

The pulse is best counted for a full minute.
Usually we shortcut by counting for 15 seconds and multiply by 4. We probably miss
much for the saving of 45 seconds. Begin the
count at 0. If the first beat of a 15-second
count is numbered 1, the pulse will be over-
estimated by four. Minor variations in
rhythm are common. The pulse increases
slightly with inspiration. Young patients may
demonstrate *sinus arrhythmia*—a mild variation in rate of no clinical importance.

When the pulse is very irregular, further
characterization is in order. The rhythm is
called *regularly irregular* if a pattern to the
irregularity is apparent. When no pattern
emerges, the rhythm is *irregularly irregular.*
When the pulse is irregular, count the rate by
auscultating the heart and by feeling the pulse
at the wrist. Heartbeats which occur very
close to the preceding beat, premature beats,
eject very little blood since the diastolic filling
period was so brief. Valve closure occurs, but

no pulse is generated. Hence, the apical pulse rate exceeds the peripheral pulse rate. The difference between the two rates is the *pulse deficit*. Premature heart beats are detected at the radial artery either as very weak early beats or *dropped beats*.

Bradycardia means slow heart rate. No clear lower limit of normal exists for pulse rate. Very slow rates may be encountered in the resting well trained athlete. If the normal atrial pacemaker of the heart is ineffective, a lower slower pacer takes over, as in complete heart block, which commonly yields rates of 40–50 a minute. Increased intracranial pressure reflexly slows the heart, sometimes to very slow rates. Bradycardia secondary to increased intracranial pressure is an ominous sign. An increase in vagal tone slows the heart rate and reduces the blood pressure as we noted in the section on blood pressure. Reduced metabolic demands produce slow heart rates, as in hypothyroidism and hypothermia.

A wide variety of conditions cause *tachycardia*. Broad categories include increased metabolic demand for oxygen, heart disease, increased circulating catecholamines, and decreased blood volume.

The most common increase in demand for oxygen occurs with exercise. Fever also results in tachycardia, and a useful rule of thumb is that an increase of 10 beats/minute can be expected for every degree increase in temperature. Thus, a temperature of 38.3°C (101°F) may generate a pulse of about 105. A much faster rate suggests an additional disease. A relatively slow pulse with a fever is seen in some enteric infections in which there is an increase in vagal tone.

Increased circulating catecholamines occur physiologically in fright and pathologically in pheochromocytomas. Many ingested substances similarly stimulate the heart rate— coffee, cigarettes, and a wide variety of drugs.

Any disease which reduces the ability of the heart to pump blood results in a compensatory tachycardia. *Tachycardia* is the first sign of congestive heart failure. Coronary artery disease, pericardial effusion, constrictive pericarditis, primary heart muscle disease, and valvular disease are common examples.

By a similar mechanism, diminished blood volume results in tachycardia. Hypovolemic shock is characterized by a low blood pressure and a rapid thready pulse. Mild degrees of hypovolemia, as in dehydration, can be detected by noting an increase in pulse rate and a fall in blood pressure when the patient assumes an upright posture.

Pulse Volume and Contour

When palpating the pulse, your fingertips detect the change in pressure from diastolic to systolic levels. Although this pulse pressure is accurately recorded by the sphygmomanometer, palpation enhances the value by allowing you to analyze the wave form.

Stroke output of the heart and peripheral vascular resistance determine the blood pressure, and alterations of either of these effect the pulse volume. *An abnormal volume is caused from too little or too much stroke output or resistance.* The combination of the pulse rate, volume, and contour usually makes it possible to say which of these disturbed mechanisms is operative.

A small volume pulsation indicates either too much resistance to flow or too little blood ejected by the heart. Increased resistance may occur at the heart, as in pericardial effusion or restriction and stenosis of the mitral valve. In these circumstances, the resistance acts to impede the stroke output, and a compensatory tachycardia will also appear. If the increased resistance is due to a stenotic aortic valve, the stroke volume may be normal but the ejection time is prolonged: *pulsus parvus et tardus*—weak and slow pulse. A localized diminution of the pulse volume indicates a local increase in resistance, most commonly partial atheromatous obstruction.

A primary reduction in stroke output likewise yields a small pulse volume. This is the hallmark of heart failure in which the pump cannot eject the full complement of blood required. Up to a point, the vascular resistance will fall to compensate but beyond that point a small pulse volume and tachycardia ensue. Blood loss and dehydration behave similarly.

Do not equate a small pulse volume with hypotension. A small pulse volume only means a small pulse pressure. A blood pressure of 160/120 will give a small pulse volume.

A large pulse volume indicates too much volume or too little resistance. Profound bradycardia of any cause calls forth a large stroke output and, thus, a large pulse volume. The stroke volume increases and the resistance falls in fever, anxiety, anemia, pregnancy, and hyperthyroidism.

Aortic valve incompetence demands an increased stroke output since a portion of the ejected volume falls back into the left ventricle. With profound aortic insufficiency, a bounding "water hammer" pulse is characteristic. An arteriovenous fistula lowers resistance by short-circuiting blood away from the small arterioles which produce the greatest resistance. Severe atherosclerosis of the aorta produces a large pulse volume. The decreased distensibility of the aorta prohibits the translation of pressure into stretch, and the systolic pressure is therefore greater. In diastole, there is no rebound contraction of the walls of the major vessels so that the diastolic level is less.These combine to give a wide pulse pressure which is further enhanced by a reduced resistance in the arteriolar bed.

Variations in pulse volume from beat to beat deserve mention. We have previously discussed arrhythmias and pulsus paradoxus as causes of variation in the pulse. In the former, variations in the cardiac output are reflected in variations in the pulse, and in the latter the variation occurs with respiration. *Pulsus alternans* is the alternation of a weak pulsation with a strong one. Although the mechanism is not understood, the sign indicates heart failure and is ominous.

Low frequency vibrations within a vessel wall are sometimes palpable. Palpable vibrations are called *thrills*. Thrills indicate a turbulent flow of blood in the underlying vessel. The turbulence may originate in the vessel being palpated or may be transmitted from a point proximal to the palpating finger. A thrill which is audible with the stethoscope is called a *bruit*. The Korotkoff sounds produced by the sphygmomanometer represent turbulence and are really bruits. Pathologically, thrills and bruits may be produced by any phenomenon which disturbs the normal laminar flow of blood.

The most profound alterations of the pulse contour result from diseases of the heart valves, and we shall consider them in detail in the section with the heart.

PRACTICE SESSION

Palpate your partner's brachial and carotid pulses and note any difference in character. See if you can detect the time difference between the carotid and dorsalis pedis, assuming you have a long reach or your partner is short.

Now, have your partner exercise briskly for a few minutes and count the pulse immediately after exercise and again 3 minutes later when it should have returned to the resting rate.

Temperature

Fever, the elevation of body temperature due to disease, is one of the oldest clinical signs. So many things influence body temperature that it is a tribute to homoeostasis that normal body temperature is kept within such narrow limits. The temperature of the human organism represents the balance between heat generated and heat lost. The temperature regulation center of the hypothalamus determines the set point. At temperatures above the set point, mechanisms for heat loss predominate—peripheral vasodilatation, sweating, and hyperventilation. If the temperature falls below the set point, heat generation is enacted—increase in the metabolic rate, tensing of the muscles, and shivering. *Abnormalities of temperature result from ab-errations of heat production, heat loss, or changes in the hypothalamic thermostat.*

Normal Temperature

Body temperature is highest at the core and falls off toward the skin. In clinical practice, we do not measure core temperature but accept instead reasonably reproducible rectal or oral measurements. In addition to graduations in temperature from the core outwards, there are variations in the temperature throughout the day. A diurnal variation occurs, with the highest temperature between 8 and 11 o'clock at night and lowest levels between 4 and 6 o'clock in the morning.

Oral temperature recording is convenient but subject to error. Normal oral temperature is $36.8 \pm 0.3°C$ ($98.3 \pm 0.5°F$). False elevations may be recorded just after smoking or drinking hot liquids. Falsely low readings may occur after drinking cool liquids or in patients who breath through their mouths. *Rectal temperature* recording is less convenient but also less subject to error. Normal rectal temperature is 37.2 ± 0.3°C (99.0 ± 0.5°F). Occasionally, it is necessary to take *axillary temperatures* which are about 0.6°C (1°F) lower than oral temperatures. Recent evidence indicates that the most reliable temperature recordings are obtained through a small thermocouple placed against the tympanic membrane. Fortunately, this degree of accuracy is only rarely required.

Fever

The physical diagnosis of fever involves more than merely recording its presence. The clinician must ask whether the cause is excess heat production, diminished heat loss, or change in the central thermostat.

Any form of hypermetabolic state may generate a fever through increased heat production. A high ambient temperature, excessive exercise, and hyperthyroidism are a few examples.

Interference with heat loss most often is interference with sweating. Many drugs with atropine-like effects are capable of producing fever. A few patients have primary neurologic defects with diminished sweating ability. Other patients with congenital absence of the sweat glands are at great risk of hyperpyrexia. Probably the most common sweating defect is dehydration. The patient with fever should always be checked for the presence of axillary sweat. A combination of diminished sweating and hypermetabolism is seen in *heat stroke*. The most common circumstance is a high ambient temperature and dehydration, the culmination of which is an extremely high fever.

A change in the central thermostat is the most common cause of fever. Infections and inflammatory diseases somehow render the hypothalamic temperature center less sensitive to heat. A characteristic sequence of events ensues. First, the patient feels an intense chill or *rigor* and may wrap himself in blankets. At this point the temperature is normal, but the thermostat calls for an elevation and shivering; the mechanism for heat generation is called on. Once the temperature rises to the new high level dictated by the

hypothalamus, shivering ceases, and the patient has the sensation of "feverishness" but is relatively comfortable. Once the pyrogen or cause for the fever is removed, the hypothalamus becomes resensitized or reset to normal temperature requirements. A profound exhausting sweat ensues as a mechanism for heat loss until the body temperature falls again to normal levels. This explains why the temperature taken at the onset of a chill is normal. Early clinicians recognized the onset of sweating as the sign that the fever was lysing. This was the "crisis passing" (Fig. 5.7).

The early clinicians, unfortunately, had little in their armamentarium to deal with fever. Much of their effort while waiting for the crisis was in documenting the type of fever. Great diagnostic significance was attached to the pattern of fever. Currently, it is rare for a febrile illness to run its course without some intervention. Nonetheless, a few patterns of fever are characteristic.

Intermittent fever is one in which periods of fever are interspersed with periods of normal temperature. Malaria is the prototype. The duration of the fever-free interval may indicate the type of falciparum responsible. A remittent fever pattern is characterized by temperature variations of at least −16.7°C (2°F) a day but never returning to normal. When the oscillations are particularly wide, with recurrent sweats and chills, the term *"hectic"* or *"septic" fever* is applied. This usually indicates the presence of an abscess or other infection which intermittently breaks into the blood stream.

Fever blisters are the vesicles of a herpes simplex virus, usually about the mouth and nose. Presumably, the temperature elevation activates a latent infection, causing it to erupt and become symptomatic.

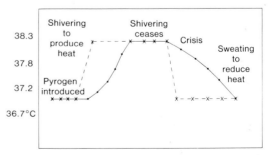

Fig. 5.7. Mechanism of chills, fever, and sweat. ●, Recorded temperature; ×, setting of hypothalamic thermostat.

Hypothermia

Temperatures below normal usually result from deficiencies in heat production or abnormalities of the thermostat. Rarely, increased heat loss, as in prolonged cold exposure, may result in hypothermia. Even in cold exposure, however, there is frequently a central depressant such as alcohol or barbiturates to explain hypothermia. The body's efforts to generate heat are normally so strong that cold exposure usually causes death through exhaustion from shivering rather than through hypothermia.

Reduced heat production occurs with hypometabolic states, such as profound myxedema or malnutrition. The physical characteristics are readily recognized. Central nervous system depression by primary brain disease, drugs, or toxins may alter the thermostat and so produce hypothermia.

Respiration

The respiratory rate is the number of inspirations per minute. Because the respiratory rate is much slower and less regular than the pulse, it should be counted for a full minute to reduce the error. As with the pulse, we are interested in more than just the rate of respiration.

Rate and Volume

The normal adult resting respiratory rate is 14–18/minute. A rapid rate is termed *tachypnea* and a slow rate is called *bradypnea.* These are two ways of increasing the amount of air exchanged. One is to increase the rate (*tachypnea*), and the other is to increase the volume (*hyperpnea*) of each breath. Both rate and depth may increase when there is an increased demand for oxygen. Recall the last time you raced up four flights of stairs. The physiologic demand may be for the elimination of carbon dioxide rather than for the supply of oxygen. In states of profound metabolic acidosis, the lungs are called on to compensate by blowing off carbon dioxide. The prototype of this very deep, rapid respiration occurs in diabetic ketoacidosis. Hyperventilation of this sort is called *Kussmaul respiration.* Acute anxiety attacks may be associated with hyperventilation which is not physiologic. The resulting lowering of the arterial carbon dioxide and subsequent alkalosis produces perioral and acral tingling. When the hyperventilation and alkalosis are profound, carpopedal spasm results.

If the volume of each breath is restricted by disease, a rapid rate results in an attempt to maintain a normal total volume exchange. The combination of rapid and shallow breathing is commonly seen in chronic lung disease, especially fibrosing diseases of the lungs. The volume may be so little that the patient frequently interrupts his sentences to take a breath.

Sighing is a normal mechanism to fully ventilate both lungs periodically. These single deep breaths prevent atelectais or collapse of poorly ventilated areas of the lung. Normally, we sigh once every several minutes. Constant repetitive sighing is a manifestation of depression.

Bradypnea, a slow respiratory rate, may be normal if the volume of each breath is great, as might occur in the experienced runner. When bradypnea coexists with shallow respirations, the term *hypoventilation* applies. Hypoventilation almost always originates from central nervous system disease, either primary disease of the brain or metabolic insults. A very slow respiratory rate is common with increased intracranial pressure.

Hippocratic respiration is a peculiar form of profound hypoventilation. Described as "fish-mouthing," this form of respiration indicates that death is imminent. The rate is usually less than 10. With each breath, the head lurches backward and the jaw drops open, much as a fish out of water. Virtually no air is exchanged.

Hypoventilation may occur in very obese patients. Presumably, the effort of breathing is just too great because of their massive bulk. Rather than increase their rate of breathing, these patients adapt to chronic hypoxia and hypercapnia by leading extremely sedentary lives. These patients have been likened to the fat boy of the Pickwick papers–hence the name *Pickwickian syndrome.*

Periodic Breathing

From time to time in clinical practice, you will see periodic or phasic respirations. *Cheyne-Stokes* respiration is the most common type. In this pattern, periods of apnea alternate with periods of hyperventilation. From the period of apnea, respirations slowly become progressively more rapid and deep than taper off again to another period of apnea. The cycle length varies but is usually on the order of 30 seconds to 2 minutes with 5–30 seconds of apnea.

Biot's respiration is less frequently encountered. Here, periods of apnea alternate abruptly with periods of constant rate and depth of breathing. In both Cheyne-Stokes and Biot's respiration, the fault lies in the feedback system controlling respiration. A sluggish response to changes in oxygen and carbon dioxide allows for wide oscillations between underbreathing and overbreathing. Virtually any central nervous system or cardiovascular disease may be operative (Fig. 5.8).

Effort of Breathing

In addition to assessing the rate and depth of respiration, the clinician notes the effort of breathing. The work required reflects the efficiency of the system.

Accessory mucles are called into play when additional work is required. Normally, the intercostal muscles raise the ribs. This, together with the falling diaphragm, expands the volume of the thorax and creates the small negative pressure within the pleural space necessary to draw air into the lungs. In emphysema, the lungs are chronically hyperexpanded. The lower ribs remain elevated and the diaphragm remains depressed. The accessory neck muscles are then used to raise the upper ribs to generate the negative inspiratory pressure. This is readily seen at the bedside.

Abdominal muscles provide additional power for expiration. Contraction of the abdominal muscles increases the intra-abdominal pressure and forces the diaphragm up. Use of these muscles indicates obstruction to the outflow of air as is seen in asthma.

Retraction of the intercostal muscles is the inward displacement of these muscles during inspiration. *Retraction always means that there is decreased compliance of the lungs.* Stiff lungs are hard to inflate. A large negative intrapleural pressure must be generated before the lungs will inflate. This large negative intrapleural pressure draws the intercostal muscles inward—retraction occurs.

Effectiveness of Gas Exchange

Now that we have looked at the rate, depth, and work of breathing, we ask "How well is the system working?" There are indicators of the supply of oxygen and the elimination of carbon dioxide.

Cyanosis is a crude index of oxgenation—crude because the factors causing cyanosis are not solely related to the lungs' ability to provide oxygen. Cyanosis occurs when the patient has 5 g% or more of deoxygenated blood. At the extremes of hemoglobin levels, cyanosis may be misleading. The patient who is profoundly anemic may have insufficient oxygen but not be cyanotic. Conversely, the patient with too much hemoglobin may have sufficient oxygen but also have 5 g% of deoxygenated blood and be cyanotic. The rate of blood flow through the skin and the skin pigmentation affect our ability to detect cyanosis. Generally, cyanosis appears first in the fingers, toes, and lips—those areas where there is a rich capillary network and relatively thin skin.

Carbon dioxide retention manifests itself in several ways. *Chronic hypercapnia is soporific.* The sensorium is dulled, and the patients frequently fall asleep in a chair or during the examination. *Asterixis* may result from hypercapnia. Asterixis is a form of tremor sometimes called "flapping." It is best elicited by having the patient extend the arms and maximally extend the wrist. Episodic sudden loss of the extensor tone causes a quick flap forward of of the hand, which then rapidly regains tone and snaps back to the extended position. Asterixis is also seen in hepatic precoma, renal failure, and other metabolic encephalopathies and so is not diagnostic of hypercapnia.

Cheyne-Stokes

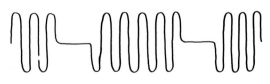

Biot

Fig. 5.8. Abnormal breathing patterns.

PRACTICE SESSION

You need to practice counting respirations without appearing to do so. If you approach a patient and declare, "I'm going to count your breathing," the normal rate will change, usually going up. During your next boring lecture, watch some of your colleagues. First, guess what the rate looks to be, then count it accurately for 1 minute. With some practice, you will get quite accurate.

A good time to check the rate and pattern is during sleep and you will have the chance to do that with many of your patients. They will appreciate the extra minute of rest and you will get good information.

chapter 6

SKIN

No discipline in medicine makes more demands on descriptive skills than dermatology. The language of dermatology is strict because the classification of skin diseases is primarily morphologic. Previous attempts to classify skin diseases according to etiology or functional relationships have resulted in confusion. The description of skin pathology, then, is exacting. Once you have mastered the skill of accurately examining and describing skin ailments, specific diagnosis is vastly facilitated. The description of skin pathology is your index to source material. Standard textbooks of dermatology divide material into such categories as "pigmented lesions," "bullous eruptions," "desquamative diseases," etc., primarily based on what the lesions look like. This is quite different from the functional or anatomic divisions of a textbook of cardiology, for instance.

The enormous scope of skin disease can be appreciated from a number of viewpoints. In terms of the impact on the population, a reliable estimate is that 1 in 20 people has a skin disease which might benefit from physician attention. Skin disease comprises the greatest segment of occupational illness and, as such, profoundly affects the economy. The cost is measured not only in terms of lost production and efficiency, but in the millions of dollars expended on proprietary skin medications.

The individual morbidity is no less profound. No other diseased system is so readily recognized by lay people. Family, friends, and all passersby quickly notice skin disease. From ancient times, skin disease has been linked to uncleanliness. The leper of biblical times was required to wear his garments in shreds and call "unclean, unclean" to all who approached. Isolationism persists today. The patient's reaction to this avoidance by others is usually a withdrawal. A simple shopping trip takes courage, and social gatherings are abhorred. The anxiety and frustration unfor-

tunately aggravate many skin conditions and initiate a vicious cycle. All of this is to say, then, that the patient who comes to you with a skin disease brings with him a complex of ills. It is imperative that you not dismiss him lightly or ignore unvoiced concerns. An optimistic, thoughtful, sympathetic approach in itself proves therapeutic.

Another way to look on the scope of skin disease is to consider the organ itself. The body surface area of 1–3 square meters must all be covered by this organ of approximately 17 kg. These dimensions make it the largest organ of the body. Functionally, the skin performs extraordinary tasks. It provides our interface with the environment affording protection from without and containment from within. It is selectively permeable. As a homeostatic mechanism, it regulates temperature within narrow ranges in spite of extremes in the environment. It has secretory, excretory, absorptive, synthetic, and sensing functions, all of which require special substructures and modifications. The basic tissue components of the skin are shared by various visceral organs—blood vessels, lymphatics, connective tissue, neural and neural-derived structures, etc. This is to point out that the variety of malfunctions which may be observed in the skin is enormous and to remind you that what you see in the skin may be the benchmark for what is occurring internally.

History

The chief complaint regarding skin disease, as we discussed earlier regarding chief complaints in general, may be pain, dysfunction, change from the steady state, or an asymptomatic observation.

Pain

Since the skin is so richly innervated, painful skin lesions are common. Inflammation and edema in the skin produce pain. Itching, or *pruritus,* is a form of pain registered only by the skin. Although common, itching is caused by so many things that its specificity in diagnosis is not great. Pruritus occurs in the obvious poison ivy and in the obscure itching associated with visceral lymphomatous diseases. Deposited bile salts of obstructive hepatobiliary disease cause intense pruritus.

The lack of pain may be diagnostically important. Skin diseases involving neurovascular bundles or nerves proper can result in anesthesia. Thus, the plaques of leprosy and the chancres of syphilis are not painful.

Dysfunction

This chief complaint primarily relates to the special structures of the skin. Excessive or diminished ability to sweat, drying, scaling, follicular plugging, and acne are examples.

Change from the Steady State and Observations

For purposes of skin disease evaluation, these two categories of chief complaints may be grouped. Because the skin is so accessible to self-examination, patients frequently consult the physician because of some change in a portion of the skin or some asymptomatic lesion which they happened to discover. The increase in the public education regarding skin cancer has helped to encourage many individuals to seek attention early.

Developing the Chief Complaint

Again, following the general outline of Chapter 1, we begin by determining the anatomic localization. This might sound redundant at first, but recall that, although the complaint may be referrable to the skin, the basic cause may be internal or systemic. For that reason, you should carefully review the associated manifestations and make a review of systems. The patient who complains of "easy bruising," "colored lines on the abdomen," and "acne" might, on close questioning, also relate the weight gain, emotional change, and diabetic polyuria and polydipsia of Cushing's syndrome, adrenocortical dysfunction.

Get the time sequence. When did the lesions first appear? Did it start as a single lesion and spread, or did all the lesions appear together? Did they appear suddenly over minutes or gradually over days or weeks? What time of the year did they start? The temporal sequence can be diagnostic. Pityriasis rosea, for instance, is a common eruption of the trunk and proximal extremities. An initial "herald patch" appears followed within 4–10 days by the development of multiple smaller patches. It most often occurs in the spring or fall.

Also important in the temporal sequence is the association of exposure to drugs, toxins, or chemicals (Fig. 6.1). The extraordinary variety of prescription and nonprescription drugs available makes it almost safe to assume that any skin disease is drug induced until proven otherwise. The skin is the end organ most frequently sited in lists of the side effects of drugs. This need not be an acute reaction to a single drug exposure but can be the result of chronic exposure, as in the skin lesions of bromide excess or the keratoses of long term remote arsenic ingestion—many of the tonics of years ago contained arsenic (Fig 6.2).

An agent not generally considered a chemical or toxin is sunlight. Sunburn is, of course, a very common dermatologic condition, but many other skin conditions are affected by sunlight—especially the ultraviolet waves of sunlight. The rare disease, porphyria, has as a hallmark photosensitivity, presumably due to the deposit of protoporphyrine in the skin. More common, but less well understood, is the photosensitivity of patients with lupus erythematosis.

Physical Examination

The skin covers all, and the examination is not complete until all is covered. Even when the complaint is localized, the entire skin must be scrutinized, preferably by completely disrobing the patient. This is no less important in the examination of the skin when the chief complaint is other than dermatologic. Some of the smallest lesions are the most devastating. If you carefully look at the skin of every patient you examine, you will at some time in your clinical career save a life by finding an early malignant melanoma. Mucous membranes must also be examined. The involvement or lack of involvement of mucous membranes is an important subdivision of your descriptive index.

The most important tool at your disposal

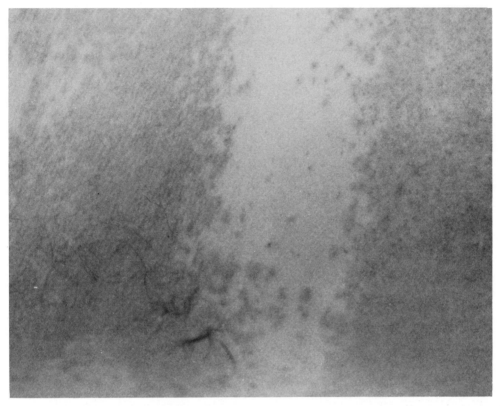

Fig. 6.1. Fixed drug eruption. The patient recently received penicillin, to which he is allergic. He developed a fever and this macular erythematous eruption, which was pruritic.

is your vision. Bright natural lighting aids considerably. Scan the skin as you would a slide under the microscope—from the low power to high power. First, get an overview of the distribution of the lesions, then assess the arrangement of the lesions, and finally analyze the individual lesions.

Special Techniques in Dermatologic Examination

Your examination of the skin can be extended by a few simple techniques. A hand lens, for instance, facilitates the scrutiny of small lesions. Manipulation of lights may be helpful. A bright lamp moved around for different angles of lighting throws various aspects of lesions into relief, clearly showing what may have only been suspected by palpation. A *Wood's lamp* generates long wave length light, so-called black light. When applied to hair or scales infected with dermatophytes, fluorescence appears. Urine containing porphyrine, as in porphyria cutanea

tarda, will also fluoresce. Skin infected with pseudomonas organisms reveals a typical metallic green color when viewed under a Wood's lamp.

Red spots in the skin are either vascular dilations or extravasated blood into the skin. *Diascopy* is the technique of pressing such a lesion with a glass slide. If the color disappears, vascular dilation is the likely etiology.

Microscopy further extends your diagnostic capability. Scales should be scraped, mixed with 10% potassium hydroxide, and examined for the mycelia of a fungus infection. Fluid from vesicles should be stained with Gram's stain for bacterial organisms and with Giemsa or Wright stain to find the multinucleated giant cells of the herpes and varicella viruses. Dark field microscopy allows you to identify the treponemes of syphilis.

Biopsies are best left to the dermatologist. On occasion, this is the only way to identify a lesion definitively. The dermatologist will select representative lesions which are not

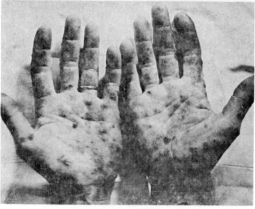

Fig. 6.2. Arsenical keratosis. The patient had taken an arsenic-containing spring tonic for many, many years. These keratotic lesions on close inspection resemble small volcanos. They are important to recognize since cancers of multiple sites are common accompaniments.

super infected or otherwise traumatized. He usually will take the biopsy from the border of a lesion, fix it immediately, and orient the block for cutting in the exact perpendicular plane.

Generalized Changes

A great deal of information concerning the general health of a patient can be obtained by examining the skin even in the absence of isolated skin abnormalities.

Skin turgor generally reflects the state of hydration. Dry, parchment-like skin is seen in dehydration and the elderly. In aged patients it reflects the loss of skin elasticity as well as a deficiency of extracellular water. Turgor is best assessed by tenting the skin of the forehead between the thumb and forefinger, rapidly releasing it, and observing how quickly it snaps back flat (Fig. 6.3). Hydra-

tion can also be checked by feeling for the presence of axillary moisture. The combination of findings of decreased skin turgor, soft sunken globes, and absent axillary sweat is diagnostic of profound dehydration.

Skin texture changes under the influence of many variables. Years of sun and wind exposure result in accentuation of skin folds and a leathery appearance. Long deep "cross feet" at the corners of the eyes, especially when extending down over the zygomatic arches, usually indicate heavy present or past cigarette smoking. A cobblestone appearance to the skin over the back of the neck has the same significance. *Thyroid hormone* profoundly affects skin texture. In *hyperthyroidism*, the skin is fine, smooth, moist, and often flushed. The hypermetabolic state hastens the maturation process so that full cornification does not occur—the cells are pushed up and out prematurely. The increased vascular flow and heat production causes skin vasodilatation and sweating to promote heat loss. When deprived of thyroid hormone, skin texture changes in one of two ways depending on the etiology of the hypothyroidism. If the cause is primary failure of the thyroid gland, the skin becomes dry, scaly, and thick for exactly the converse of the reasons described for hyperthyroidism. If, however, the hypothyroidism is secondary to pituitary failure and thus is accompanied by gonadal and adrenal failure, the skin becomes pale, dry, smooth, and silky. The skin of panhypopituitarism has been called "alabaster skin." The reason is not known, but it is a helpful diagnostic point in differentiating primary from secondary hypothyroidism.

Skin color is likewise affected by many variables. Melanin disturbances may be generalized or local. Complete absence of pigmentation identifies the *albino*. Increase in melanin pigmentation occurs in some systemic diseases. Addison's disease, primary failure of the adrenals, may be suggested by the presence of darkening of the knuckles, elbows, and palmar creases and patches of hyperpigmentation in the buccal mucosa (Fig. 6.4). The cause is the increase in ACTH, which has melanocyte-stimulating effects. When adrenal failure is the result of pituitary failure, there is, of course, no increase in ACTH and consequently no increase in pigmentation. Abnormal skin pigment is seen in *hemochromatosis*. This disease, a disorder of iron metabolism, is characterized by a bronze

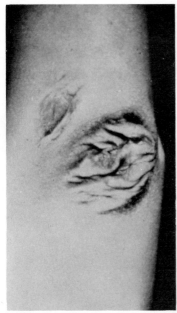

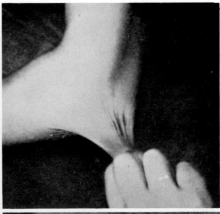

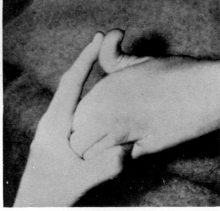

Fig. 6.3. Ehlers-Danlos syndrome. The abnormal connective tissue is here demonstrated by the loose folds of skin over the elbow and the hyperextensible joints.

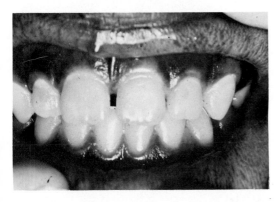

Fig. 6.4. Addison's disease. The buccal hyperpigmentation is especially well shown. The speckled dark spots over the lips and on the gingivae can be seen. This is assumed to be related to the high levels of circulating ACTH, which has melanocyte-stimulating properties (Courtesy Endocrine Unit, Massachusetts General Hospital.)

pigmentation of the skin, diabetes mellitus, and cirrhosis. All three are the result of iron deposition into the tissues and subsequent reaction to the iron. *Wilson's disease,* hepa-

tolenticular degeneration, is a disturbance of copper metabolism. The excess copper is visible in the eyes as Kayser-Fleischer rings and may be seen under the nails as blue lunulae. *Argyria* is the term used for silver intoxication. In argyria, the skin takes on a slate-grey hue as a result of silver accumulation in the skin, usually as a result of chronic abuse of silver nitrate-containing nose drops. Skin color changes from cyanosis and jaundice have already been mentioned.

Localized Changes

We go now from the low power overview to the medium power scan and evaluate the distribution of lesions.

Distribution of Skin Lesions

Skin composition varies considerably from area to area. Lesions which appear in only certain areas suggest that the affliction is related to some peculiarity of the makeup of the skin in those areas. Intertrigo, for instance, appears in areas which are moist and in which one skin surface opposes and macerates another skin surface. Such areas include the axillae, groins, and infra- mammary folds. Psoriasis appears in areas with rich keratinization, especially those subject to trauma as the elbows, scalp, and knees (Fig 6.5). The involvement of mucous membranes in some skin diseases implies that the keratin layer is not necessary in the production of the lesion. As was previously mentioned, lesions distributed in sun-exposed areas carry a special connotation. Secondary syphilis and Rocky Mountain spotted fever characteristically include palmar lesions (Fig. 6.6). The pathophysiology of this distribution is not known. *Petechiae*, which are small red spots of blood extravasated into the skin, are likely to appear first on the lower legs. The reason is that the hydrostatic pressure within the vascular tree is greatest in the dependent areas. Deficiencies in platelets or in the walls of blood vessels will show first on the lower legs even though the disorder is generalized.

Arrangement of Lesions

The medium power scan also asks, "What is the pattern of the lesions?" Skin lesions

distributed along a dermatome indicate a neurogenic basis. Herpes zoster is the prototype (Fig 6.7). Here, the vesicular lesions map out quite precisely the distribution of the infected nerve. When the disease process involves the epidermis only, the lesions fre-

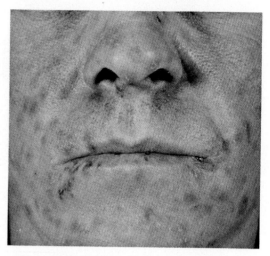

Fig. 6.6.   Secondary syphilis. Copper-colored macules involving the entire skin. Importantly, they are seen on the palms and soles and are infectious.

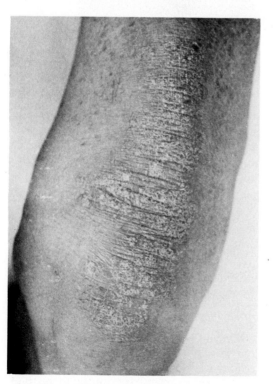

Fig. 6.5.   Psoriasis. The typical silver scales and lichenification are well demonstrated on this elbow.

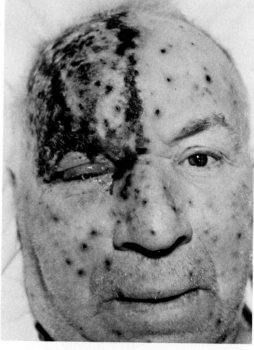

Fig. 6.7.   Herpes zoster. The affliction in this patient involves the first branch of trigeminal nerve as well as a few focal lesions elsewhere.

quently line up along the axis of the fine skin creases.

Linearity is an important observation. Lesions streaked along the long axis of a limb mean one of three things. Scratching by the patient is the most common cause of linear lesions. The eruption of poison ivy takes this form as the irritant is spread by scratching up and down. Blood vessels and lymphatics are generally arranged along the long axis of the limb and inflammatory diseases of either may produce linear lesions.

Satellite lesions, that is, a large central lesion surrounded by two or more similar smaller lesions, indicate an origin and spread from it, as is seen in malignant melanoma. Grouping is similar except no central original defect is apparent.

Types of Skin Lesions

Turning to the high power examination of the skin, we include the palpation of individual lesions. First, decide if the lesion is above, in the plane of, or below the normal skin.

Lesions above Normal Plane

Papules are solid elevated lesions (Fig. 6.8). Generally less than a centimeter in diameter, a papule may result from hyperplasia of one of the normal skin elements, or by infiltration of a foreign element. When a papular lesion consists of many small excrescences, it is termed a *vegetation* (Fig. 6.9). The typical filiform wart is a vegetation.

Plaques are larger elevations than papules, extend deeper into the dermis, and may have sloped edges. A plaque may result from the coalescence of many papules. *Pigskinning* appears if the features of the skin are accentuated. Frequently, because of the size and elevation of a plaque, there is proliferation of the keratin layer producing *lichenification*.

Nodules differ from papules in that they extend deeper into the dermis. They are round, oval, or elliptical. A nodule may originate in the subcutaneous tissue and only secondarily elevate normal overlying skin. Nodules must be taken seriously, as they may indicate systemic disease such as lymphoma, metastatic carcinoma, and late syphilis (Figs. 6.10–6.12).

Wheals are special elevated skin lesions caused by edema fluid in the upper dermis. They are pale red in color and evanescent.

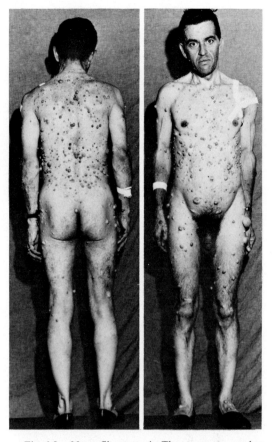

Fig. 6.8. Neurofibromatosis. The enormous number of skin neurofibromas apparent here is not ususual in this disease.

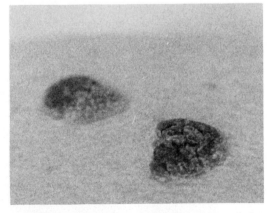

Fig. 6.9. Senile keratosis. This is a verrucous lesion which is prevalent in the elderly. It is benign.

They may be small or large and are frequently pruritic as with insect bites. Huge disfiguring wheals occurring spontaneously in the lips, eyes, ears, epiglottis, and intestine are seen in *hereditary angioneurotic edema.* This life-threatening wheal reaction is caused by an inherited disorder of the complement system. Wheals can be raised in 25% of normal people by stroking the skin with a blunt object like a tongue blade. This reaction to trauma is called *dermatographia* (Fig. 6.13). Darier's sign is the wheal raised when stroking the brown macule of urticaria pigmentosa. Malignant histiocytes, infiltrating the skin in this disease, release histamine when traumatized and produce the wheal.

Vesicles and *bullae* are circumscribed, fluid-filled elevations. The fluid contained may be lymph, blood, or serum. Bullae are large vesicles, the dividing line arbitrarily put at 0.5 cm. When examining vesicles or bullae, it is important to try to ascertain the depth of the fluid. To produce such a lesion, the skin must split to accommodate the fluid. When the fracture occurs high in the epidermis, the roof is very thin and may be translucent.

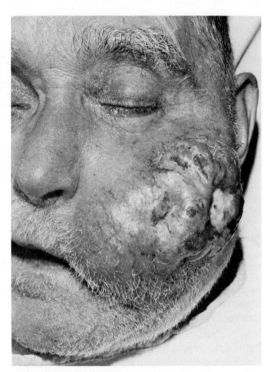

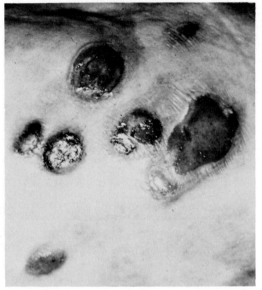

Fig. 6.10. Mycosis fungoides. This peculiar skin lymphoma is distinguished by these plaque-like lesions with superficial ulcerations. In other areas they can be detected as subcutaneous tumors.

Fig. 6.12. Extensive squamous cell carcinoma of the cheek. The patient denied the existence of this lesion until it was far advanced.

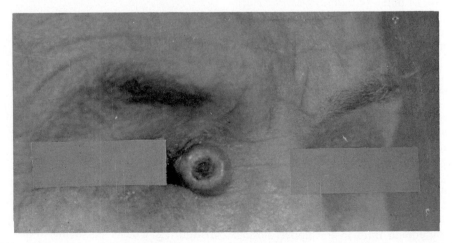

Fig. 6.11. Epidermoid carcinoma. This lesion was painless. Note the waxy, rolled, heaped up outer margins and central necrosis. The lesion has a tendency to local invasion.

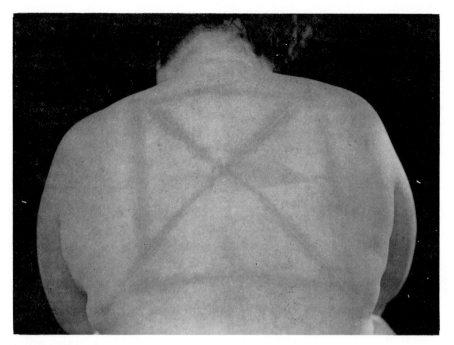

Fig. 6.13. Dermatographia. This patient was stroked with a tongue blade 5 minutes before this photograph was taken. Wheal and flare reaction was apparent. The persistent flare can be easily seen. Many normal individuals demonstrate this phenomenon.

Deeper lesions have a thicker roof. The cleavage plane may have to be determined histologically. A small difference in this level helps to distinguish pemphigoid, a relatively benign condition, from its frequently fatal look-alike, pemphigus. Vesicles with a small central depression, *umbilicated vesicles*, are characteristic of viral diseases such as herpes, variola, and varicella (Fig. 6.14). Vesicles and bullae rupture easily. The appearance of unroofed or healed lesions should be noted. When the base is composed of the basal layer of epidermis, it is called an *erosion*. Erosions weep clear fluid and do not scar. Deeper lesions into the papillary dermis and below are called ulcers. Ulcers may bleed and scar.

A *pustule* is a vesicle filled with purulent debris. The color of a pustule is usually creamy but may be green or yellow. Any vesicular lesion may become secondarily infected to form a pustule. Follicular pustules arise in hair follicles, are conical, and have a hair projecting from the center. These heal without scarring. When follicular pustules coalesce and form necrotizing *carbuncles*, scarring may result.

Lesions in Plane of Skin

Since these are not palpable, their hallmark is a color change. Skin color is influenced by

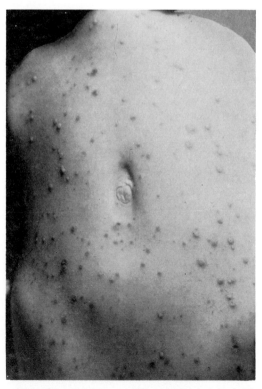

Fig. 6.14. Chicken pox. This eruption is 4 days old. Lesions are in various stages of development. Some umbilicated vesicles can be noted. In addition, there are small ulcer-like lesions. (Courtesy Dr. Louis Weinstein.)

several factors. Melanocytes, endogenous to the skin, are responsible for basic skin color. Absence of melanocytes in albinism causes the skin to take on a pale pink color of the underlying vascular bed. Even in normal individuals, vascular tint contributes greatly to skin color. The thickness of the overlying skin and the richness of the vascular plexus allow for many shadings. The lips are very vascular and the skin very thin, hence the pink-red color. The degree of oxygenation of the blood in the surface vascular further modifies the hue.

Exogenous pigments deposited in the skin are another source of color change as, for instance, in jaundice. Local abnormalities, then, of melanocytes, vascularity, or pigment all may produce isolated lesions in the plane of the skin.

An isolated pigmented lesion in the plane of the skin is called a *macule* (Fig. 6.15). Macules vary in size from a millimeter to several centimeters and in color from white to jet black.

Macules caused by disturbances of melanocytes are commonly encountered in the physical examination (Fig. 6.16). Freckles

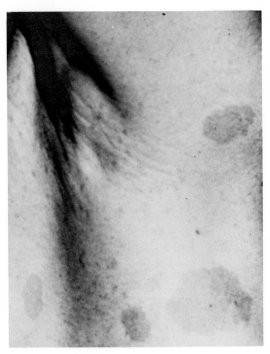

Fig. 6.16.   Neurofibromatosis. The café au lait spots are apparent. Also note the presence of axillary freckles, which is characteristic of this disease.

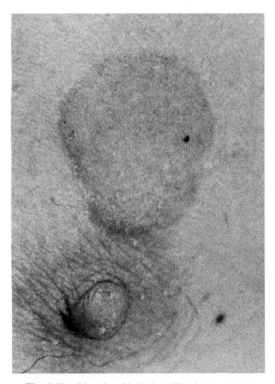

Fig. 6.15.   Macular skin lesion. This lesion is distinguished only by its color and is not palpable.

and nevi are local increases in the number of melanocytes or the number of melanin granules contained in the melanocytes. Localized absence of melanocytes also qualifies as a macule, in the case of a white macule. *Vitiligo* is the term applied to patchy, geographically widely distributed areas of amelanotic skin. Vitiligo, which may be inherited, is seen also in many autoimmune diseases such as idiopathic hypoparathyroidism, Addison's disease, hyperthyroidism, and pernicious anemia. *Malignant melanoma* may be in the plane of the skin or above it. This disease, frequently aggressively malignant, merits special comment. Composed of malignant melanocytes, the color of a melanoma may vary from pale white, to pink, to jet black. A helpful rule of thumb is to remember "red, white, and blue" as combinations of these colors are almost always present in a melanoma. Painless, these lesions may be quite large as in the superficial spreading type or small as in the nodular type. Satellite lesions are common. *Peutz-Jegher* syndrome is an interesting melanin disturbance. Dark melanin macules are found about the lips, perineum, oral mucosa, hands, and face. These are associated with multiple intestinal polyps, particularly of the upper small intestine.

Macules may be caused by vascular anomalies. When blood is extravasated into the skin, the term *petechiae* or *purpura* is applied. Petechiae are small and discrete while purpura are large and more diffuse. Diascopy helps distinguish vascular dilatation from ex-

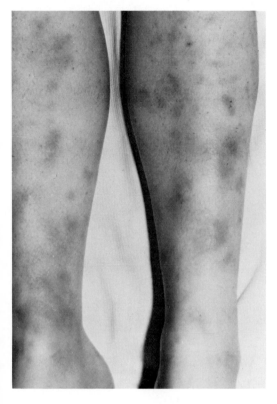

Fig. 6.17. Erythema nodosum. These lesions begin as painful spots on the lower extremities, which subsequently become indurated, raised, and erythematous. Resolution takes the form of a spreading bruise. It is associated with a wide variety of diseases.

travasation of red cells. Seepage of red cells into the skin implies trauma from without or abnormalities within the vessel. Clotting factor or platelet deficiencies, inflammation of the vessel walls, or emboli account for most petechial lesions. When petechiae are palpable (no longer macules), a *vasculitis* is almost surely the cause. The vessel wall not only leaks red cells but, because of the inflammation, accumulates a surrounding zone of polymorphonuclear leukocytes and edema fluid.

Vascular anomalies may be in the plane of the skin or above it (Fig. 6.17). Any component of the vascular tree may be, or become, malformed. *Spider angiomas* are highly branched stellate arterial lesions which pulsate and blanch on pressure (Fig. 6.18). Pregnancy and hepatobiliary diseases are the most common conditions associated with spider angiomas. For some unknown reason, spider angiomas rarely appear below the umbilicus. *Hemangiomas* are benign lesions derived from or composed of endothelial-lined vascular spaces. They may be capillary or venous in origin. *Senile hemangiomas* (cherry spots) can be found on the trunk of practically every adult patient. They are only a few millimeters in diameter and compress with difficulty. Cavernous hemangiomas may elevate the overlying skin. Frequently large, these hemangiomas are deep purple and boggy to palpation. The borders are poorly defined and they fill quickly after compression.

Telangiectasia are non-neoplastic vascular dilations which should arouse your interest promptly. They frequently benchmark systemic disease—particularly the collagen vascular diseases. Although they may occur anywhere, they are readily recognized at the borders of the upper eyelids and at the base

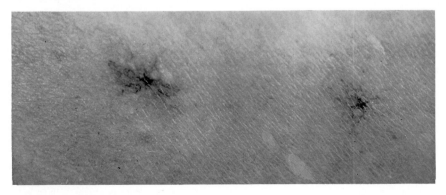

Fig. 6.18. Spider angiomas in a case of alcoholic cirrhosis of the liver. These lesions compress readily and are also seen in pregnancy.

of the nails (Fig. 6.19). They are lace-like and
are intertwining dilatations of normal vascular elements. Dermatomyositis, systemic lupus erythematosis, metastatic carcinoid, and
hereditary hemorrhagic telangiectasia should
be considered when you find these lesions.

Nevus flammeus (port wine stain) is a common deep burgundy vascular macule often
distributed along the course of a peripheral
nerve. Nevus flammeus occasionally indicates its counterpart in the viscera beneath.
When nevus is seen above the line between
the eye and ear a similar malformation may
be present in the meninges and the syndrome
of *Sturge-Weber* is present.

Lesions beneath Plane of Skin

The common denominator of lesions below
the normal plane is a loss of skin substance.
This atrophy of skin may occur in a variety
of ways.

Vascular disease accounts for many such
lesions. Occlusion of the end arteries of the
skin results in loss of skin substance. The
degree of loss depends on the degree of oxygen deprivation. Anoxia damages structures
in direct relationship to their degree of specialization. Fibroblasts, being very primitive,
can tolerate far lower oxygen concentrations
than can neurons, for instance. This basic
rule applies to the skin. Skin appendages such
as hair follicles and sweat glands atrophy
first. This accounts for the shiny, dry, hairless

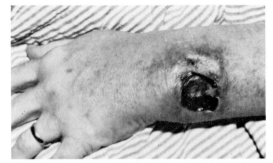

Fig. 6.20.   Ichthyma gangrenosum. This deep ulcer
crater is characteristic of pseudomonas septicemia.

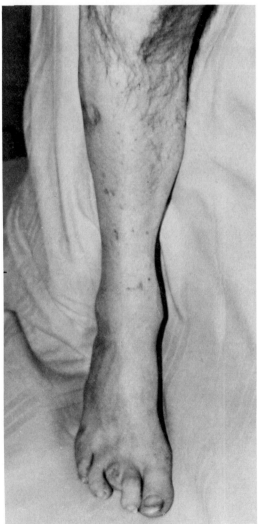

Fig. 6.21.   Diabetes mellitus. Many of the changes of
diabetes are visible on this photograph. Note first that
the patient has had a toe amputated because of small
vessel disease. There is absent hair growth below the
midcalf. The depressed area on the midcalf is an area of
necrobiosis diabeticorum. Smaller similar lesions are also
present.

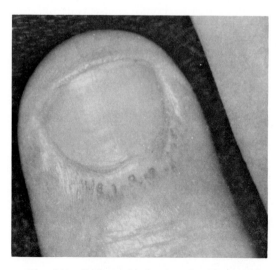

Fig. 6.19.   Periungual telangiectasia. The hairpin
loops of dilated blood vessels are readily apparent along
the nail margin. These lesions suggest the presence of a
collagen vascular disease—dermatomyositis or systemic
lupus erythematosis. (Courtesy Dr. Howard Baden.)

legs of the patient with the peripheral vascular disease, as we will see in a later chapter. If vascular occlusion happens suddenly, an infarct will result. *Skin infarcts* appear first as erythematous macules which become purpuric and darken in color until black. A bulla forms, followed by an ulcer, followed by scarring (Fig. 6.20). When multiple skin infarcts are found, septicemia and vasculitis should be considered. Gradual occlusion results in slow scarring without blatant necrosis. The prototype of this kind of lesion is seen in diabetes mellitus. The diabetic capillary angiopathy produces *necrobiosis lipoidica diabeticorum* in the skin (Fig. 6.21). These lesions are found on the anterior lower legs as shiny depressed areas with an orange to brown hue bordered by tiny telangiectasia.

Loss of skin substance also results from abnormalities of the connective tissue of the skin. The *striae* of *Cushing's disease*, when carefully examined, are below the plane of the skin. Small connective tissue hernias in the skin come about under the influence of excess corticosteroids. The pigmentation of these striae is secondary to the increased vascularity showing through these dermal rents. Stretch marks from rapid weight change or after pregnancy are similar in formation but are not pigmented as the vascularity of the skin is not increased. *Scleroderma* may be a disorder of connective tissue although some authorities believe it to be a vascular disease. Connective tissue, in any event, is profoundly altered in the skin. It becomes "hide-bound" especially over the fingers and face. The border of the atrophy is indistinct. The skin becomes shiny, smooth, and relatively immobile over the underlying structures. Digital ulcers and gangrene appear in severe cases.

PRACTICE SESSION

While toweling off after your next shower or bath, go over your own skin carefully. You should be easily able to find four different skin lesions. With your new descriptive skills, define each by its location, color, place in the normal plane, blanching characteristics, and exact size. Find the classmate who has psoriasis (there is bound to be at least one) and ask him or her to tell you what the condition has been like for them so that you might understand what it means. If they are courteous and you are genuine in your interest, you will likely be allowed to examine representative plaques.

chapter 7

HEAD, EYES, EARS, NOSE, AND THROAT

The head, eyes, ears, nose, and throat are considered together because of the obvious anatomical advantage which makes the examination of these systems conveniently performed in sequence. Discussion of the cranial nerves is deferred to the chapter on the nervous system.

Head

History

Perhaps no chief complaint is more frequently borne to the physician than *headache*. Because it is so common and is so commonly unassociated with significant disease, the import may be missed to the detriment of the patient. In developing the chief complaint of headache it is important to characterize the pain precisely. Is it the throbbing pain of vascular dilatation? Is it the dull deep toothache pain of deep-seated tumor? Is it the lightning-like stabs of pain caused by neuralgia? Is it the pulling ache of tension?

The temporal sequence gives valuable clues. Generally, headache present in the morning or waking a patient from sleep is more significant than the pain occurring toward the end of the day. Hypertension and tumor both may give morning headaches. Afternoon and evening headaches commonly are tension related or are secondary to eye strain from minor astigmatism. Similarly, headache experienced for many years is unlikely to be related to neoplasm.

Have the patient be specific about the location of the pain. Pain from intracranial disease is referred according to the location of the disease. Mass lesions above the tentorium cerebelli refer pain to the branches of the fifth cranial nerve and pain is detected somewhere anterior to the ears. Lesions below the tentorium are reflected in the distribution of the 9th and 10th cranial nerves behind the ears. Migraine always occurs uni-

laterally and has an aura or warning, usually visual. Tension headaches are almost always occipital as a result of spasm of the paraspinous muscles of the neck. Inflamed sinuses produce pain over the frontal or maxillary sinuses if they are diseased, but may be referred to the occiput if the sphenoid sinus is afflicted.

The dural arteries, arteries at the base of the brain, and the great venous sinuses register pain when in spasm or when especially turgid. Mechanisms distant from the head which alter vascular tone may then produce headache, as in fever, carbon dioxide or carbon monoxide retention, or hypoxia.

Increased intracranial pressure stretches the meninges as well as the vascular structures. This traction and displacement produces pain. This is probably the mechanism operative in brain tumor. When tumors in the brain cause headache, they may also cause *projectile vomiting*—projectile because it is unassociated with nausea. Coughing or straining at stool further increases intracranial pressure and aggravates this form of headache.

Examination

Phrenologists took pride in analyzing the skull and interpreting the mind and character of the subject. Perhaps the only valid premise of phrenology is the fact that the skull, during growth, enlarges in response to the pressure from within. There are several determinants of the shape of the skull.

Hydrocephalus and *microcephaly* are dramatic examples of the response in growth of the skull to the brain. In childhood hydrocephalus, the cranial plates are forced widely apart, and suture closure may never occur. Inspection shows normal features up to the ears, above which the head may be grotesquely enlarged. When severe, the patient's eyes are forced downward so that he appears to be staring at the floor. A bright light applied to the head in a dark room will *transilluminate* brightly. Percussion with the tip of the middle finger against the side of the skull returns a note like a *cracked pot*. When the brain fails to develop, the skull will not enlarge and microcephaly results.

Craniosynostosis is the premature closure of a cranial suture. Normal skull proportions are dependent on the proper sequential closing of the sutures. If a suture closes prematurely, an abnormal head shape will result. By inspecting the head of such a patient you can deduce which suture was at fault by remembering that growth always occurs in a direction perpendicular to the line of the suture. When one suture closes, compensatory growth occurs at the remaining open sutures. When the sagittal suture closes prematurely, lateral growth ceases and anteroposterior growth compensates, producing a long narrow head, *scaphocephaly*. Should the coronal suture fuse, anteroposterior growth halts and the skull assumes short, broad, conical proportions called *acrocephaly*. If one half of the coronal suture closes prematurely, that side becomes flattened and the orbit on that side is elevated, *plagiocephaly*.

Frontal bossing is the prominence of the frontal bones which makes the forehead appear prominent and the hairline appear recessed. It is seen in Paget's disease with exuberant bone growth or in the periostitis of congenital syphilis or rickets. The high blood flow in the bone of Paget's disease occasionally produces a bruit heard on auscultation of the skull. Similarly, breath sounds may be audible in the pagetic skull.

The skull must always be palpated when the patient complains of headache, especially when the headache is localized. Isolated swellings should be taken seriously, as the scalp and skull are common sites of metastatic disease.

Scalp hair is modified by several genetic and hormonal influences. *Frontal recession* of the hairline occurs under the influence of testosterone. It is a helpful clue to distinguish *virilization* from simple *hirsutism*. Virilization of the female, in addition to producing increased hairgrowth or hirsutism, also produces temporal recession of the hairline and enlargement of the clitoris. Many structural genetic defects also produce a low hairline posteriorly. Balding is an inherited trait and

can occur in the female. When balding occurs in an unusual pattern or as a result of a disease, it is called *alopecia*. An extraordinary number of diseases, drugs, and toxins may cause damage to hair follicles and lead to alopecia.

Eyes

Gram-for-gram, the eye holds more diagnostic information than any other organ available for physical diagnosis. The vascularity alone permits the diagnosis of anemia, diabetes, hypertension, hyperviscosity states, and arteriosclerosis. Endocrinopathies such as hypo- or hyperthyroidism may be suggested by eye findings. Neuro-ophthalmology is sufficiently complex and sophisticated to merit the devotion of entire clinical careers. Six of the 10 cranial nerves influence the eye, as do sympathetic and parasympathetic pathways. The pathology may be distant and the effect visual and visible, as for instance, in the ptosis, miosis, and anhidrosis of *Horner's syndrome*. This syndrome results from interference with the sympathetic pathways to the eye and may be the only finding in carcinoma of the apex of the lung. Even if these windows on systemic disease were not present, the eye would deserve careful examination. Sight is so precious that the failure to avert preventable blindness is cruel beyond words.

History in Eye Disease

When the chief complaint refers to the eyes, a careful history may tell you what form of pathology to look for. Eye *pain* must be taken seriously. Inflammatory diseases of the eye may cause pain. *Iritis* or *iridocyclitis*, inflammation of the anterior chamber, is commonly painful. *Conjunctivitis*, on the other hand, usually does not cause pain. Of the two, iritis is much more dangerous. The cornea is richly innervated, and tiny abrasions of the corneal epithelium are quite painful.

The eye is a hollow viscus, and like all such structures becomes painful when distended. *Glaucoma* is the abnormal elevation pf pressure within the globe secondary to obstruction to the egress of vitreous humor. When the pressure rises abruptly, as in acute glaucoma, the pain may be so intense that nausea and vomiting ensue.

Photophobia is the aversion to light usually because light exposure causes pain. Very bright light will produce pain in normal persons. Albinos, lacking the protective screening pigment in the iris and retina, avoid light. Systemic infections, particularly viral and rickettsial diseases, may be associated with photophobia. Here, the pain is a kind of ciliary body myalgia. The position of rest for the iris is when it is dilated. Light forces the muscles of the ciliary body to contract, and this gives pain.

Asthenopia, eyestrain, generally indicates focusing problems. Refractive errors such as *myopia* (nearsightedness), *hyperopia* (farsightedness), and especially *astigmatism* (variations in corneal curvature) and *phoria* (muscle imbalance) are common faults.

Dysfunction, change in visual ability, frightens most patients. Usually, visual loss leads to a prompt call for help. When the loss is partial, as with a field cut or a *scotoma* (blind spot), or unilateral, it may go unnoticed. The patient, for example, who develops tunnel vision or bitemporal hemianopsia may not seek aid until several garage doors have been eliminated. Loss of vision, per se, only indicates difficulty somewhere between the cornea and the occipital lobe of the brain. Historically, defining it in terms of the specific character of the loss, onset, duration, and associated problems frequently allows a diagnosis prior to examination.

When the images falling on the two retinas do not correspond, *diplopia* or double vision is said to be present. The delicate coordination of both sets of extraocular muscles can be disturbed by diseases of the orbit or diseases of the cranial nerves controlling these muscles. The history should include: onset, gradual or sudden; the direction in which the degree of diplopia is the greatest and that in which it is the least; and whether the two images are side by side or above and below. Determine if the problem is remittent and question for associated disorders, diabetes being a common cause, for instance.

Having spots before the eye is a common complaint. Any foreign body between the cornea and retina which blocks light will be registered as a spot. The history may tell you exactly where the obstruction is before you examine the patient. Foreign bodies in front of the lens move in the direction of gaze, i.e. when the patient looks to the left the spot moves to the left. When the obstruction is behind the lens the movement is opposite, i.e.

when the patient looks left the spot moves right. *Free floating debris* in the humor is just that, free floating, tending to drift in and out of the field of vision. A constant spot is usually a fixed defect like a corneal opacity.

Red eyes have confronted many of us in the morning mirror. Conjunctival infection, while alarming, is usually painless. Similarly, a subconjunctival hemorrhage is more frightening than symptomatic, and the etiologic cough or sneeze has been forgotten.

Puffy eyes should prompt you to ask about allergic conditions, edema elsewhere, renal disease, or other source of protein loss.

Examination

Again, proceeding in a logical reproducible fashion, the wealth of clinical information hidden in the eyes becomes available for the clinician. Several special techniques are required and should be mastered.

Orbit and Adnexa. The hairs of the eyebrows grow very slowly. Loss of the lateral third of the eyebrows is occasionally observed in myxedema.

The globe rests within the confines of the bony orbit and is embraced by the extraocular muscles on the sides, the optic nerve and a fat pad posteriorly, and the conjunctiva anteriorly. Abnormalities of any of these structures may alter the position of the eyeball in the orbit. *Exophthalmos* is the abnormal protrusion of the eye from the orbit (Fig. 7.1). This position can be measured accurately by placing a rule in the edge of the lateral orbital angle and sighting from the side across the front edge of the cornea. Exophthalmos is present when this distance exceeds 16 mm (Fig. 7.2). Hyperthyroidism, tumors in the orbit, and arteriovenous fistulas can cause exophthalmos. Unilateral exophthalmos usually means local disease, but it is worth noting that some patients with hyperthyroidism will have unilateral changes. Auscultation of the globe may reveal the typical continuous murmur of an arteriovenous fistula as far back as the carotid syphon (Fig. 7.3). To listen over the globe, have the patient close his lids lightly and gently apply the bell of the stethoscope to the eye.

The complex structure of the lids leads to a variety of maladies. Each lid has a central *tarsal plate* of relatively rigid tissue. Through injury or advancing age, this plate may be-

Fig. 7.1. Unilateral exophthalmos. The left eye appears more prominent than the right. Direct measurement confirmed proptosis of the left eye secondary to a retro-orbital tumor.

The position of the lid relative to the eye is determined by three sets of muscles: the orbicularis oculi muscle innervated by the seventh cranial nerve, the levator muscle innervated by the oculomotor cranial nerve, and the smooth muscle of Müller innervated by the sympathetic and parasympathetic nerves. Normally, the upper lid passes over the cornea at a level just touching the iris. *Ptosis*, drooping of the lid, occurs with interruption of any of the nerve pathways to the lids, that is, the oculomotor nerve, the facial nerve, or the parasympathetics. *Stare* is the abnormal elevation of the upper lid so that the white sclera appears between the lid margin and the iris. *Lid lag* can be demonstrated by having the patient look up and slowly move the eyes downward. When this sign is positive, the lid *lags* behind and, again, the white sclera appears above the iris. Stare and lid lag occur in hyperthyroidism probably because of the sympathetic hyperresponsiveness of the disease affecting the smooth muscle of Müller.

The *lacrimal apparatus* consists of the lacrimal gland in the upper outer wall of the anterior orbit and the upper and lower puncta

come lax and *ectropion* (eversion) or *entropion* (inversion) of the lid occurs (Fig. 7.4).

Sebaceous glands invest the roots of the eyelashes. A *hordeolum* (stye) is an infection of one of the glands. To examine the underside of the lid, place a cotton-tipped applicator stick against the upper lid about one-third of the way toward the lashes. Then pull down and outward on the lashes and quickly snap the lid over the applicator to reveal the moist palpebral conjunctiva (Fig. 7.5). Thin linear vertical streaks are the Meibomian glands which exist all along the lid edge. *Meibomitis* is infection in one of these glands. Occasionally, cysts develop in these glands; they are called *chalazions*.

The angle formed medially and laterally by the junction of the upper and lower lids is called the *canthus*. A fold of extra skin draped over this angle is called an *epicanthal fold*. Epicanthal folds distinguish the eyes of a child with Down's syndrome (mongolism). When seen in nonorientals, epicanthal folds should raise your index of suspicion for congenital anomalies elsewhere.

The distance between the eyes also may vary in congenital disease. The distance between the two inner canthi should be no more than 40 mm, between the pupils no more than 75 mm, and between the outer canthi no more than 95 mm. Ocular hypertelorism is present whenever these limits are exceeded.

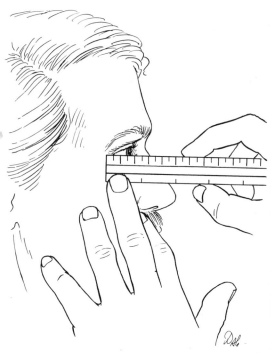

Fig. 7.2. Measuring the protrusion of the globe. The ruler is applied to the angle of the orbit and a sighting taken directly across the foremost edge of the cornea. Normal eyes do not exceed 16 mm. One eye should be compared with the other.

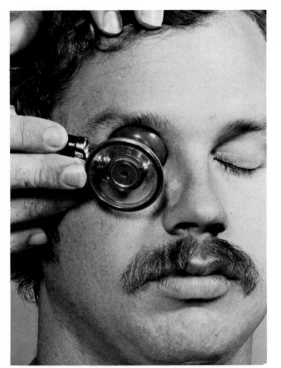

Fig. 7.3. Listening for an ocular bruit. The bell of the stethoscope is applied to form a tight seal around the orbit.

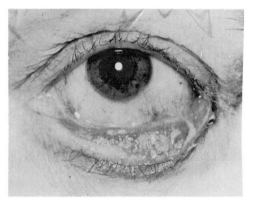

Fig. 7.4. Ectropion. The lower lid is everted because of degeneration of the tarsal plate. The exposed conjunctiva has become secondarily infected and shows a mucopurulent discharge. This is a common condition in elderly people.

which drain from the upper and lower inner lid margins. The adequacy of the lacrimal apparatus can be checked by *Schirmer's* test. A strip of filter paper 5 mm in width and 2 cm in length is used. Bend a few millimeters of the strip and place it in the conjunctival sac over the lower lid. After 5 minutes, the normal lacrimal gland will produce enough tears to saturate 15 mm or more of the strip.

Conjunctivitis is the inflammation of the conjunctiva, either bulbar, palpebral, or both. It must be distinguished from *episcleritis*, which is inflammation of the scleral vessels. Both of these conditions make the eye look red. The conjunctival vessels run toward the limbus (margin of the iris) while the scleral vessels originate at the limbus and run outward. Conjunctivitis usually spares a thin rim around the limbus, and episcleritis is most intense at the limbus. When in doubt, take the lower lid and press firmly against the cornea and then draw the lid downward; conjunctivitis will blanch and episcleritis will not (Fig. 7.6).

Extraocular Muscles. Delicate coordination

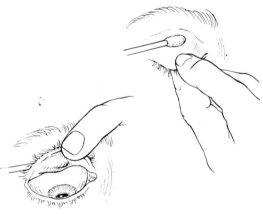

Fig. 7.5. Everting the lid for examination. (Reproduced by permission from Rosen, P., and Sternbach, G. L. *Atlas of Emergency Medicine,* p. 112. Williams & Wilkins, Baltimore, 1979.)

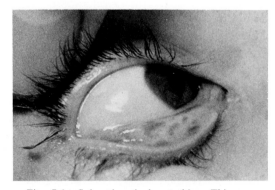

Fig. 7.6. Subconjunctival petechiae. This young woman had rheumatic heart disease and developed subacute bacterial endocarditis. These hemorrhagic lesions were the result of emboli from the infected heart valves.

of the extraocular muscles is required for normal binocular vision. Anything interfering with this balance produces *heterophoria*. *Esophoria* is the tendency for one eye to turn nasally, and *exophoria* is the tendency for one eye to deviate outwards. The patient with heterophoria may not see double if he can correct with an extra effort. The extra effort, however, is likely to produce asthenopia. To detect heterophoria, we employ the *cover test*. With the patient fixing on a light, cover first one eye and then the other. The eye under cover will deviate to its position of rest. When the cover is shifted to the other eye, the previously resting eye will fix again and move centrally. In exophoria, for instance, the eye will move inward when uncovered.

Strabismus, or *squint*, exists when the eyes are not aligned in the same direction and diplopia results. Severe heterophoria which cannot be compensated for will result in strabismus. When long standing, strabismus results in the suppression of vision of one eye— a condition known as *amblyopia*. Subtle strabismus can be detected by shining a small penlight at the patient from a distance of several feet. The reflected light should appear just slightly nasally from the center and on the same point on each globe. The cover test is also useful here. If the fixing eye is covered, the deviated eye will assume fixation and move centrally.

Inability to move the eye in a given direction indicates muscle incompetence either through primary disease of the muscle or through disease of the innervation of the muscles. We will consider nerve paresis in a later chapter.

Ophthalmoscope. Before going any farther, we should consider the ophthalmoscope, a useful tool for the remainder of the eye examination. Simply stated, the ophthalmoscope is a light source fitted with a series of lenses to allow focusing at varying distances. Different lenses are required because the focal length is limited, that is, the limits between which an object is seen clearly. Most ophthalmoscopes are calibrated in two scales—black for positive diopters (unit of measurement for focal length) and red for negative diopters. With high black numbers, close objects come into focus. The higher the red number, the more far objects come into focus. To start with objects close to you, begin with high black, move down the scale to 0, and then up the red scale as you focus on progressively

more distant objects. The cornea, for instance, is seen best with a black 15 or 10 lens, the eye lens with a black 6 lens, and the retina with a red 1 to 3 lens.

Many ophthalmoscopes also provide different lenses over the light source. These will produce a round white dot, a white dot with a grid, a round green dot, or a longitudinal slit of light. The uses of these different light sources will be discussed shortly.

Use the ophthalmoscope in a darkened room to allow for maximum pupillary dilatation. Have the patient fix his gaze on an object 8–12 feet away. (Recall, that to view the retina you must look through the patient's own lens, which can vary your focus.) To examine the right eye, hold the ophthalmoscope in your right hand and view with your right eye. Your right index finger should rest on the focusing wheel. Hold the instrument so that the light falls on the pupil and you can see a *red reflex*, a bright red reflection from the retina. Now move the instrument toward the eye until the knuckle of your middle finger touches the patient's zygomatic arch. Now look through the viewer beginning with a black (+) 15 lens to focus first on the cornea, then the anterior chamber, then the lens, then the vitreous, and finally the retina by changing the lenses to lower black numbers to 0 and then up the red scale (Fig. 7.7).

Cornea. The cornea may be examined directly or with the aid of the ophthalmoscope. The cornea is entirely avascular and richly innervated. The delicate corneal epithelium is easily damaged with serious consequences. An aid to the diagnosis of interruption of the corneal epithelium is a fluorescein-impregnated strip of filter paper. The tip of the strip is gently placed against the conjunctiva of the lower lid. The fluorescein will diffuse rapidly over the globe and brightly stain any breaks in the epithelium.

Substances normally foreign to the cornea may be deposited within it under abnormal circumstances. The most common corneal deposit consists of lipoids and is called *arcus senilis* (Fig. 7.8). Arcus senilis appears as a thin band of milk-white material arranged around the edge of the cornea usually denser above and below than at the sides but forming a complete ring. Arcus may be a normal finding in the elderly, especially blacks, but when it is seen in young patients it may indicate a hyperlipoproteinemia.

Band keratopathy is the deposition of cal-

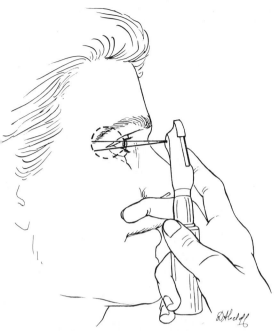

Fig. 7.7. Using the ophthalmoscope. The middle finger rests on the cheek and the index finger on the focusing wheel. Note that the beam of light illuminates only a small portion of the total retina.

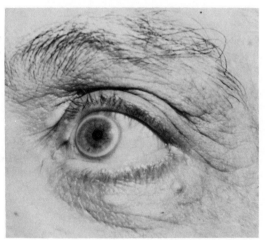

Fig. 7.8. Arcus senilis. The brilliant white band surrounding the iris is readily apparent. If you look carefully, you can also note two xanthelasma, one in the inner canthus and one below the lower lid. This patient was hospitalized for an acute myocardial infarction.

cium salts in the cornea. This occurs in hypercalcemia of long standing. Here, the deposit is denser along the exposed band of cornea, presumably because the exposure to air allows carbon dioxide diffusion, a consequent higher pH, and, therefore, a milieu more conducive to calcium precipitation.

Kayser-Fleischer rings are golden brown deposits of copper in the cornea. These are almost pathognomonic of Wilson's disease (hepatolenticular degeneration) and occur in approximately 80% of affected individuals. They should be looked for in anyone with chronic liver disease.

Uvea—Anterior Chamber, Iris, and Ciliary Body. The aqueous humor of the anterior chamber is normally clear. When debris accumulates, it may first be seen as small dots of reflected light much as sun reflects on dust particles in a room. When blood extravasates into the anterior chamber, it will layer out at the bottom as a *hyphema*. A similar layering out of pus is called a *hypopyon*.

The iris should be round and symmetrical. An irregular pupil indicates disease of the iris, usually old uveitis. Scar tissue or adhesions between the cornea and the lens called *synechiae* tether the iris in a fixed position at a single point or points, producing an irregular pupil.

Change in the size of the pupil occurs through the influence of the third cranial nerve and sympathetic parasympathetic pathways. The pupillary reaction must be tested in several ways. First, shine a light quickly and directly into one eye and observe for the normal contraction. There is commonly a brief overreaction followed by a slight dilatation. Next, shine the light in the eye and observe the reaction in the other eye, the *consensual reflex*. These two maneuvers establish the adequacy of the arc from receptors to effectors in both the tested and contralateral eye. Contraction also occurs when the eye accommodates to near vision. Have the patient focus on a far object then focus quickly on your finger held 8–12 inches in front of his face and observe for the contraction. The *Argyll Robertson* pupil is one which will not react to light, but which constricts with accommodation. It most commonly occurs in central nervous system syphilis but is also seen in other encephalitides.

Anisocoria, unequal pupils, or any abnormality of the light reactions, infers pathology anywhere from the reception of light through the optic nerves to the brain stem, the third cranial nerve, sympathetic, or parasympathetic pathways.

Egress of the vitreous humor is afforded by the canal of Schlemm in the anterior chamber in the angle between the iris and the cornea. Any obstruction to this outflow results in glaucoma. *Narrow angle glaucoma* occurs when the anterior chamber is shallow. When the pupil dilates, the contracted iris may jam up the canal of Schlemm. Using the ophthalmoscope and the slit of light on the front of the cornea, look from the side. You will see two strips of reflected light. One curves over the front of the cornea and the other over the iris. A crude index of the depth of the anterior chamber is the distance between these two reflections. *Open angle glaucoma* results from degenerative changes which obstruct flow. The anterior chamber is normal in depth.

Obstruction leads to increased intraocular pressure and progressive destruction of vision. The pressure in the globe can be measured indirectly. Simple palpation of the eye tells much to the experienced examiner. *Impression tonometry* measures the depth of the indentation produced in the eyeball by a plunger of defined shape, diameter, and weight. The Schiotz tonometer is such an instrument. The cornea is anesthetized with the patient supine. The instrument is gently applied directly vertical to the cornea. The reading on the scale is converted to millimeters of mercury from a table according to the weight applied to the plunger. The normal range is 10–22 mm Hg (Fig. 7.9).

Lens. A black (+) 6 lens on the ophthalmoscope will bring the eye lens into focus. The lens, like the cornea, is normally crystal clear and completely avascular. Unlike the cornea, the lens has no nerve supply. Any opacity in the lens is termed a *cataract*. Although the etiologies vary considerably, the same complaint of failing vision results. Viewed through the ophthalmoscope, cataracts appear as black spots against a red background since they block the reflected light from the retina. The most common form of cataract obstructs light centrally more than peripherally. Cataracts never completely obliterate light perception. If loss of light perception is the chief complaint, the finding of a cataract is insufficient explanation and the examination must be pursued further.

The lens may *dislocate* if the suspensory ligaments are torn or are incompetent. If the dislocation is partial, the curvilinear edge of the lens can be seen intersecting the pupil. Patients with hereditary disorders of connective tissue such as *Marfan's syndrome* and *homocystinuria* are prone to dislocation of the lens.

Fig. 7.9. Tonometry. The cornea has been anesthetized. The lids are held open and the tonometer applied in a vertical position. (Reproduced by permission from Rosen, P., and Sternbach, G. L. *Atlas of Emergency Medicine*, p. 113. Williams & Wilkins, Baltimore, 1979.)

Retina (color Plates 7.1–7.4). The retina is the bank, according to Sutton's law. (Willie Sutton, when asked why he always robbed banks, is alleged to have replied, "cause that's where the money is.") Extraordinarily complex, the retina affords an unmatched opportunity to visualize directly a cranial nerve and end arterioles.

The central retinal artery is a branch of the ophthalmic artery of the internal carotid artery. A 0 to red (−) 6 lens brings the retina into focus (Plate 7.1*A*). The central retinal artery enters the fundus in the middle of the optic disc and immediately branches into a superior and inferior papillary artery. At the edge of the disc, these each divide into a nasal and temporal branch. The retinal veins maintain a close association. Normally, these vessels are extremely thin and translucent. What you see with the ophthalmoscope consists of a column of blood as a red streak. The wall in itself is displayed only as a thin brilliant streak of reflected light. The width of the entire arteriole compared with the width of the white streak is an important indicator of disease. Arteriolar narrowing when localized indicates spasm, and when generalized indicates a vasculopathy for any of a variety of

PLATE 7.1

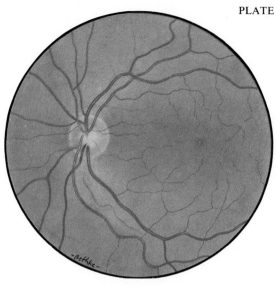

A. Normal fundus. Note the relationship between the arteries which are thinner and slightly paler to the veins. The macula is clearly seen with a normal surrounding halo of darker coloring. The temporal margin of the disc is slightly less well defined than the nasal margin.

B. Multiple cotton wool spots. Note that these spots obscure the underlying blood vessels and are, therefore, superficial in the retina.

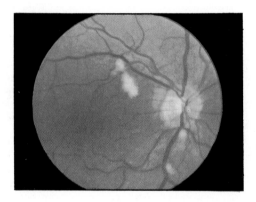

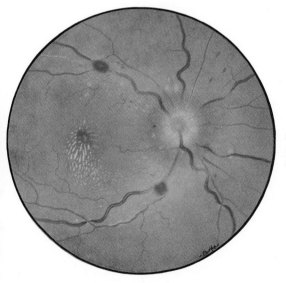

C. Retinopathy of hypertension. Cotton wool spots, flame hemorrhages, and arteriolar tortuosity are apparent. In addition, the disc margin is slightly edematous.

PLATE 7.2

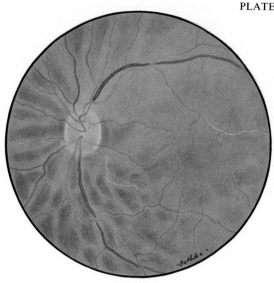

A. Retinal artery plaques. This patient had evidence elsewhere of arteriosclerosis. Note the silver arterioles and marked arteriovenous nicking.

B. Angioid streaks. These ruptures in Bruch's membrane radiate from the disc and have the appearance of vessels. They are, in fact, linear accumulations of pigment along the ruptures. This patient had Paget's disease.

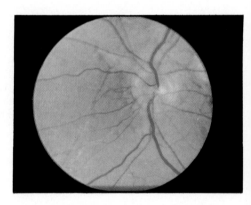

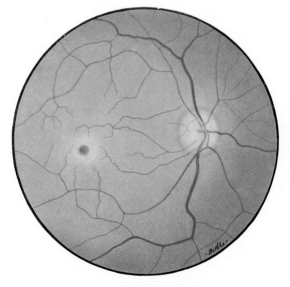

C. Central retinal occlusion. Sudden blindness in the right eye prompted this patient to seek medical attention. Note the generalized oligemia of the retina with the cherry red macula.

PLATE 7.3

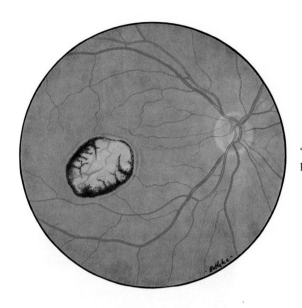

A. Chorioretinitis. This is the characteristic "bomb cluster" of old chorioretinitis. This patient presumably had toxoplasmosis.

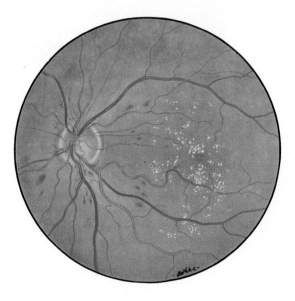

B. Severe diabetic retinopathy. Extensive hemorrhages, exudates, and edema are readily apparent. Microaneurysms are not apparent on this slide.

PLATE 7.4

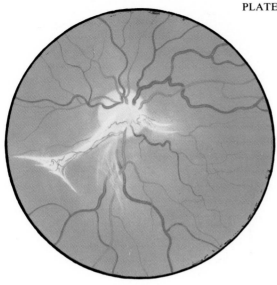

A.   Neovascularization. A plexis of new blood vessels can be seen at 7 o'clock in this fundus.

B.   Papilledema. The retinal vessels emerge and rise over the rim of the enormously swollen optic disc. The fundus is hyperemic. The patient suffered a brain tumor with markedly elevated cerebrospinal fluid pressure.

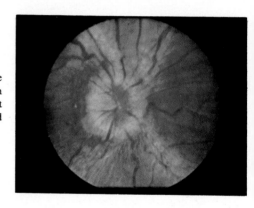

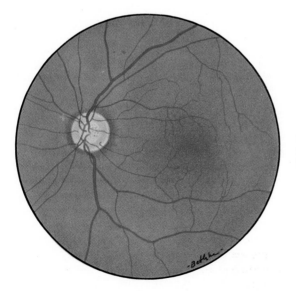

C.   Optic atrophy. Note how strikingly white the disc appears. The retinal vessels are small but otherwise normal. This patient was completely blind in this eye.

reasons. Sclerosis of the vessel wall reduces its transparency so that the central light streak becomes broader and shinier, *copper* or *silver wiring* (Plate 7.2*A*). When severe, sclerosis causes *sheathing*, which is white lines embracing the vessel. The appearance of sheathing relates to light striking the thick vessel on the tangent above and below. The ratio of the width of he arteriole to its accompanying vein (A-V) is another important measurement. Normally, this A-V ratio is 3:4 or 2:3.

The central retinal artery may be occluded by an embolus with resultant sudden blindness (Plate 7.2*C*). The usual source of the embolus is thought to be atheromatous debris from the aortic arch.

When an arteriole crosses a vein, they share a common adventitial sheath. These anchor points bear careful scrutiny. Sclerosis or spasm of the artery may impinge on the vein, leading to A-V *nicking*. The proximal vein may then appear engorged relative to the distal vein. Foreshortening of the arterioles accompanies sclerosis and produces another change at these crossing points. As the arteriole shortens, it drags the vein with it so that the vein angles at a crossing point or appears to change direction after coming out from under an arteriole.

Microaneurysms (Plate 7.3*B*) are small saccular dilatations of the arterioles close to branching points. Virtually pathognomonic of diabetes mellitus, they may portend a similar vasculopathy in the kidneys, Kimmelstiel-Wilson nephropathy. Microaneurysms are more easily visualized with the green filter over the light source. *Neovascularization* (Plate 7.4*A*) is another change of diabetes. This is recognized as a patch of tortuous new vessels which occasionally seem to be growing straight out into the humor.

Hemorrhages into the retina (Plate 7.1*C*) attend many diseases. Typically, they appear as red blotches with a flame shape, i.e. a narrow point at the origin spreading in a wedge shape. Hemorrhages are particularly important findings in hypertension, diabetes, and profound hypoxia. *Exudates* take different shapes depending on the layer of the retina in which they occur and their etiology. *Hard white exudates* are sharply defined and consist of lipophages in response to localized hypoxia. *Cotton wool* exudates (Plate 7.1*B*) are less sharply defined and usually not as bright. These result from ischemic infarcts of the retina.

The venous tree also has a tale to tell. The most dramatic abnormality is *central vein occlusion* in which the entire venous system becomes engorged and tortuous. *Venous pulsations* are normal but difficult to see. Best described as the sensation that you just saw something, the presence of venous pulsations excludes increased intracranial pressure. *Hyperviscosity syndromes* cause venous sludging of blood. Such veins appear as a string of box cars or linked sausage.

The *optic nerve* is not a true peripheral nerve but rather an extension of the brain carrying with it the dura mater right up to its disc at the retina. The fibers radiate from the retina to the disc. Through the ophthalmoscope, the normal disc has sharp edges frequently outlined by dark pigmentation, a pale pink center, and a gentle depression. The depression, or *optic cup*, is markedly accentuated in long standing glaucoma. *Papilledema* (Plate 7.4*B*) indicates increased intracranial pressure. The pressure cone transmitted under the dura lifts the optic disc and everts it into the posterior chamber. The margins become edematous and blurred. The vessels become draped over the elevated edges and the venous system becomes engorged. It may take a change of several diopters with the ophthalmoscope to focus on the top of the disc and then the bottom of the rim.

Optic atrophy (Plate 7.4*C*) is the end result of damage to the nonregenerative nerve. The disc in optic atrophy assumes a dirty gray to bright white hue depending on the loss of blood vessels and the amount of glial proliferation.

The pigment of the retina differs widely among normal individuals. Generally, retinal pigmentation matches the skin pigmentation. Fair-skinned individuals display pale retinas against which the vessels appear prominent. *Retinitis pigmentosa* is a characteristic pigmentary abnormality occurring usually as an isolated entity. Described as "bone spicules," these small discrete areas of hyperpigmentation begin peripherally and later extend near the disc. *Chorioretinitis* (Plate 7.3*A*) affects any ares of the retina and when healed shows as a small bomb cluster of hyper- and hypopigmentation standing as testimony to a preceding inflammatory, often infectious, disease such as toxoplasmosis.

Since the retinal pigment is melanin, it is well to remember that *malignant melanoma* has a propensity for retinal origin. A very dense black spot, usually slightly elevated, is characteristic.

Angioid streaks (Plate 7.2*B*) appear at first glance to be abnormal vessels. Close inspection shows them to consist of radiating streaks of gray or brown pigment which do not reach the optic disc. Microscopically, they are ruptures in Bruch's membrane which distort the pigment. Angioid streaks are seen in some normal eyes, some patients with Paget's disease of bone, and in patients with pseudoxanthoma elasticum.

Retinal detachment is a profound event symptomatically and pathologically. A rupture between the retina proper and the pigment epithelium allows fluid to accumulate in the rent. The retina is lifed up and vision in that area is impaired. Once the rupture begins, it tends to progressively dissect free.

Visual Acuity

The end result of proper functioning of all of these structures is, of course, vision. Any defect to this point may impair the ability to see. The final pathway of the optic nerve to the occipital lobes completes the complex mechanism.

Normally, the human eye detects objects which form an angle of 5 minutes. The *Snellen chart* consists of letters of decreasing size (Fig. 7.10). The code number of each row of letters equals the distance in feet at which the letters will subtend an angle of 5 minutes; i.e. "200" means that at a distance of 200 feet the letters can still be read by the normal eye. Testing is performed at 20 feet. Visual acuity is designated as follows:

Distance of patient from the test chart
—————————————————————————————
Distance at which a normal person
could see the letter

If a patient can see only the large 200 letters at 20 feet, his visual acuity is expressed as 20/200. Small cards appropriately scaled down are available for testing at arm's length and are useful at the bedside.

Measure the acuity first with the unaided eye then with glasses if the patient wears them. Finally, if the acuity is impaired, repeat the measurement with the patient looking through a pinhole. The latter test corrects for refractive errors.

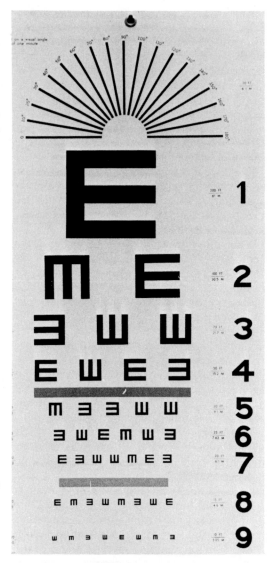

Fig. 7.10.   Snellen chart. This simplified visual acuity chart does not require that the patient understand letters. Instead he must merely indicate which direction the three prongs of the **E** point.

Color vision is most commonly tested by the *Ishihara* cards. Each card has a background of colored dots into which is blended a figure composed of dots of color likely to be confused with that of the background. The person with normal color vision will be able to see the superimposed figure, while the color-blind patient will not. A sex-linked characteristic, color blindness occurs in 4% of men and 0.3% of women.

PRACTICE SESSION

Begin by taking a good look at the external features of your partner's eyes. Look carefully at the hair appendages. Now, measure the distance between the inner canthi, the pupils, and the outer canthi. Next, take a cross sight measurement of the protrusion of the globe. Rest the ruler carefully against the bony corner of the orbit and sight tangentially across the cornea.

Now, listen over the globe. Have your partner close his eye lightly and gently apply the bell of the stethoscope. You may hear some sounds of muscle tensing or lid movement but should not hear a to-and-fro swishing sound of a vascular malformation.

Now, practice everting the lid over a cotton-tipped applicator as in Fig. 7.5. Look at the Meibomian glands and look at the lacrimal gland tucked under the upper outer edge and the lacrimal duct at the inner canthus. Get a good idea of the size of the gland for future reference.

If appropriate Schirmer filter paper is available, perform the test of lacrimal flow. Gently place the end of the strip under the lower lid and have the subject close the eye and leave the strip hanging out for 5 minutes. Remove the strip and measure the length of moisture.

Now, try the cover test to assure that no heterophoria is present. Watch the fixation of the eyes as you have your partner look at a point and alternatively cover and uncover each eye.

Now, partially darken the room and with your penlight shine at the eyes from about 3–4 feet away. Notice carefully where the reflected spot is on each eye. Now, have him follow your light as you shine from different angles and confirm that the reflection remains equal on both eyes in all directions of gaze.

With the room still darkened, try the ophthalmoscope. First, make sure that both of you are comfortable so that you will be free to take your time and have your attention only on the eye, not the kink in your back from bending way over.

First, with the ophthalmoscope about 4 inches from your partner's eye, and the same distance from your eye, confirm the red reflex through the pupil. This is the same red reflex you've seen in home color snapshots.

Now, with the ophthalmoscope lens set at a black 20, put the scope to your eye and move in toward your partner until your middle finger rests on his zygomatic arch. Remember to use your right eye when looking into his right eye. Move the finger wheel until the cornea comes into clear view. Now, move a few more clicks of the wheel and take a moment to marvel at the beauty of the iris with its distinctive pattern and colors. Confirm that the pupil is absolutely round.

A black 5 or 0 on the ophthalmoscope should put you onto the eye lens. What, you don't see anything? Good, that is normal. Remember the lens is normally crystal clear and should be all but invisible.

Now take a break, for both of you. Set your ophthalmoscope lens at a red 2 and come into position again. Don't panic, nobody sees the retina well the first time. You will find that you get quick glimpses and then everything goes out of focus again. Make certain that your partner is focused on a distant point. Look more toward the nasal side of the retina and try several times for short intervals. Gradually you will get the knack. Look now at the disc. See how it blends into the retina. Notice that it appears to be sharper in its border on the nasal side than on the temporal side. Now look at the vessels in the disc. You can see how they arise down deep and drape slightly over the rim of the disc. You should have the distinct impression of a depression in the center of disc. Pay attention to the apparent depth of this cup in your normal subject so that you can compare it to diseased eyes later.

Now, follow one of the major vessels coming out of the disc. Pick one of the smaller lighter ones, as that will be an artery, and follow it out as far as you can through several branch points. Find a place where an artery and a vein are close together and compare their relative widths and colors. Look closely at points where arteries and veins cross. Look very closely at one large vein and see if you can detect the minute pulsations within it.

You will get the impression that the upper vessels and the lower vessels tend to deviate around the macula. The macula is probably the hardest point to find. It is relatively free of major vessels and when you have your light directly on the macula you will experience the brightest back reflection. A little patience and you will see this little spot of increased color and sheen. Really sharp vision is centered around the macula so that lesions here are particularly devastating.

To get used to it, try the green filter. At first you will want to give up on it since it looks so dark. Persist, however, and you will see the vessels a new way and also get a sense of the sheen of the retina better. Now, try using the slit as suggested by holding the ophthalmoscope in proper position against your partner but move your head to the side and note the reflection of the light first on the cornea and then on the iris to get an idea of the depth of the anterior chamber. Now give your partner a lollipop for putting up with you.

Ears

History

Unlike visual loss, hearing loss is likely to be more symptomatic to the patient's family and friends then to the patient himself, "He's always got the TV on so loud," "I can't talk to her on the phone," "I always have to repeat myself." When eliciting the history of hearing loss, it is well to question the family. The patient may volunteer a preference for one ear when on the telephone. Careful questioning may reveal the range of frequency loss. The patient may admit that he hears men better than women and women better than children, thus indicating a high frequency hearing loss.

Tinnitus represents another form of dysfunction. Tinnitus is a buzzing, clicking, or ringing, constant or intermittent, unilateral or bilateral. Tinnitus originates somewhere proximal to the oval window and has many etiologies.

Pain in the ear will cause the patient to seek medical attention. Pain on manipulation of the tragus means inflammation of the external canal. Pain behind the ear points to mastoid inflammation. Deep-seated pain, aggravated by bending over, means disease of the middle ear. When the eustachian tube becomes closed because of edema, the middle ear becomes, in effect, an obstructed hollow viscus. Distention with air or pus under pressure produces severe pain. A chronic discharge from the ear most commonly originates from the external canal. A sudden burst of material followed by relief from pain is the rule when the drum spontaneously perforates.

Vertigo is a special form of dizziness. The patient with true vertigo describes the sensation of the room spinning about him. Many abnormalities can cause this irritative phenomenon of the labyrinth mechanism.

Eliciting associated complaints in ear disease is particularly important. Are the symptoms of otitis media accompanied by symptoms of a brain abscess? Is the vertigo associated with hearing loss; i.e. are both functions of the eighth cranial nerve affected? Has the patient with hearing loss been taking any medications?

Examination

The ear may give you valuable clues to the presence of kidney disease. There are many occasions in clinical medicine when this oto-renal axis becomes apparent. Congenital malformations of the external ear frequently accompany anomalies of the urinary tract, particularly the collecting system. Congenital deafness with nephritis is a well recognized entity. Many drugs share both otic and renal toxicity. Gouty deposits of uric acid are common in the helix of the ear as well as in the kidney.

External Ear. Note the position of the ear on the head. The origin of the helix should be on a horizontal line with the corner of the eye. Low set ears often accompany congenital malformations elsewhere.

The cartilage of the ear should be stiff but not rigid. This cartilage *calcifies* in response to remote frostbite, occasionally in Addison's disease, and in the rare metabolic disease ochronosis. Gouty deposits are hard white discrete nodules along the upper helix (Fig. 7.11). The markedly deformed "cauliflower ear" is the result of trauma, hemorrhage, and/or infection. The ear of the adult generally does not exceed 12 cm in vertical span. Very large ears are said to be seen in pernicious anemia.

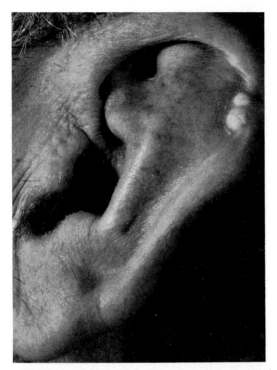

Fig. 7.11. Gouty tophus. The two white lesions on the helix of this patient's ear were hard and nontender. In addition, he had marked tophaceous gout involving the hands, feet, and elbows.

The mastoid air cells communicate with the middle ear. In the preantibiotic era, mastoiditis as a result of the extension of otitis media was common. Mastoid tenderness is common in acute otitis media, but true mastoiditis is now rare. With mastoiditis, the ear projects farther away from the head than its normal counterpart. Tenderness over the mastoid antrum can always be elicited in mastoiditis. The antrum can be approached by bending the ear forward and insinuating the index finger into the triangular depression between the mastoid prominence and the attachment of the ear (Fig. 7.12).

External Meatus and Ear Drum. The *otoscope* is a simple illuminating device attached to a speculum to which is added a magnifying lens (Fig. 7.13). Specula of varying sizes are available to fit children and adults. Disposable covers insure against cross contamination. Most instruments also provide for the attachment of a rubber tube and bulb for air insufflation.

The skin of the external ear, being closely applied to the bone, is relatively unyielding

and painful if traumatized. Arnold's nerve, a branch of the vagus nerve, runs in the external canal. Irritation of Arnold's nerve accounts for the coughing induced in some patients when you introduce the speculum into the canal.

The external canal in children is straight. The adult canal angulates, and it becomes necessary to pull upward and backward on the auricle to visualize the drum. To examine the right ear, take the auricle between your left thumb and foreginger and spread the other fingers over the occiput. You will then be able to maintain traction to straighten the canal and move the head at the same time while holding the otoscope with the other hand. Examine the external canal for erythema, debris, foreign bodies, discharge, and patency.

Next, examine the *ear drum*. Normal landmarks should be identified or their absence noted. The drum angulates so that the superior aspect is closer to you than the inferior portion (Fig. 7.14).

The normal drum is pearly gray and concave. A bright triangular light reflex appears with the apex pointing toward the center of the drum and its base directed toward the jaw. A localized bulge close to the apex of this light is the tip of the malleus. The arm of the malleus runs up and forward toward the eye to the short process. Two folds in the

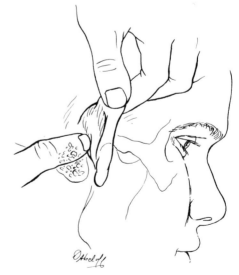

Fig. 7.12. Palpating the antrum of the mastoid. The index finger is pressed firmly behind the tragus in the position shown.

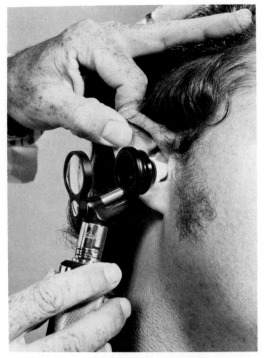

Fig. 7.13. Using the otoscope. The thumb and index finger of the left hand grasp the pinna while the other fingers of the left hand are rested on the head for control of position. The otoscope is gently introduced into the external auditory canal.

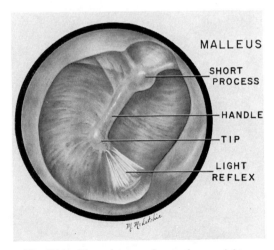

MALLEUS

SHORT PROCESS

HANDLE

TIP

LIGHT REFLEX

Fig. 7.14. Normal tympanic membrane, right ear. The patient's nose would be to the right of this diagram and his occiput to the left. The normal landmarks are indicated.

drum radiate from the short process and run forward and backward.

Acute inflammation of the drum appears as erythema, and small blood vessels may course over its surface. *Suppurative otitis media* causes inflammation and bulging of the drum with loss of the normal landmarks. *Serous otitis media* appears as bubbles of air in fluid behind the drum with little inflammatory change. Perforation may happen anywhere in the drum. The end result of untreated infection is scarring. A *scarred drum* becomes white, thick, and retracted. The retraction makes the malleus appear more prominent. Repeated infections or a chronic inflammatory process may produce a *cholesteatoma*. This mass of squamous epithelium and debris looks like a chalky granulated nodule anywhere along the rim of the drum. It is not a true tumor but may expand into adjacent structures.

The mobility of the drum is lost with scarring. If the drum appears normal, have the patient pinch his nose and attempt to blow against closed lips. A normal drum will bulge outward in response to the increased pressure transmitted up the eustachian tube. Do not do this test when there is obvious disease in the ear. Mobility may also be tested by gentle insufflation of air from the otoscope with the tube and bulb attached.

Hearing

Hearing loss is crudely classified into two types which have different etiologies and different treatments. *Conductive hearing loss* results from interference with the translation of air vibrations delivered to the inner ear. *Sensorineural hearing loss* infers disease somewhere from the organ of Corti to the brain. Combination defects are common. The two types can be distinguished by a tuning fork, although the precise nature of the defect requires sophisticated audiometric testing.

A tuning fork with 512 vibrations/second is the most reliable. Rinne's test detects conduction deafness. The normal state of affairs exists when sound transmitted through the air is heard better than sound transmitted through bone. If disease prevents normal conduction of sound waves, then transmission through bone will bypass the difficulty. (Beethoven used this phenomenon when his hearing began to fail. He would put his teeth against the piano while playing in order to "hear" his compositions.) To perform the Rinne test, strike the tuning fork and apply the handle to the mastoid prominence. When the patient indicates that he can no longer hear the sound, immediately place the vi-

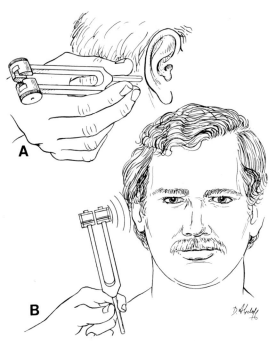

Fig. 7.15. Rinne test. *A*, The vibrating tuning fork is first applied to the mastoid prominence until the patient no longer detects sound. *B*, The tuning fork is then held close to the ear, and the patient is asked if he can hear the sound once again. Normally, air conduction is better than bone conduction.

brating heads close to the external canal. Normally, the sound is again heard and the test is *positive*. A negative test means that conduction deafness exists in the tested ear (Fig. 7.15).

Weber's test will confirm conduction deafness or indicate the presence of sensorineural deafness. Strike the tuning fork and apply it to the vertex of the head and ask the patient if he hears the sound better in one ear than the other. Normally, the sound appears to originate in the middle. If there is a conduction deafness in one ear, the sound will appear louder in that ear. If there is a nerve deafness in one ear, the sound will appear closer to the normal ear (Fig. 7.16).

Schwabach's test confirms sensorineural deafness by comparing the patient's hearing against the normal examiner. Strike the tuning fork and apply it to the mastoid of the patient. When the sound is no longer heard, transfer the fork to your own mastoid. If you can hear the vibration, the patient has a sensorineural defect. If you cannot hear the vibration, ask a colleague to test you.

With these three tests, the two types of

hearing loss and their various combinations can be diagnosed fairly accurately. Vestibular testing will be discussed in the neurologic chapter.

Nose and Sinuses

History

Dysfunction of the nose registers as a loss of smell, inability to filter or cleanse the air, or problems with humidification of the inspired air.

Loss of smell is a physiologic event of aging. It is not usually noted as such but, rather, is likely to be called a loss of taste since these functions are intimately associated.

Filtering and cleansing is accomplished by the cilia of the respiratory epithelium and the mucus-secreting cells. Irritation calls forth increased mucus which if overwhelming or associated with inadequate ciliary function bothers the patient as *rhinorrhea*, "clogged nose," or postnasal drip. The nasal epithelium is a rich source of plasma cells responsible for

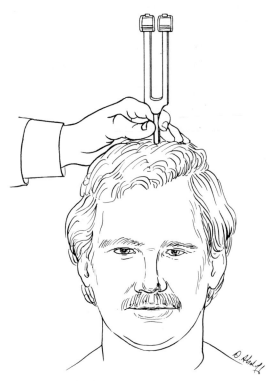

Fig. 7.16. Weber's test. The vibrating tuning fork is placed on the middle of the head behind the frontal sinuses. The patient is asked whether of not he hears the sound more strongly in one ear than the other.

secretory IgA and allergic IgE antibodies—hence, the rhinitis of hay fever and sneezing of allergic reactions. In this context, a seasonal or exposure history is helpful.

The free extrusion of fluid from the mucous membranes warms and humidifies the inspired air. The patient who complains of painful breathing in cold air may be telling you that his nasal mucosa is dry.

Pain as a symptom usually relates to the sinuses. Inflammation with obstruction leads to severe pain since secretion continues but egress is prevented. The maxillary and frontal sinuses localize pain over their locations. Sphenoid sinusitis presents as occipital pain. Sinus pain is aggravated by anything which increases the pressure within the sinus. Thus, the patient will avoid bending over, coughing, sneezing, or blowing the nose.

Epistaxis is a startling, usually painless observation. Elicit any history of minor trauma, upper respiratory infection, or irritation as a local cause but do not neglect a history of hypertension or bleeding disorders as more ominous etiologies. *Snoring* is an observation frequently more symptomatic to the spouse than the patient. This frequently indicates nasal obstruction and should not be dismissed out of hand, especially in the child. The nose is one place from which spinal fluid may find free drainage to the outside. Basilar skull fractures with small ruptures in the cribriform plate allow free flow of cerebrospinal fluid. A clear discharge is the symptom and recurrent meningitis may result. Another observation pointing to the nose is a change in the timber of the voice. The nasal passages provide a resonating chamber for speech. Obstruction prevents normal reverberations of sound.

Examination of Nose and Sinuses

The nose is best examined with a nasal speculum and bright light source directed to a head mirror. Alternately, the large short speculum on the otoscope will suffice. Remember that the axis of the nasal passage is at a right angle to the face, not parallel with the bridge of the nose (Fig. 7.17). To obtain adequate visualization, lift up on the tip of the nose and introduce the speculum straight in.

The nasal septum divides the passage approximately, rarely exactly, into two equal chambers. Inspect the mucous membrane of the nose for variations in the normal pink-red

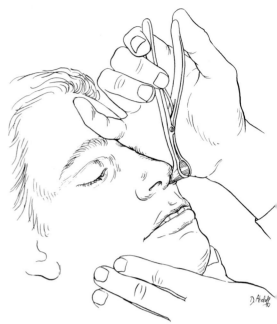

Fig. 7.17. Using the nasal speculum. The hand is rested on the bridge of the nose and the speculum controlled between the thumb and index finger. Note the axis of the speculum which is perpendicular to the face and not parallel to the nose.

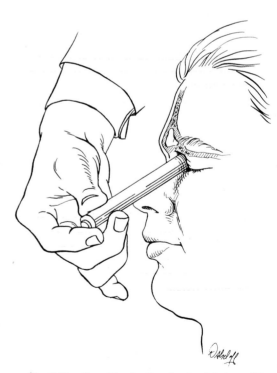

Fig. 7.18. Transilluminating the frontal sinus. The room should be darkened and a penlight firmly applied in the position shown. One side is compared with the other.

color. The septum is a common source of epistaxis. *Perforations* of the septum may be the result of chronic irritation or trauma or may indicate the gummatous destruction of syphilis. Marked deviation of the septum will obstruct one passage.

Directing the light laterally, several bulbous structures will come into view. Inferiorly, the smooth rounded anterior end of the inferior turbinate is seen. Above this, the anterior end of the middle turbinate is seen. The cleft-like recess between these turbinates is the middle meatus. The superior turbinate cannot be seen anteriorly. The meatus between adjacent turbinates is the point of sinus drainage. Purulent discharge from a meatus indicates sinusitis. The narrow dark cleft between the middle turbinate and the septum is the olfactory sulcus.

Examine for mass lesions. Swollen hypertrophied turbinates may appear as masses. They will be seen, however, to be fixed and sensitive to manipulation. *Nasal polyps*, common in atopic patients, appear as grape-like, pale pink, relatively mobile masses. Malignancies look grey-white, are friable and relatively insensitive.

The nasopharynx is examined with the aid of a mirror directed up and behind the uvula. This technique will be discussed shortly in the section on the larynx.

The frontal and maxillary sinuses can be examined indirectly. With the room darkened, have the patient put a bright light source in his mouth to *transilluminate* the maxillary sinuses. A bright flashlight will suffice. The normal air-containing sinuses will light up symmetrically. If one contains pus, secretions, or blood, it will appear darker than its mate. Similarly, a small bright penlight pressed under the superior orbital ridge will transilluminate the frontal sinus (Fig. 7.18).

Direct percussion over a sinus which is acutely inflamed elicits exquisite pain. The patient with acute sinusitis usually tolerates this only once.

PRACTICE SESSION

Look carefully at your partner's ear. Measure the length of the ear and note the lobe configuration. Bend it forward and find the triangular depression of the antrum of the mastoid.

Now, with the otoscope in your right hand, pull the right ear slightly backward with your left hand. Use your thumb and index finger to do this and rest your other fingers on the head so that you can maintain traction on the ear to straighten the canal. Gently introduce the speculum. Note the color of the canal and the presence of any wax. Now, check the landmarks of the drum against the picture and confirm each major feature. Have your partner blow gently against closed nostrils and watch the drum move toward you.

Compare your hearing with your partner by measuring the distance from your ear where you can just no longer hear a ticking wrist watch. Then perform the Weber and Rinne tests.

Now, with the nasal speculum, examine the nostrils. Remember to use a bright light and to introduce the speculum perpendicular to the plane of the face. Note the color of the normal mucosa and identify the middle and inferior turbinates and the middle meatus between them. Which way is the septum deviated? (It always is to a certain extent.)

Now, transilluminate the sinuses. First, darken the room. Use your penlight and shine it into the upper inner corner of the eye as shown in Figure 7.18 to transilluminate the frontal sinus. The maxillary sinuses can be illuminated from inside or out. Place the light on the zygoma close to the nose and have your partner open his mouth wide. Look at the illumination of the hard palate on that side and compare its luminosity to the other side. Alternatively, have your partner close his mouth over a standard flashlight and look at the illumination on the outside.

Mouth and Throat

Again, we find a compact area with much clinical information, frequently of systemic importance.

History

Pain in the mouth is common and not surprising considering the various foodstuffs, liquids, and gases we assault the mouth with.

Toothache is the most common mouth pain and usually of obvious etiology. Tooth pain may be referred pain, as with a myocardial infarction. _Sore throat_ must be distinguished from pain with swallowing. The former usually means infection while the latter implies a mechanical association. Clearly, a severe sore throat makes swallowing painful, but when the complaint is pain just with swallowing consider a mass lesion. The complaint of a lump in the throat poses a difficult problem. This complaint most commonly is of hysterical origin—_globus hystericus_—but a mass lesion must be excluded.

The many tasks of the pharynx lead to a variety of dysfunctional complaints. _Trismus_ is spasm of the masticatory muscles which impedes the ability to open the mouth. A relatively early complaint in tetanus, it is also seen in acute ethmoid or sphenoid sinusitis and is occasionally a form of hysteria. _Dysphagia_ is difficulty swallowing. This complaint must always be taken seriously. Ask the patient if the difficulty is greater with solids or liquids. Dysphagia with solids means mechanical obstruction. Dysphagia with liquids is neurologic. It takes much more sophisticated motor control of the pharyngeal muscles to prepare a bolus of liquids than solids. The place that the patient indicates food "sticks" is a fairly reliable indicator of the point of obstruction.

Loss of taste may, as already noted, be in reality loss of smell. If the taste loss is unilateral, there is interference with the lingual branch of the seventh cranial nerve. Bilateral loss of taste frequently accompanies severe systemic illnesses of any form. The patient with hepatitis notes a loss of taste for cigarettes. A metallic taste bothers patients with a variety of metabolic derangements—uremia, for instance.

Mouth breathers develop a dry mouth which may be troublesome. Obstruction in the nose or dyspnea may be the cause. _Xerostomia_ also results from insufficient saliva. Inflammatory disease of the salivary glands impairs their secretory ability. _Sjögren's syndrome_ is xerostomia and keratoconjunctivitis sicca (similar loss of tears) seen in a variety of immunologic diseases such as rheumatoid arthritis.

Hoarseness is another dysfunctional complaint demanding thorough investigation. Throat pain and hoarseness makes local la-ryngeal disease suspect. Painless hoarseness is more ominous. Diseases as remote as carcinoma of the lung with involvement of the recurrent laryngeal nerve will produce hoarseness. The history of heavy use of the voice as in professional singers, auctioneers, and teachers may correlate with the finding of vocal cord nodules or hyperkeratosis.

A few historical observations which might come out in the systems review are pertinent. Family and friends may volunteer this information before the patient.

Halitosis, bad breath, may originate anywhere in the pharynx or lower respiratory tree. Dental caries is the usual source. The sweet musty odor of severe liver disease, fetor hepaticus, the metallic odor of uremia, and the putrid smell of lung abscess can all be noted.

A change in the voice may be commented on by the patient or family. A difference in resonance, as noted, occurs in nasal obstruction. Paralysis of a vocal cord yields a breathy, not truly hoarse, voice because of air wasting. As vocal cords become thicker, the timber of the voice gets lower. This explains the deep voice of myxedema and acromegaly or testosterone excess in the woman. Difficulty articulating is called _dysarthria_. When transient, it may be the hallmark of cerebrovascular disease—the patient sounds intoxicated. _Scanning speech_ is a peculiar form of dysarthria common in multiple sclerosis. It is an alternating hesitant and sliding speech. _Dysphonia_ is dysarthria-caused interference with the motor control of phonation. The dysphonic patient may be unaware of his deficit.

Examination

Examine the lips for angular stomatitis, also called _cheilosis_. This fissuring and cracking at the corners of the lips is caused by poor nutrition or ill fitting dentures. _Rhagades_ are white linear scars at the corners of the mouth and should prompt a search for other stigmata of late syphilis. The pigmentation of the lips is altered in some interesting illnesses. _Addison's disease_ causes a blotchy hyperpigmentation of the lips and oral mucosa. The telangiectasia of _hereditary hemorrhagic telangiectasia_ syndrome are prominent on the lips. Punctate dark macules found on the lips are characteristic of _Peutz-Jeghers syndrome_ or intestinal polyposis.

Carcinoma of the lip is the most common oral neoplasm. Ninety-five percent of the carcinomas of the lip occur on the lower lip. The characteristic lesion has heaped up hard margins with a central ulceration or depression. Primary chancres of syphilis on the lip may look like carcinoma but are most often on the upper lip. This lesion also has hard borders with a central ulcer but is usually painless.

After inspecting the lips, evert them to inspect the fornix around the mandible and maxilla. *Mucous retention* cysts may be found anywhere along these fornices. They are nodular and transilluminate brightly.

Leukoplakia is a patchy white lesion found on any of the mucous membranes of the mouth. Generally the result of chronic irritation, the white color is the result of an abortive attempt at keratinization. This lesion may be premalignant. It must be distinguished from *thrush* or *Candida* stomatitis. In monilia, the white patches can be lifted and peeled away.

Healthy gums are bright pink. The junction of teeth and gums is sharp and firm. If this junction recedes, infection and *alveolar pyorrhea* results. The teeth appear abnormally long, are loose, and discolored. *Hypertrophied gums* appear as bulging tissue encroaching on the teeth. Chronic therapy with diphenylhydantoin for seizure disorders, scurvy, and leukemic infiltrations causes this condition.

Pigmentation along the gum line occurs in heavy metal intoxications. Lead, bismuth, and silver are characterized by a series of tiny grey-black dots not quite at the dental gum line.

If the patient wears dentures, have him remove them as the chronic irritation is a common precursor of leukopakia and dental granuloma—the latter is a firm pink fleshy swelling covered with normal mucosa. Carefully inspect intact teeth for a source of occult infection. Check the occlusion. Acquired malocclusion may be the first sign of a neoplasm of the jaw. Chronic jaw clenching and teeth grinding is a nervous habit and may result in the loss of a significant portion of the occlusive surfaces. A blue-grey tooth is a dead tooth and probably an infected tooth. Notched incisors, *Hutchinson's teeth*, and the *mulberry* or five-pointed bicuspid—both signify congenital syphilis (Fig. 7.19).

The *tongue* as a muscular organ is covered with mucosa so altered as to provide a sensation of taste. First examine the tongue in its relaxed position on the floor of the mouth. Have the patient roll his tongue from side to side (Fig. 7.20). Lateral indentations along the adjacent teeth indicate *macroglossia*. Macroglossia is usually an insignificant finding but may indicate swelling from edema fluid, lingual malignancy or, rarely, infiltration by amyloid or myxedema. Now, have the patient

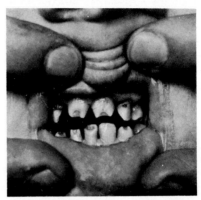

Fig. 7.19. Hutchinsonian teeth. Congenital syphilis can be implied when these notched teeth are seen. An additional lesion not shown here is the mulberry molar. Such molars have five points, four circumferential and one central.

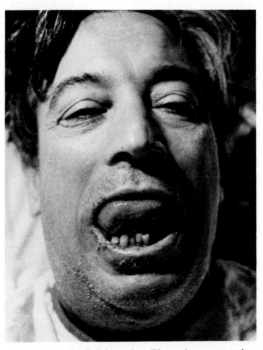

Fig. 7.20. Ludwig's angina. The patient cannot close his mouth because of the extensive cellulitis of the tongue and mucous membranes of the mouth.

fully extend the tongue. Inability to do so in the adult commonly indicates a malignancy of the tongue (Fig. 7.21). In the child, this disability may be the result of a short frenulum. Such a child has difficulty speaking—difficulty particularly with words which require elevation of the tongue. Deviation of the protruded tongue to one side means either a malignancy of that side of the tongue or damage to the hypoglossal nerve on the same side.

Examine the surface of the tongue. The papillae vary in shape ranging from the delicate filiforms anteriorly, to the midportion conical papillae, to the posterior large circumvallate papillae. The latter form a "V" with the apex posteriorly pointing to the foramen cecum, which is the remnant of the thyroglossal duct. Rarely, a lingual thyroid island can be seen in this foramen. Denudation of the papillae occurs in children in a geographic pattern. This is harmless and self-limited. Complete denudation in the adult gives the *beefy red tongue* which is characteristic of malnutrition and pernicious anemia. *Fissures* of the tongue are commonly encountered. Congenital fissures run across the tongue while the fissures of syphilis or malnutrition (especially ariboflavinosis) are longitudinal (Fig. 7.22).

Have the patient touch the roof of the mouth with the tip of the tongue and examine the undersurface. The membranous frenulum should be loose and pliable. Flanking both sides of the base of the frenulum are the ducts of the sublingual salivary glands. The glands can easily be palpated with the gloved index

finger under the tongue and the other hand pressing up from below the jaw. Its consistency should be lobular, firm, and uniform.

The posterolateral border of the tongue beneath the tonsillar pillars can also be palpated. To view this area, a common site of neoplasia, grasp the tongue with a gauze pad and pull outward and to the opposite side.

Stensen's duct, the orifice of the parotid glands, is examined by retracting the cheek away from the upper molars where it can be seen as a small papilla. It will be inflamed in parotitis and may drain purulent material.

Fig. 7.22.   Congenitally fissured tongue. The daughter and granddaughter of this patient have similar appearing tongues. (Courtesy of Dr. David Weisberger.)

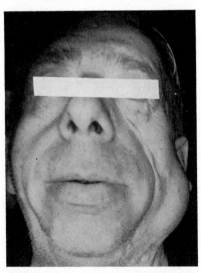

Fig. 7.23.   Mixed tumor of the parotid gland. The swelling extends over the angle of the jaw conforming to the anatomic position of the parotid gland. Surgical excision was accomplished and the microscopic evaluation was that of a typical mixed tumor. (Courtesy Dr. David Weisberger.)

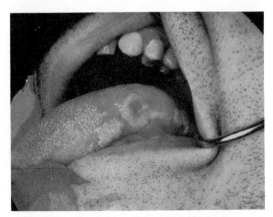

Fig. 7.21.   Carcinoma of the tongue. Note the round lesion with a central depression located on the side of the tongue. It is painless. Anterior to it is a small area of leukoplakia. (Courtesy Dr. David Weisberger.)

Koplik's spots, which are virtually pathognomonic for measles, are tiny grey-white spots with a rim of erythema. These spots are seen around or just below Stensen's ducts. A short portion of the duct can be palpated between the two fingers, one in the mouth and the other outside just below the zygomatic arch. In this fashion, an obstructing stone may be detected (Fig. 7.23).

As noted above, there are many indications for inserting the examining finger into the mouth. A protective maneuver for the examiner is to push a fold of the cheek in between the teeth with the other hand. This is especially helpful in children or patients with a hyperactive gag reflex (Fig. 7.24).

The *palate* has transverse rugal folds over its hard portion, and the soft portion is smooth. A high arched *gothic palate* accompanies many congenital malformations and is also seen in the mouth-breathing patient with a nasal obstruction. A midline palatal mass is a normal bony variant called *torus palatinus.* Any masses not in the midline are suspect for neoplasia. The palate may give the first hint of jaundice, the yellow color being most apparent in bright natural light. The palate is also a common place to find the microemboli of subacute bacterial endocarditis. These appear as small discrete petechiae. Small vesicles with surrounding erythema accompany viral pharyngitis (usually Coxsackie A, herpangina). A hole in the soft palate results from a syphilitic gumma.

The tonsils rest between two folds of tissue, the tonsillar pillars. Unless the tongue is fully

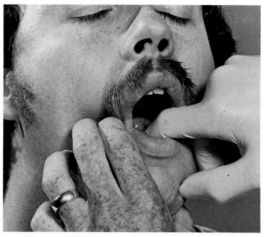

Fig. 7.24. Palpating the floor of the mouth. The gloved finger palpates beneath the tongue and against the thumb of the opposite hand. Note that the cheek has been pressed between the teeth with the left hand. This avoids having the patient bite your finger, an especially helpful maneuver with children.

extended and the tongue blade inserted deeply, this area will not be adequately visualized. Too often, the timid examiner presses on the midtongue, making an obscuring mound of posterior tissue. The tonsils naturally enlarge to age 7 and thereafter shrink in size unless chronically inflamed. Acute tonsillitis appears as red swollen tonsils, which may be so large as to meet under the uvula. Most infections cause the crypts to appear prominent, and they may contain pus. The bacterial infections call forth more purulence than the viral infections.

PRACTICE SESSION

Begin by inspecting the lips and the vermillion border. Now evert the lower lip and look carefully at the fornix and the junction of the gum and teeth. Do the same with the upper lip.

Now, using a tongue blade, push the lateral cheeks away from the teeth and complete this inspection. Find the entrance of the ducts from the parotid salivary glands on each side.

Now, with the tongue elevated to touch the palate, note the frenulum and the ducts of the submandibular glands. With the tongue extended to its maximum, look at the different papillae going all the way back to the remnant of the thyroglossal duct.

The soft palate should elevate when your partner phonates with his mouth open. Check to see if he has been previously relieved of his tonsils. If not, get an idea of how big they are and note the multiple crevices of this tissue, and the two sets of pillars surrounding them.

Practice depressing the posterior tongue with the tongue blade. You must get back far enough to truly depress the tongue but not so far as to make him gag excessively. Your success in doing this is as much a matter of your slow and deliberate technique as it is the sensitivity of the patient to a gag reflex.

Now, put on a glove and palpate the submandibular salivary gland and the terminal portion of the duct from the parotid gland.

Nasopharynx and Hypopharynx

These areas cannot be directly visualized without special equipment. A small additional effort is required to examine them indirectly. The tools include a head mirror, bright light source, and a dental mirror. The examination is gentle and performed with care.

Sit in front of the patient or to his right with the light source shining at you from the patient's left. Bring the head mirror down over your right eye so that you see clearly through the central hole with your right eye and around the edge of the mirror with your left eye. The curvature of the mirror determines the focal length at which the reflected beam is brightest. By moving your head back and forth you can vary this point of maximum illumination. Now, have the patient lean toward you slightly with his neck slightly flexed and chin up. Have him protrude his tongue fully and wrap it with a gauze pad (Fig. 7.25). Put your thumb under the tongue and your index finger on top. Apply traction and rotation so that your thumb is now on top. By so turning the tongue, you protect the delicate undersurface from injury against the teeth and you aid maximum protrusion. Warm the mirror in hot water or over an alcohol lamp, test the temperature against the back of your hand and gently insert the mirror from the corner of the mouth until it touches the base of the uvula. Continue by lifting the uvula up and back. Focus the beam of light on the mirror and the larynx will come into view. To view the nasopharynx, the mirror is introduced under and behind the uvula. The patient must breath slowly and easily or the soft palate will elevate and obstruct your vision. Recall that the reflection in the mirror will be upside down. When viewing the larynx, the epiglottis will appear at the top and the posterior pharyngeal wall at the bottom of the mirror.

When examining the nasopharynx, first find the *eustachian tubes* laterally. The upper posterior wall of the nasopharynx is the site of *adenoidal* tissue which, when hypertrophied, may obscure and obstruct the posterior aspect of the nasal passages. The *choanae* are separated by the white pillar of the posterior nasal septum. The posterior ends of the turbinates come into view in each choanal cavity. Neoplasms, inflammation, bleeding, and polyps may all be detected here.

Reversing the mirror, to view the hypopharynx, the *epiglottis* appears most prominent. The anterior surface leads down to the reflection of the epiglottis onto the tongue with two trough-like spaces laterally, the *valleculae*. If the epiglottis obscures the larynx,

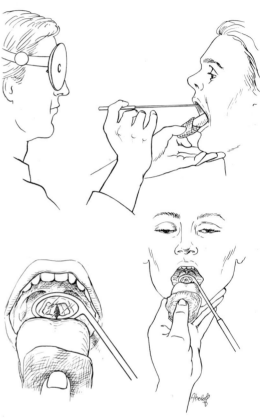

Fig. 7.25. Indirect laryngoscopy. The tongue is wrapped in a gauze pad and pulled forward. The fingers under the jaw allow you to manipulate the head. Introduce the mirror without touching the soft palate to avoid a gag response and focus the light with your mirror.

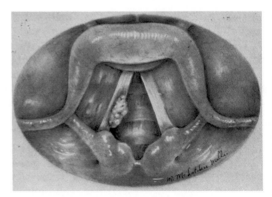

Fig. 7.26. Early carcinoma of the right vocal cord as seen with a mirror. This small lesion would prevent the cords from coming together in the midline and result in hoarseness.

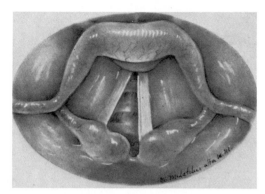

Fig. 7.27. Paralysis of the left vocal cord as seen with a mirror. The left vocal cord cannot be moved away from the midline—abductor paralysis. This would be the appearance of the vocal cords in a patient with a carcinoma in the left lung involving the recurrent laryngeal nerve.

apply gentle twisting traction on the tongue and have the patient say "eh."

The lateral edges of the epiglottis sweep posteriorly, embracing the larynx as the *aryepiglottic* folds just over the posterior arytenoids. The *true vocal cords* show as pearly white strips running anteroposteriorly. They should be smooth, moist, and symmetrical. Any irregularity, nodularity, erythema, or secretion is abnormal (Fig. 7.26).

Normally, the vocal cords move outward with inspiration, and the greater the inspiratory force the greater the motion. On expiration, the cords move toward each other but do not meet in the midline. One cord may appear to move more than the other. This is more apparent than real and is caused by tilting the mirror. During phonation, the cords vibrate against one another, slowly with low pitches and rapidly with high pitches.

Paralysis of a cord may be of the abductors or adductors (Fig. 7.27). Unilateral abductor paralysis causes few symptoms since the affected cord will lie in the midline and phonation will be near normal. This is *incomplete paralysis*. With progression to *complete paralysis*, the adductor fails and the cord moves outward. The normal cord will come across the midline for phonation and a slightly breathy quality to the voice is apparent. *Bilateral adductor paralysis*, most commonly hysterical in origin, prohibits phonation since the cords will not come together. The hysterical etiology can be proved by having the patient cough—during which the cords will close normally. Bilateral adductor paralysis causes stridor and severe difficulty. Neither cord moves away from the midline and the airway becomes compromised.

chapter 8

NECK

With the exception of the thyroid and parathyroids, the neck is an area of transit. Anatomically we are concerned with lymph chains draining separate regions, blood vessels to and from the head and neck, the esophagus, trachea, and cervical spine.

History

Of the many observations a patient can make about his own body, the most frightening to him is the discovery of a *lump*. In our zeal, we have taught the lay public that any lump is cancer until proven otherwise. Recognize, then that such a patient presenting with a chief complaint of a lump in the neck is an anxious patient. A temporal history is important. A slowly enlarging mass is more significant than one which has been present for years. The association of an upper respiratory infection with a lump in the neck usually connotes normally functioning lymph nodes. A mass which appears after eating or drinking is virtually a diagnostic history of a pharyngeal pouch or high esophageal diverticulum. Such a patient will also report regurgitation of liquid or masticated food when supine.

Pain in the neck is as varied as the structures present. *Tension headache* almost always includes neck pain, worse toward the end of the day. The patient usually will rub the back of his neck as he tells you of the discomfort. Pain produced by moving the head in a certain direction originates from the musculoskeletal structures. *Thyroiditis* causes pain in the anterior neck which is aggravated by touching the area, constricting it with clothing, or swallowing.

A number of remote complaints signal the need for a careful neck examination. *Thyroid diseases*, because of their profound metabolic effects, may present as complaints as distant as diarrhea, oligomenorrhea, weight loss, fatigue, or tremor in hyperthyroidism, and as constipation, fatigue, lassitude, menorrhagia, or weight gain in hypothyroidism. *Hypercalcemia* from a parathyroid neoplasm insidiously disables a patient. Constipation and fatigue are common. The loss of renal concentrating ability from hypercalcemia produces polyuria and secondary polydipsia. Weight loss with abdominal complaints suggesting a neoplasm should prompt a search for the left supraclavicular *"Virchow's node"* close to the drainage point of the thoracic duct. Similarly, supraclavicular nodes in the patient with pulmonary complaints might afford tissue for the diagnosis of cancer, sarcoid, or tuberculosis. The patient who complains of episodes of unilateral blindness (*amaurosis fugax*) or episodes of dizziness, nausea, lightheadedness, or transient hemiparesis might be telling you of critical *compromise of the carotid artery* which if uncorrected will progress to a complete stroke. Encroachment on the neural foramina of the cervical spine by the bony overgrowths of osteoarthritis commonly occurs in the elderly. The complaint, however, may be not of neck pain but rather of shoulder and arm pain with weakness. Most of the time, wheezing means asthma. More than one physician, however, has been chagrined by neglecting obstruction of the upper airway as a cause. This area of transit then may be diseased with effects at either end of the respective conduits.

Examination

The neck is artificially divided into a number of components. Delineating the midline is the notch of the thyroid cartilage and the suprasternal notch. The *submandibular triangle* is defined by the mandible and the two bellies of the digastric muscle. The prominent sternocleidomastoid defines the lateral border of the *anterior cervical triangle*. The *supraclavicular fossa* is found between the clavicle, the insertion of the sternocleidomastoid, and the insertion of the trapezius on the lateral clavicle. Superiorly, this space becomes the *posterior cervical triangle* with the trapezius and the sternocleidomastoid as the major markers. Although these areas are artificial, the accurate localization of a mass lesion may permit swift diagnosis (Fig. 8.1).

Inspect the neck for obvious swellings. These are sometimes better seen than felt. Have the patient swallow and note the motion of the thyroid cartilage and the motion or lack of motion of any detectable masses (Fig. 8.2). Notice how the patient carries his head.

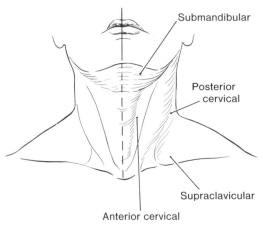

Fig. 8.1.   The triangles of the neck.

Spasm of the neck muscles forces a patient to hold his head canted and fixed so that the offending muscle or muscle group is on minimum stretch. Such a patient will turn his trunk to turn his field of vision. Severe osteoarthritis of the cervical spine likewise causes a patient to rotate at the shoulders rather than the neck. In the latter, motion is limited by bone, not by pain.

Palpate a mass to determine its *location, consistency, size,* and *mobility.* Lymph nodes may be discrete or matted, rubbery or stony hard, free or fixed, painless or tender. Physiologic enlargement of lymph nodes in response to an active filtering function usually causes discrete enlargement with firm nonfixed characteristics. If the node contains polymorphonuclear leukocytes (lymphadenitis), tenderness will be present. Metastatic disease usually imparts a stony hard sensation to the palpating fingers. Lymphomatous disease is usually rubbery. Lymphoma, and less so carcinoma, will mat nodes since a pericapsular extension of the disease is the rule. Cystic masses in the neck are fluctuant to a greater or lesser extent dependent on the thickness of the cyst wall, the viscosity of the entrapped fluid, and the pressure within the cyst. Transillumination is similarly variable (Figs. 8.3 and 8.4).

Palpation is performed from both the front

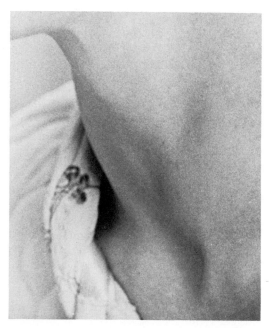

Fig. 8.2.   Diffuse thyromegaly. This young lady had Grave's disease. A smooth swelling is apparent on both sides beneath the sternocleidomastoids.

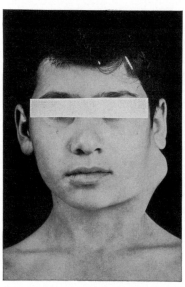

Fig. 8.3.   Tuberculous lymphadenitis. This young man had been aware of a painless swelling at the angle of the jaw for several months. Untreated, the natural history would be spontaneous drainage with a resultant sinus tract. This is a "cold" abscess.

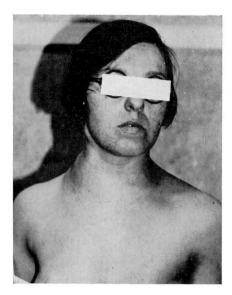

Fig. 8.4. Lymphoma involving the cervical lymph nodes. This young woman had painless enlargement of lymph nodes for 4 weeks. On palpation, they were rubbery and matted. Biopsy subsequently confirmed the diagnosis of Hodgkin's lymphoma.

left main stem bronchus will now transmit its pulsations up the trachea to your fingers.

To palpate the *carotid artery*, lay three fingers alongside the thyroid cartilage and gently displace the sternocleidomastoid laterally and the thyroid cartilage medially (Fig. 8.5). Use the right hand to feel the left carotid and vice versa. Palpate gently and only on one side at a time to avoid carotid sinus stimulation and reflex bradycardia. The carotid artery is a favorite site of atherosclerosis. The atheromatous material projecting into the lumen disturbs lamellar flow and sets up turbulence which is audible as a *systolic bruit*. This sound, which is high pitched, is best heard with the diaphragm of the stethoscope while the patient holds his breath (Fig. 8.6).

In addition to the carotid artery, the ante-

and the back of the patient. From the front, first palpate the posterior cervical space beginning at the mastoid process and run your fingers down along the border of the trapezius. A chain of lymph nodes is present here which drain the posterior scalp and retropharyngeal space. They may be selectively enlarged in German measles and are frequently involved in infectious mononucleosis and other viral illnesses. Just anterior to this chain of nodes, you can feel the lateral processes of the cervical vertebrae and then the posterior edge of the sternocleidomastoid.

The thyroid cartilage is the most prominent structure in the midline. It can be manipulated freely from side-to-side but not up and down. Lateral movement should produce palpable *crepitus* as it slides across the anterior cervical spine. Lack of such crepitus may signify a mass in the retropharyngeal space.

The hyoid cartilage lies above the thyroid cartilage. In the elderly, the lateral horns of the hyoid may ossify and feel like stony hard lymph nodes. Both cartilages rise with swallowing. A *tracheal tug* can sometimes be felt in patients with aneurysms of the thoracic aorta. To detect this sign, have the patient swallow and apply firm pressure on the thyroid cartilage. Trap the cartilage at its highest point of ascension to stretch the entire trachea. The dilated aorta as it curves over the

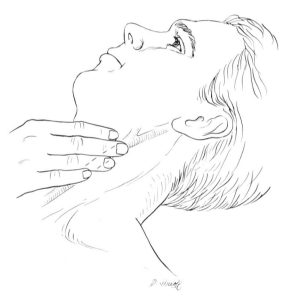

Fig. 8.5. Palpating the carotid artery. The sternocleidomastoid is displaced laterally and the trachea displaced medially.

Fig. 8.6. Listening for carotid bruits. The patient holds his breath and the carotid artery is ausculted over the bifurcation.

rior cervical triangle contains an important string of lymph nodes. Pharyngitis and tonsillitis are the most common causes of enlargement of these nodes. Laryngeal, pharyngeal, and thyroid cancers all tend to spread to these nodes.

With the exception of a single node just above the thyroid, midline swellings are never of lymph node origin. The thyroid gland during its embryologic descent from the hypopharynx may leave a *thyroglossal duct remnant* anywhere along its midline trail. If the duct maintains a connection with the foramen cecum, it will move upward as the tongue is protruded. If the cystic remnant is attached to the thyroid, it will ascend with swallowing.

Lateral nonlymph node swellings include *carotid artery aneurysms, cystic hygromas,* and *branchial cleft cysts.* Aneurysms will pulsate, cystic hygromas are seen primarily in children and transilluminate brightly, and branchial cleft cysts usually appear in adult life as firm cystic swellings beneath the sternocleidomastoid near the angle of the jaw (Figs. 8.7 and 8.8).

Inspect the thyroid gland from in front of the patient and palpate both from in front and from behind. Thyroid enlargement may be diffuse or asymmetrical (Fig. 8.9). Palpate first with the fingers across the trachea, then

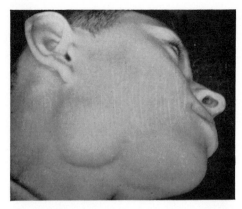

Fig. 8.8. Branchial cleft cyst. This is the typical location of such a lesion. This is really a congenital lesion and corresponds to the location of the branchial clefts or embryologic gills. They generally do not appear until early adult life.

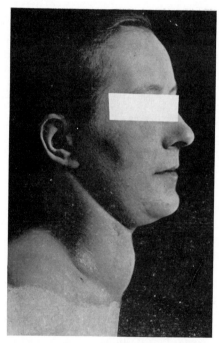

Fig. 8.9. Goiter. The pronounced enlargement of this patient's thyroid gland had been present for many years. Examination and biochemical tests confirmed that this patient was euthyroid—nontoxic goiter.

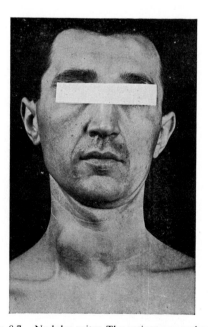

Fig. 8.7. Nodular goiter. The patient was euthyroid. A similar nodule had previously been removed from the left lobe of the thyroid.

seek to outline both lateral lobes (Figs. 8.10–8.12). Put the tips of your fingers against the lateral border of the sternocleidomastoid and the tip of your thumb on the midline just above the sternal notch and have the patient swallow. If the thyroid is enlarged, it will be palpable during swallowing as the tissue rises

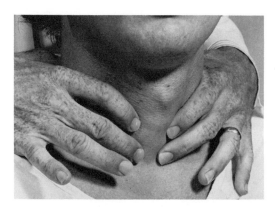

Fig. 8.10. Palpating the thyroid from behind. The fingers are placed on either side of the tracheal cartilages while the patient is asked to swallow. An enlarged thyroid will be felt to rise underneath the fingers with swallowing.

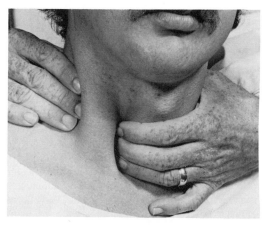

Fig. 8.11. Palpating the right lobe of the thyroid. The fingers are so positioned to trap the right lobe of the thyroid beneath the bulk of the sternocleidomastoid.

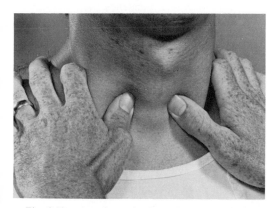

Fig. 8.12. Anterior palpation of the thyroid gland. The thumbs are so positioned to locate the lateral lobes of the gland on each side of the tracheal cartilages.

under your fingers. The normal thyroid cannot be easily felt. In thyroid diseases which cause diffuse enlargement, the pyramidal midline lobe is easily felt. Occasionally the enlargement extends down beneath the sternum as a *substernal goiter*, detectable only by hyperextending the neck and having the patient swallow (Fig. 8.13). A hyperfunctioning thyroid gland is also hypervascular. The enlarged vessels may impart a thrill to the hand and a bruit to the ausculting ear. Now, moving behind the patient, again feel for thyroid enlargement. From this position, both hands can palpate simultaneously (Fig. 8.14).

The posterior approach also affords a good opportunity to palpate the submental and submandibular lymph nodes. Next, direct your attention to the supraclavicular fossae. If the fossae are deep, palpation for masses can be facilitated by having the patient perform a Valsalva maneuver (attempted expiration against a closed glottis). This forces the apex of the lung and the structures in the supraclavicular fossae upward against your fingers (Fig. 8.15).

A *cervical rib* is an accessory rib or fibrous band from the seventh cervical vertebrae. Most commonly an incidental finding, it sometimes causes symptoms. The first thoracic nerve and the subclavian artery must angulate sharply to cross such a rib with the production of vascular or neural symptoms

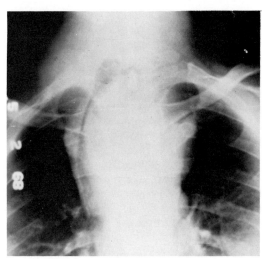

Fig. 8.13. Substernal goiter. The marked enlargement of the thyroid gland can be seen in the anterior mediastinum below the heads of the clavicle. The thin lucency which transcribes a gentle arc to the right is the compressed and displaced trachea.

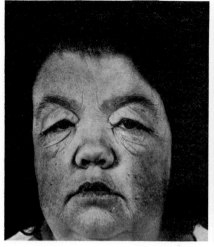

Fig. 8.14. Myxedema (*left*). The coarse features, dull expression, and baggy skin are characteristic. Thyrotoxicosis (*right*). The widened palpebral fissures give the patient the appearance of having been startled. The white sclera is apparent all the way around the cornea. The hyperresponsiveness to catecholamines affects the sympathetic levators of the eyelid. They respond by producing this stare.

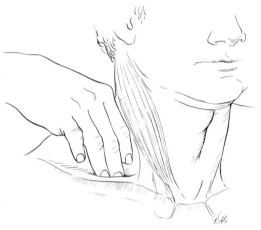

Fig. 8.15. Palpating the supraclavicular fossa. The fingertips are pressed deeply in and behind the clavicle.

The cervical spine is the cause of great discomfort in many people, particularly the elderly. Inspect all motions of the neck. Head nodding and rotation involve primarily the allanto-occipital articulation. Lateral flexion to either side requires a freely moving midsection of the cervical spine while flexion and extension call the lower cervical vertebrae into action. With the head in maximum flexion, run a finger down from the inion of the skull to the vertebral prominence, the seventh cervical vertebrae. A tough ligament, the ligamenta nuchae, can be felt. This band is partially ruptured in *whiplash* injuries and may avulse the spinous process of the seventh cervical vertebra. Tenderness to percussion over a spinous process is fairly hard evidence of pathology within that vertebra. Spasm of paraspinous muscles can be caused by direct muscle injury or spine disease. A muscle in spastic contraction becomes firm and tender to touch. The patient will assume a position which allows maximum shortening or relaxation of the affected muscle or muscle group.

Meningitis causes spasm and pain which are aggravated by stretching the meninges. Flexion of the head, attempting to put the chin on the chest, will produce severe pain in meningitis or will be prohibited by intense spasm of the neck muscles. The spasm can be so intense that in your effort to bend the head forward the entire trunk will come up from the bed.

in the arm and hand. Pallor or cyanosis of the hand and pain in the ulnar nerve distribution with wasting of the thenar eminence are common. *Adson's maneuver* helps suggest the diagnosis. With the arm hanging straight down, palpate the radial pulse. Then, have the patient turn his head toward the side of the symptoms and take a deep breath. This puts the scalenes on tension; raises the first, second, and accessory ribs; and may obliterate the pulse and produce pain.

The subclavian and vertebral arteries may produce audible bruits in the supraclavicular fossae and may become aneurysmally dilated.

PRACTICE SESSION

Let's start by identifying all the landmarks of the neck. Facing your partner, identify first the thyroid cartilage with its prominent notch. Move this cartilage from side-to-side and note the fine crepitus as it slides over the vertibrae. Follow the trachea down to the suprasternal notch, just lateral to which are parts of the insertion of the sternocleidomastoid muscles on each side. Follow the forward edge of that muscle on each side to define the anterior cervical triangles. Gently palpate the carotid artery on one side then the other, as shown in Figure 8.5, using three fingers to push the trachea and muscle away from each other. High in the triangle on each side, you may well be able to palpate the posterior ends of the submandibular salivary glands.

Now, go to the mastoid prominence and palpate down the posterior border of the sternocleidomastoid. This is the posterior cervical triangle, the posterior border of which is the trapezius muscle. There are very likely to be a few small lymph nodes here in the normal individual. Also, palpate deeply in this triangle to feel the lateral spines of the cervical vertebrae.

Now, begin at the inion of the skull and feel the tense ligament which extends from there to the vertebral prominence of the seventh cervical vertebra. This is best felt with the patient's head flexed.

Palpate the supraclavicular fossa on both sides. Do this first with your partner in a neutral position, then do it with his shoulders elevated to lift the clavicles up.

Now, examine the thyroid from the front and from behind as shown in Figures 8.8–8.12. A glass of water for your partner makes this easier. Watch carefully as he swallows. Note how the thyroid cartilage rises and falls and watch carefully for a swelling between it and the suprasternal notch. Someone in your class will have a palpable thyroid. This is *not* necessarily abnormal. The gland is frequently palpable in thin-necked people, in premenstrual women, and just as a variant.

chapter 9

BREASTS, AXILLA, AND LYMPHATICS

The discovery of a lump is the most common chief complaint of breast disease. Usually, the consultation with the physician is sought early but, tragically, it occasionally takes months for the woman to decide that ignoring a mass will not negate its existence. Denial plays a potent role in the death rate of breast carcinoma.

The usual questions are used to probe the history of the present illness—the temporal sequence, pain or its absence, enlarging, constant, or fluctuating in size, and associated symptomatology such as fever or weight loss. There are, in addition, a few specific questions which help sort out breast diseases. A lactating woman with a painful breast *most likely* has a *galactocele*. A mass which becomes tender during the menstrual period is *probably* a *fibroadenoma*. A tender mass arising following *trauma* may be a *hematoma*. Note the equivocations "most likely," "is probably," and "may be," to emphasize the treacherous nature of breast cancer.

Since early detection is the best weapon against breast cancer, we must question every woman carefully during the interview. We are coming to realize that there are factors which put some women at great risk for breast cancer. A family history of the disease is one such factor. The patient whose mother or sister has had breast carcinoma is far more at risk for the disease herself. If the patient has had a previous mastectomy for malignancy, she is at great risk to develop it in the remaining breast. Advancing age becomes a risk since the incidence of carcinoma of the breast increases directly with age. Dispute continues about the factors of parity, nursing history, ethnic background, and breast size. Nonetheless, you should record such information so that future students will have the data you lack.

Examination of Breasts and Axilla

Inspect the breasts with the patient disrobed to the waist and sitting with her back straight. The left breast generally is slightly lower than the right and the nipples point down and outward.

Nipple retraction when bilateral and lifelong is of no significance but when unilateral and recent indicates withdrawal by inflammatory or malignant tissue (Fig 9.1). A dry eczematous lesion of the nipple is characteristic of *Paget's disease*, a slowly growing carcinoma superficial to the breast.

The areolae contain specialized sebaceous glands subject to the same problems as such glands elsewhere—*retention cysts* and suppuration. The areolae of the nullipara are pink and become brown with the first pregnancy and remain so thereafter. *Addison's disease* causes hyperpigmentation of the areolae as well as the palmar creases and knuckles.

Now, inspect the skin over the breasts for deviations in the normal contour (Fig. 9.2). Several maneuvers are helpful if a question arises. Have the patient put her hands on the back of her head or have her lean forward until the breasts hang in a dependent position. *Peau d'orange* (orange peel) almost always means an underlying malignancy (Fig. 9.3). This peculiar accentuation of skin markings is caused by direct invasion of the skin with mitotic figures or by obstruction of the skin lymphatics. *Dilated superficial veins* over both breasts will be seen in the pregnant or lactating breast. This is an ominous sign when unilateral. Carcinomas demand an increased blood flow—hence, the dilated venous channels and the use of infrared photography and thermography as diagnostic techniques. Finally, look just below the insertion of the

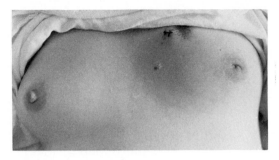

Fig. 9.1. Inverted nipple. The left nipple is inverted, secondary to a carcinoma in the upper inner quadrant. The small incision is a site of biopsy.

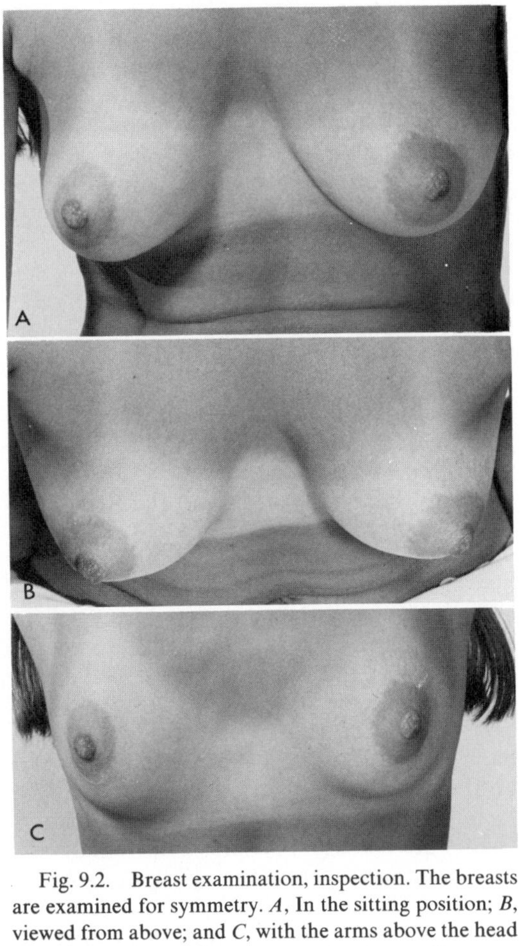

tors—four quadrants and a tail. The four quadrants are defined by imaginary horizontal and perpendicular lines drawn through the nipple. The tail extends from the upper outer quadrant toward the axilla. Now describe the *size, consistency*, and presence or absence of *pain*. A fixed mass is more likely malignant than not. Test *fixation* of the skin by immobilizing the mass with one hand and pinching up the overlying skin with the other hand. If the skin is fixed, you will be unable to move it freely over the mass (Fig. 9.6). A cancer may also extend into the deep pectoralis major. Have the patient put her hands on her waist and push inward. This tenses the

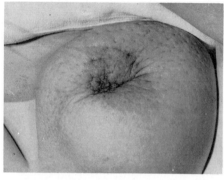

Fig. 9.3. *Peau d'orange.* The orange peel appearance of the skin is caused by an extensive carcinoma beneath the nipple of this breast. Involvement of the skin lymphatics leads to this appearance.

Fig. 9.2. Breast examination, inspection. The breasts are examined for symmetry. *A*, In the sitting position; *B*, viewed from above; and *C*, with the arms above the head to flatten the breast tissue against the chest wall.

pectoralis major for any obvious lymph node enlargement (Figs. 9.4 and 9.5).

If the patient complains of a lump, have her find it for you first—she will be expert at this. Next palpate it and describe its *location*. The breast is artifically divided into five sec-

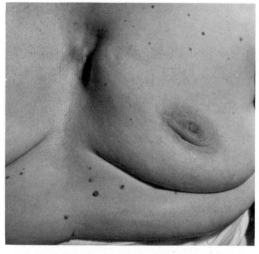

Fig. 9.4. Carcinoma of the breast. Note the mass lesion in the upper inner quadrant of the left breast. This was fixed to the underlying bone on examination and proved to be a scirrhous carcinoma.

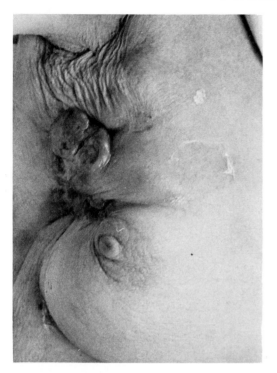

Fig. 9.5. Fungating carcinoma of the breast. This extensive carcinoma had ulcerated through the chest wall and fixed the skin and underlying ribs.

pectoralis, and a mass invading it will become immobile.

Complete the examination as you would with a normal patient. First, palpate the entire breast with the flat of your hand, pressing and moving in a circular fashion (Fig. 9.7). A very large breast is best examined with the patient supine, arm over her head and no pillow under the head. In order to drape the tissue over the chest wall you might have her roll slightly to bring the examined breast uppermost. Next, systematically examine each quadrant with both hands, finishing with the tail of the breast (Fig. 9.8). Palpate beneath the areolae and nipple and watch carefully for any discharge. Finally, examine both axillas. Have the patient raise her arm over her head. Lay the tips of your right hand fingers in the left axilla (vice versa for the other side). Now bring her arm down and rest her forearm on your forearm. This allows deep insinuation of the palpating hand. As you press against the chest wall and slowly move your hand downward, lymph nodes will be easily detected. Now put your thumb in the axilla and press against the head of the humerus to detect other nodes (Fig. 9.9).

What does the normal breast feel like? A question frequently asked and hard to answer. The breast is glandular and supported by fibrous stroma admixed with fat. The glandular tissue is bosselated and rubbery while the fibrous stroma is firm. The relative proportions of these elements dictate the consistency of the individual breast.

Cancer of the breast is hard. Often, the margins mold into normal tissue (cancer meaning crablike is apt here). A small early breast cancer can be simulated by palpating a marble on a hard table through a feather pillow or down quilt.

Fibrocystic disease occurs most commonly in women 30–50 years old and may be localized or diffuse. It is caused by a disconjugate proliferation of glandular and fibrous tissue and can feel very much like a malignancy. *Fibroadenomas* are small and very firm and occur in young women. Their most characteristic quality is their elusiveness. Just when you feel it between your fingers it slips away.

As with other exocrine glandular tissue, obstruction of a duct will cause cystic retention of secretions. Such a *breast cyst* fluctuates and will transilluminate.

Always note a nipple discharge. If the patient complains of this, it is well to have a few slides handy to prepare smears for Gram's stain and for cytological examination. With a careful examination, you should be able to detect from which quadrant the discharge emanates. Since the ducts are spokewheeled, a discharge from the right breast at 10 o'clock indicates trouble in the upper inner quadrant, and so forth. An *intraductal* papilloma fre-

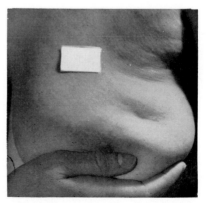

Fig. 9.6. Skin dimple of an underlying carcinoma. Manual elevation of the left breast on this patient clearly shows the fixation to the skin of the underlying mass lesion.

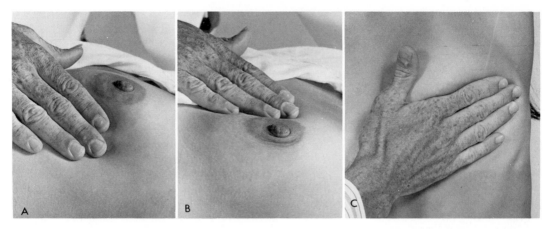

Fig. 9.7.   Manual palpation of the breast. All four quadrants are examined with the flat of the hand and tips of the fingers.

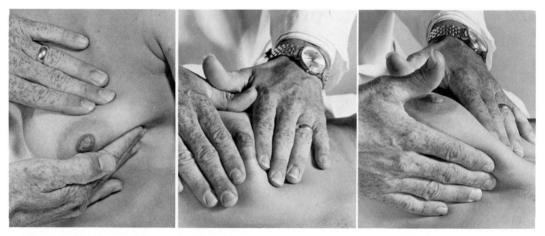

Fig. 9.8.   Bimanual palpation of the breast. Again, all areas of the breast are examined between both hands. This is especially useful for examining the subareolar area and tail of the breast.

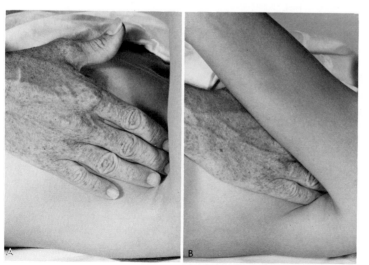

Fig. 9.9.   Examining the axilla. A, The fingertips are first placed deep into the axilla; B, next, the arm is brought down over the examiner's forearm, and the node-bearing regions are examined by firm pressure against the chest wall.

quently exudes bright red blood, a *carcinoma* crystal clear fluid, an *abscess* purulent material, a *galactocele* opalescent fluid, and a *retention cyst* yellow serous discharge.

Supernumerary nipples are very common in both men and women. They appear anywhere along the mammary line which extends from the midaxilla down and inward toward the umbilicus. They are of cosmetic significance only (Fig. 9.10).

The male breast is not immune to disease. *Gynecomastia* results from a variety of disorders, all of which share a hormonal alteration. In obese men, apparent gynecomastia is simulated by fat. Careful palpation will distinguish adipose tissue from true glandular hypertrophy. The administration of *estrogens* to a male causes gynecomaatia (Fig. 9.11). Certain *drugs* chemically resembling estrogens yield gynecomastia in some men—digoxin and spironolactones, for instance. *Choriocarcinoma* of the testis with the overproduction of chorionic gonadotropin for some reason leads to gynecomastia. Cirrhosis of the liver with faulty degradation of normally circulating estrogens allows a relatively higher concentration to stimulate the breast. Unilateral gynecomastia is commonly seen in pubertal boys. Gynecomastia is a feature of *Klinefelter's syndrome*.

When you discover gynecomastia in the male, the examination is not complete until you carefully review drug exposure, examine the liver, and examine the testes.

The male breast contains the same basic elements as the female breast and, similarly, can develop a carcinoma.

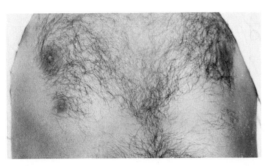

Fig. 9.10. Supernumerary nipples. The accessory nipples are apparent below the normal nipples along the mammary line. This is a dominant characteristic found in families.

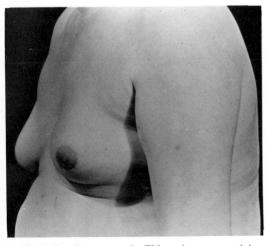

Fig. 9.11. Gynecomastia. This patient was receiving estrogen therapy for carcinoma of the prostate.The gynecomastia is apparent.

Lymphatics

We have already discussed some of the lymphatics and will do so again when we review other specific anatomic areas, however, it is of value to review some general features of the examination of lymphatics and the common diseases which affect them.

There are several node-bearing regions which are easily examined. These include the nodes in the neck and the groin. It is usual that small, nonfixed, soft lymph nodes can be found in the triangles of the neck and in almost every adult groin. Both regions drain structures frequently infected or are responsible for drainage of large areas of tissue. The axillae, as just described, also frequently harbor small palpable nodes. The difference between nodes enlarged through normal functioning and those enlarged because of ominous pathology is difficult to describe. Generally speaking, nodes which are painful are filled with polymorphonuclear leukocytes of infection. Very hard, stoney, nodes are caused by infiltration with malignant cells. Matted coalescing groups of nodes are the rule in lymphomas. Recall from your pathology that lymphomatous tissues characteristically extend beyond the capsule of the node, thus they tend to "run together."

The supratrochlear nodes are worth looking for. To find these, start by grasping the medial condyle of the humerus between your thumb and forefinger then move up the tissue proximal to this point. If the node is enalrged, it will be felt somewhere in the first 5 cm of tissue.

Another site of occasionally enlarged nodes is over the sacroiliac joints. Four or five nodes can sometimes be felt here on both sides.

The lymphatic vessels may be diseased anywhere. The entire skin contains these and, when inflamed, they declare their presence as a bright red streak. The streak is always linear and will be parallel to a limb or follow the obvious drainage pattern. The streak may be hot to touch and quite tender in "lymphangitis." If a major drainage trunk is obstructed, tissue fluid builds up proximal to the obstruction with marked edema. When acute, this may be somewhat painful, more like an ache and, when chronic, it is painless. There is a congenital form of absent lymphatic drainage called "Milroy's disease" which affects one or both legs and is seen mostly in women. It usually becomes most marked at about puberty.

Although in this text, for the sake of simplicity, we consider the lymphatics according to anatomic region, you must consider them as a system if you find one part of it to be abnormal. When, therefore, you detect lymphadenopathy anywhere, be particularly careful to examine all node-bearing areas.

chapter 10

HANDS

Hands are expressive, individual, and accessible, and they impart much clinical information to the interested physician. The changes we discussed earlier about the skin—hormonal effects, oxygenation, and blood volume status—are all reflected in the skin of the hands. In addition, however, to systemic changes involving the hands as a part of the whole, there are systemic diseases which are reflected exclusively or predominantly in the hands. Many of these elude adequate pathophysiologic explanations and will be discussed somewhat randomly or, at best, categorized to the extent we do understand them.

History

A few *pain* syndromes involving the hands are virtually pathognomonic. Irritation of the ulnar, median, or radial nerves produces pain in the hand. When intrinsic hand weakness is combined with pain, the radial nerve is excluded since it has only sensory functions in the hand. Pain radiation to the ulnar side of the hand and ring and little fingers point to the ulnar nerve. The radial side of the palm and the thenar muscles depend on the median nerve. The pathology may arise anywhere from the cervical spine to the distal nerve. The *carpal tunnel syndrome* results from entrapment of the median nerve between the bony and fibrous elements of its canal at the wrist. The pain has characteristic median nerve radiation, and the patient may volunteer that hyperextension of the wrist aggravates the pain.

Many of the arthritides affect the hands. The pain pattern often suggests the kind of arthritis present. Morning stiffness is the rule in *rheumatoid arthritis*. Such patients, usually young women, relate that the pain improves with activity. The older woman with pain aggravated by activity more likely has *osteoarthritis*.

Raynaud's phenomenon is a bizarre event which follows exposure to cold in the affected individual. The patient relates that initially the distal portions of the fingers become chalky white with a very sharp demarcating line between the affected skin and the normal. Aching pain ensues. Following warming, the affected area becomes blue, then purple and red. Most patients with Raynaud's, on careful questioning, also admit to dysphagia since disordered esophageal motility is a common accompaniment. This presumed vascular instability may occur alone or as a part of a generalized collagen vascular disease such as systemic lupus erythematosis or systemic sclerosis.

Pain with cold exposure can also be caused by circulating *cold agglutinins*. These globulins complex as a result of cold exposure and presumably plug the microvasculature.

Hand *weakness* may be a chief complaint. Characterize the weakness as you take the history. *Myasthenia gravis* may affect the intrinsic muscles of the hand. The usual story is that of weakness progressing to paralysis with repetitive muscle contractions. *Amyotonia congenita* prohibits a patient from releasing his grip. This rare inherited disease consists of this muscle abnormality, frontal baldness, premature cataracts, nasal speech, and hypogonadism. The neuropathy of *diabetes mellitus* causes hand weakness far more often than is generally recognized.

Fasciculations, small involuntary muscle twitching, may be observed by the patient in his hands. Fasciculations always mean lower motor neuron disease. When they occur in the hands and are associated with weakness, consider *amyotrophic lateral sclerosis*.

The *absence* of hand weakness may be important in the patient who complains of weakness elsewhere. Selective proximal weakness sparing the hands, forearms, and lower legs is characteristic of the myopathy of endocrinopathies such as hypo- or hyper-thyroidism, Addison's and Cushing's diseases, and internal malignancies of many sites.

Examination

No special techniques or equipment are required. The small hand lens used for skin examinations will be useful.

The examination begins with the greeting handshake. Sweaty nervous palm? Coarse skin? Firm grip? Does his hand fit yours?

Vascularity

The radial and ulnar arteries feed the hand. Both are normally palpable. The ulnar artery is somewhat deeper than the radial and firmer pressure is required to feel it. With patent palmar arches, each artery is capable of supplying the entire hand should the other become occluded. The patency of these arches is tested by firmly obstructing both arteries at the wrist, raising the hand above the shoulder until it blanches, then releasing one artery. The spreading normal red color will be confined to only one-half of the hand if the arches are nonfunctional. (see Fig. 13.11 in Chapter 13).

Raynaud's phenomenon can be produced in the susceptible individual by immersing the hand in ice water.

Palmar erythema is a curious finding in several unrelated conditions. It presents as a deep red erythema most prominent over the heel of the palm spreading laterally over the hypothena and thenar eminences. In flagrant examples, it involves the fingertips as well. Careful inspection shows that the erythema is the result of deep telangiectases. Cirrhosis, pregnancy, and rheumatic heart disease are the most common identifiable causes.

Bony and Tendinous Structures

The bones of the hand are not immune to generalized changes affecting bones elsewhere. The hands enlarge in acromegaly as bones do elsewhere. A few systemic diseases, however, have specific changes in the hands.

Marfan's disease of connective tissue is also called arachnodactyly to emphasize the spider-like long fingers characteristic of this disease. That the fingers are long and lax can be shown by having such a patient close his fist over the thumb. The tip of the thumb will protrude from the ulnar side of the palm,

unlike a normal hand. The hyper extensibility can also be shown with the maneuver in Fig. 10.1.

Pseudohypoparathyroidism, a congenital end organ unresponsiveness to parathormone, can be suspected by examination of the hands. Curiously, in this disease, the fourth and fifth metacarpals are markedly shortened with resultant apparent shortening of these two fingers. When such a patient makes a fist, the fourth and fifth knuckles are not seen (Figs. 10.2 and 10.3).

Sex influences finger length. The ring finger of a male is longer than or the same length as the index finger. Females, however, possess an index finger which is longer than or the same length as the ring finger.

A common finding in many congenital diseases is an incurved fifth finger.

Osteoarthritis results in bony overgrowths at joints. In the hand, this phenomenon particularly affects the distal interphalangeal joints. These bony excrescences which appear

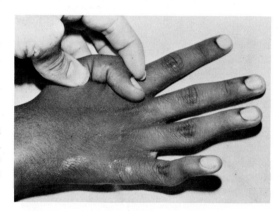

Fig. 10.1.   Hyperextensible joints. This patient demonstrates marked laxity of the joints. He also suffered from acquired scoliosis and, in addition, had the "click-murmur" syndrome.

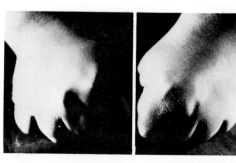

Fig. 10.2.   Pseudohypoparathyroidism. The clenched fists demonstrate quite clearly the shortened metacarpals. The fourth and fifth knuckles are not apparent.

on the lateral side of the joints are called *Heberden's nodes.* Although occasionally destructive, osteoarthritis is rarely inflammatory so that heat, redness, and soft tissue swelling are uncommon findings (Fig. 10.4).

Rheumatoid arthritis is a more inflammatory, destructive, joint disease. The greatest morbidity of this disease results from hand involvement. Rheumatoid arthritis has a propensity for the *proximal* interphalangeal joints and the metacarpophalangeal joints. The swelling which occurs is not bone but rather synovium and soft tissues filled with edema fluid. The swelling varies throughout the day and characteristically assumes a fusiform, or sausage, configuration. The proliferating pannus of the synovium destroys the joints. Palpation reveals warmth and, frequently, fluid within the joint. As the process proceeds in the metacarpophalangeal joints,

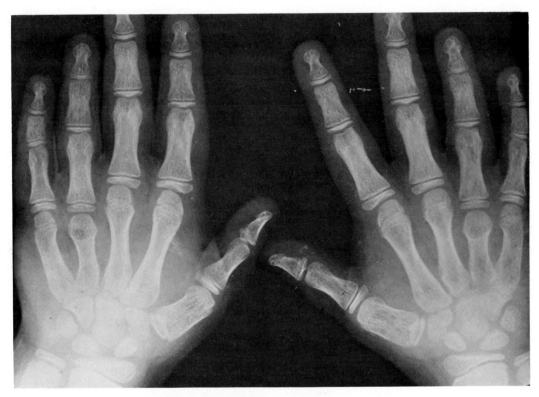

Fig. 10.3. Pseudohypoparathyroidism. This radiograph shows clearly the shortened fourth and fifth metacarpals.

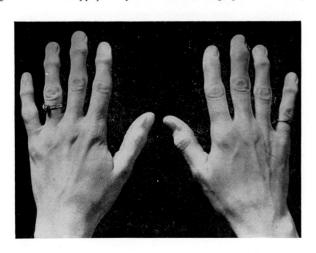

Fig. 10.4. Osteoarthritis. The distal interphalangeal joints are most prominently involved. The bulges on the sides of these joints are called Heberden's nodes. On palpation, they will be noted to be bony and not soft tissue. The metacarpophalangeal joints are spared, and deformity is minimal.

ulnar drift results—the fingers rather than straight extensions of the hand point toward the ulnar side of the wrist. *Subluxation* causes the heads of the metacarpals to ride up and over the heads of the phalanges so that the knuckles appear very prominent. The disease in the proximal interphalangeal joints involves the extensor and flexor tendons by proximity. The extensor tendons which pass over the top sides of the joint may be eroded and slip to the bottom side of the joint. In this position, they become flexor in action. The result is that the joint is constantly flexed—the so-called *boutonnière* deformity (Figs. 10.5–10.7).

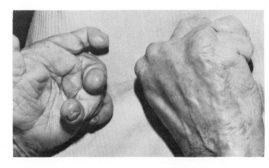

Fig. 10.7.   Psoriatic arthritis. This is the mutilating variety of psoriatic arthritis. The digits are soft and telescoped. On palpation very little bone can be felt.

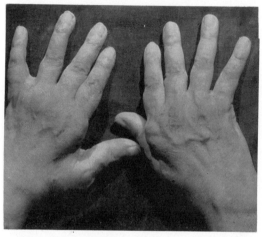

Fig. 10.5.   Rheumatoid arthritis. The metacarpophalangeal joints are especially involved in this patient. Diffuse swelling is obvious. The right hand shows the typical ulnar drift of the fingers. Fusiform swelling of the proximal interphalangeal joints is also apparent. The landmarks of the wrist are obscured by similar soft tissue swelling.

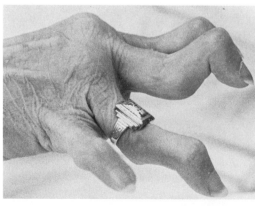

Fig. 10.6.   Rheumatoid arthritis, marked. The marked deformity of the metacarpophalangeal joints and proximal interphalangeal joints is apparent.

Localized *tendon nodules* are also common in rheumatoid arthritis. The flexor tendons are particularly prone to this abnormality. When these nodules become large enough, they have difficulty sliding under the retinacula of the palm. The nodule snaps back and forth under this fibrous sheath, producing a *trigger finger*. The nodules can be felt by palpating the palm and having the patient flex and extend the fingers.

The flexor tendons are also involved in *Dupuytren's contracture*. Here, fascial proliferation tethers the flexor tendons of the fourth and less so the fifth and third fingers, holding them in a fixed flexion at the metacarpal phalangeal joints. The fascial overgrowth is easily felt. The cause of this progressive deformity is unknown, but it seems to be an accompaniment of advancing age, of alcoholism, and of chronic disease.

The extensor tendons are favored as a site for fatty deposits in certain *hyperlipidemias*. *Tendinous xanthomas*, as they are called, are fleshy swellings which glide back and forth under the skin as the tendons move. Palpating the back of the hand while the patient alternately flexes and extends the fingers permits detection of small xanthomas. This is a maneuver worth repeating on all patients with premature coronary artery disease (Fig. 10.8).

Diffuse xanthomata of the hands are multiple yellow skin plaques frequently seen at skin creases, and they occur in some diabetic patients and patients with advanced biliary cirrhosis and very high plasma triglycerides.

Tophaceous gout, an increasingly rare disease, may appear in the hands as hard tumors around the joints. Careful examination shows that these deposits are not bone and sometimes can be moved slightly over the joint (Figs. 10.9 and 10.10).

Fingernails

Many systemic disorders reflect themselves in the fingernails. Since a nail takes approximately 6 months to grow from base to tip, a calendar effect exists. A normal nail springs from the nail bed embraced by a pale semicircular *lunula* and rises off the finger tip distally at a sharp border.

Clubbing of the nails was recognized cen-

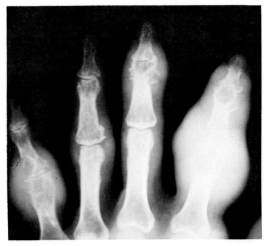

Fig. 10.10.   X-ray of gout. The soft tissue swelling is apparent on the index finger. The typical gouty lesions are seen as punched out holes in the ends of the phalanges.

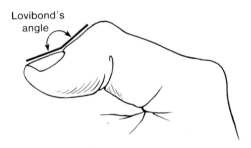

Fig. 10.11.   Lovibond's angle. This is the angle which is disturbed in clubbing of the nail. The drawing shows the normal angle which is usually about 150° and does not exceed 180°.

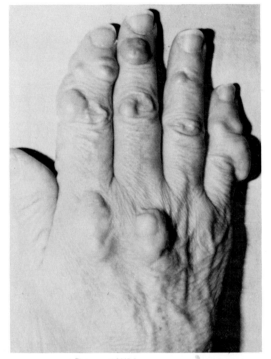

Fig. 10.8.   Tuberous xanthomas. These soft, fleshy nodules are readily apparent on the extensor tendon surfaces of the hand. They are associated with a systemic lipid disorder.

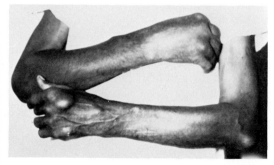

Fig. 10.9.   Tophaceous gout. Note the large accumulations of urate located over the head of the first metacarpal and over the olecranon of the left elbow. With adequate therapy, this lesion is becoming increasingly rare.

turies ago but is still poorly understood. Inspect a normal nail from the side and you will note that the nail forms an angle with the nail bed skin. This angle normally does not exceed 160° (Fig. 10.11). Clubbing results when the nail bed overgrows, pushing the proximal nail upward and obliterating the normal angle. The nail also assumes an hourglass or beaked appearance as well. Place your finger on top of the patient's finger pointing in the same direction and push down on the nail bed. The patient who has clubbing will have a spongy or floating nail. Clubbing occurs in a variety of disorders including cyanotic congenital heart disease, carcinoma of the lung or pleura, bronchiectasis, hyperthyroidism, and cirrhosis. Hypoxia and/or increased digital blood flow seem to be common denominators. Clubbing may also be a congenital familial disorder of no significance (Fig. 10.12).

The lunula appears most prominently in the thumb and progressively less so toward the fifth finger. Profound anemia causes apparent loss of the lunula since the distal nail bed becomes as pale as the lunula. *Half-and-half* nails call attention to the possibility of renal disease. A ground glass, dull white proximal half of the nail abruptly meets a red or reddish brown distal half of the nail. This finding may antedate symptomatic renal disease (Fig. 10.13).

Beau's line and *Mee's line* attest to the

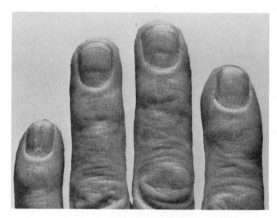

Fig. 10.14.   Beau's lines. This patient suffered a myocardial infarction approximately 10 weeks prior to this photograph. The horizontal ridgelike depression corresponds to this insult. Recall that the normal nail takes approximately 6 months to grow from nail bed to tip. (Courtesy Dr. Howard Baden.)

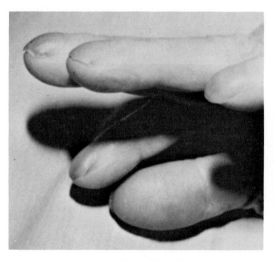

Fig. 10.12.   Clubbing. Lovibond's angle is markedly distorted. Instead of being 170° or less, in this photograph it can be seen to be in excess of 180°. Palpation of the nail bed reveals a spongy quality.

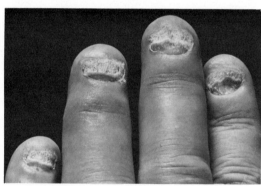

Fig. 10.15.   Psoriasis of the nails. This patient has bitten her nails considerably and reveals the underlying psoriatic debris which leads to the onycholysis.

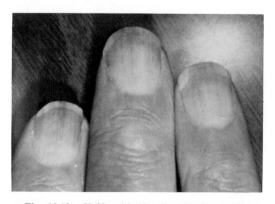

Fig. 10.13.   Half-and-half nails. The lunula is not apparent. A soft white color of the nail becomes a dark rim of red-brown distally. Although frequently seen in chronic renal disease, this finding is not specific. (Courtesy Dr. Howard Baden.)

chronologic aspect of the nail exam. Beau's line is a transverse groove in the nail while Mee's line is a similar transverse white line. Either of these may result from a severe physical or emotional shock such as overwhelming infection, myocardial infarction, or death of a loved one. Both of these lines grow out from the nail after the insult. Beau's line in the midnail, for instance tells you that some catastrophic event occurred approximately 3 months earlier (Fig. 10.14).

Irregular separation of the nail distally is termed *onycholysis.* The most common cause is psoriasis, where the accumulation of keratin debris beneath the nail is responsible. *Hyperthyroidism* may cause the same irregular separation, probably from accelerated irregular growth of the nail. *Pitting*, tiny peck marks on the surface of the nail, is another hallmark of psoriasis and helps differentiate it from hyperthyroidism (Figs. 10.15 and 10.16).

Hypoalbuminemia, when profound from any cause, may result in the appearance of two thin parallel white transverse bands (Fig. 10.17).

Splinter hemorrhages appear as longitudinal red-to-brown hemorrhages beneath the nail. They can be seen anywhere under the nail but most often appear distally. Trichinosis is the most common worldwide cause of splinter hemorrhages, but in the United States subacute bacterial endocarditis ranks first. The mechanism may be hemorrhage from tiny bits of embolic material or a peculiar form of vasculitis known to accompany bacterial endocarditis (Fig. 10.18).

Periungual telangiectases are tiny hairpin vascular anomalies around the border of the nail. They are important clues to the presence of collagen vascular diseases such as lupus

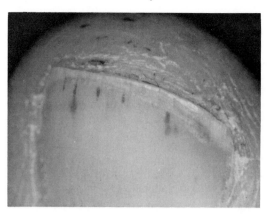

Fig. 10.18.   Splinter hemorrhages. This patient with subacute bacterial endocarditis demonstrated characteristic splinter hemorrhages beneath the tip of the nail and on the distal portion of the finger. In addition, there were subconjunctival hemorrhages characteristic of embolic phenomenon. (Courtesy Dr. Howard Baden.)

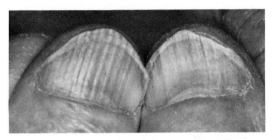

Fig. 10.19.   Koilonychia. These spoon-shaped nails were found on an elderly woman with iron deficiency anemia. They are also longitudinally ridged. (Courtesy Dr. Howard Baden.)

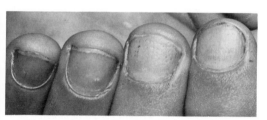

Fig. 10.16.   Psoriatic pitting. This is a characteristic lesion of psoriasis. Note the pin-point crypts located on the surface of the nails. It is very unusual to find psoriatic arthritis without the presence of pitting of the nails. (Courtesy Dr. Howard Baden.)

erythematosis, dermatomyositis, and scleroderma.

Scleroderma affects the digits early in the disease. The digital pads become atrophic and the skin over the fingers shiny and "hidebound." Raynaud's phenomenon frequently accompanies scleroderma, and ulcers on the fingertips may result.

Longitudinal ridging (Fig. 10.19) and white patches, leukonychia, (Fig. 10.20) are common findings of unknown significance.

Dermatoglyphics

Here, we take a leaf from law enforcement notebooks and examine fingerprints. Our sleuthing is of a somewhat different nature. Instead of using fingerprints as individual characteristics, we are looking for common patterns reflecting disease. Since fingerprint development is completed by the end of the

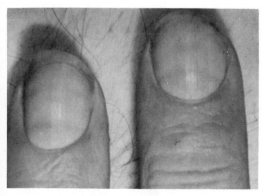

Fig. 10.17.   Muercke's lines. Two parallel white bands are separated by a dark red strip of nail. When you see this lesion, suspect hypoalbuminemia. (Courtesy of Dr. Howard Baden.)

first trimester of pregnancy, diseases we look for are congenital and frequently inherited.

The description of fingerprints may be qualitative or quantitative. There are a few easily recognizable patterns. In addition to the patterns on each individual finger we note certain palmar characteristics. Do not confuse the coarse palmar creases with the delicate parallel ridges of true finger and palm prints.

An *arch* is the simplest finger print. The *loop* is a hairpin turn. The ridges of a loop turn 180°. The open end of the loop may point to either side of the hand and hence the designation of radial and ulnar loops. A *whorl*, or double loop, has ridges that turn 360°. The geometry of print ridges dictates the need for *triradii*. A triradius is the junction of three sets of parallel ridges forming a Y. For each 180° turn of ridges, one triradius is required to compensate. Therefore, a loop always has one associated triradius and a whorl has two triradii. The quantitative aspect of dermatoglyphics comes in counting the number of ridges between a triradius and the center of the loop or whorl (Fig. 10.21).

There is a triradius at the base of each finger and one at the heel of the hand. The *digital* triradii are lettered "*a*" through "*d*," index-to-little finger by convention. The triradius at the heel is called the axial triradius and is lettered "*t*."

The angle formed by the triradius of the index finger, the axial triradius, and the tri-

Fig. 10.21. Typical fingerprints. *Upper left*, simple arch; *upper right*, a loop; *lower left*, double whorl, note the two triradii; *lower right*, another simple loop.

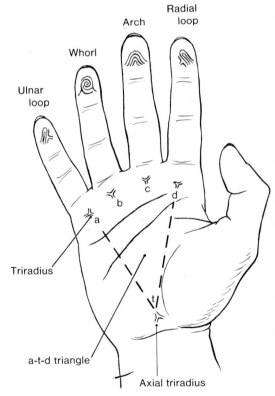

Fig. 10.22. Dermatoglyphics. Typical fingerprints are represented. The digital triradii are lettered.

Fig. 10.20. Leukonychia. These frequently observed white blotches scattered over the fingernails have no known clinical significance. (Courtesy of Dr. Howard Baden.)

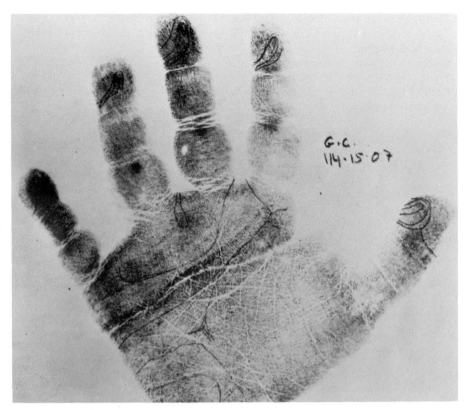

Fig. 10.23. Hand print of a patient with Down's syndrome. The high axial triradius can be seen above the hypothenar eminence. It has been penciled in. In addition, note the wide space between the fourth and fifth fingers and the single Simian crease across the palm. (Courtesy Endocrine Unit, Massachusetts General Hospital.)

radius of the little finger is designated the a-t-d angle. Many diseases cause the axial triradius to be displaced upward in the palm, and the a-t-d angle therefore enlarges. In normals, the angle does not exceed 60° (Figs. 10.22 and 10.23).

A variety of influences disturb the pattern of ridges. Intrauterine infections such as German measles, and chromosomal alterations such as Down's syndrome may yield peculiar fingerprints.

Radial loops rarely appear on the fourth or fifth fingers in normals but are very common in Down's syndrome. Ulnar loops on all ten fingers are seen in 4% of normal people but in 33% of patients with Down's syndrome. Also in this disease, the a-t-d angle is very

wide, and there may be only a single transverse palmar crease, the so-called *Simian fold.*

The number of ridges in a loop or whorl is somewhat determined by sex chromosomes—specifically, the number of X chromosomes. The more X's, the fewer ridges, so that women tend to have a lower ridge count than men. Patients with multiple X chromosomes have progressively fewer ridges.

The variety of patterns that have been documented is far too great for our purposes here. Suffice it to say that when an inherited or congenital disease is suspected, dermatoglyphics may help elucidate the nature of the problem. Pay particular attention to the digital loops, ulnar or radial, and the position of the axial triradius.

PRACTICE SESSION

Check the length of the digits and confirm the relative lengths of the index and ring fingers as a function of sex. Examine the nails and note the presence of the lunulae and measure

Lovibond's angle. Note the smooth separation of the nail from the bed at the distal end. There is likely someone in your class with psoriasis who could show you onycholysis and perhaps some nail pitting.

Now, describe the basic pattern of fingerprints on each finger. Find the triradius of the loops and whorls. Next, find each digital triradius and the axial triradius. Take your time, they are hard to find at first. Use a good light and a small hand lens and don't be fooled by the large creases. Measure the a-t-d angle.

chapter 11

THORAX

History

The wide variety of vital structures contained in the thorax, because of their varied functions and anatomy, present a broad spectrum of diseases. The historical correlations are similarly broad. A careful characterization of the patient's complaints generally allows discrimination of the clinical possibilities.

Chest pain to the layman means heart disease and, regardless of the age of the patient, you must consider that fear-generating connotation. The chest wall, pleura, heart, great vessels, trachea, and esophagus are the structures within the chest which can produce pain. Note that the lungs are painless. Lung disease causing pain does so only through involvement of adjacent structures, usually the pleura. As with pain anywhere, the history should include the duration, temporal relationships, character, aggravating and alleviating factors, and associated symptoms.

The chest wall, containing musculoskeletal elements, is prone to diseases of these elements elsewhere. The patient with chest wall pain will relate that he can press on the area and produce pain and, often, that supporting the area with the flat of his hand reduces the pain. Chest wall motion, coughing, sneezing, and just breathing aggravate the pain. A history of trauma simplifies the diagnosis, but the trauma may be indirect, as in the fractured rib from violent coughing. The ribs have a true articulation with the spine and may be involved in generalized arthritis, especially *ankylosing spondylitis*, and become painful. The junction of the bone of the rib and its cartilage extending to the sternum is a site of inflammation in Tietze's syndrome, *costochondritis*. The ribs are also a favored site for *metastatic malignant deposits*, probably because of the rich vascularity secondary to marrow activity which persists even into advanced years. When metastases expand the rib and involve the periosteum, pain results.

Pleural pain has a different character. This lancinating pain or "catch," as the patient will call it, results from the loss of the normal lubricating function of this serous membrane. The pleural space, a potential space, is lined on both surfaces by the delicate pleural membrane. Irritation from any source calls forth inflammatory debris of neutrophils and fibrin. The grating of these two raw surfaces during respiration is painful. Primary pleural diseases are exemplified by Coxsackie B virus, *pleurodynia*, or the pleurisy of systemic lupus erythematosis. When lung disease causes pleurisy, it is a valuable anatomic marker indicating which lung is diseased and tells you that the process involves peripheral lung tissue. If pneumonia, it must be a consolidating process and, if a pulmonary embolus is responsible, it must have caused infarction. The pleura is also sensitive to sudden changes in tension. *Spontaneous pneumothorax*, the sudden rupture of a lung, produces dyspnea and severe chest pain (Fig. 11.1). The disease affects young men with no previous lung disease and older patients with known bullous diseases of the lung. As the air escapes into the pleural space, the affected lung collapses to accommodate the gas, and the visceral pleura is decompressed.

Esophageal pain may resemble cardiac pain. A careful history usually links the pain with function. The esophagus registers mucosal irritation as pain. Acid reflux from the stomach through an incompetent gastroesophageal sphincter produces a burning substernal discomfort. *Hiatus hernia* frequently predisposes to incompetence of this sphincter. The patient, usually obese, admits to aggravation of the pain by bending over or lying down after eating and relief of the pain with antacids. Pain associated with dysphagia is an obvious road sign to the esophagus.

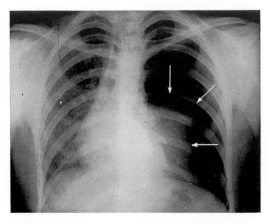

Fig. 11.1. Tension pneumothorax. This young man suffered the sudden onset of acute pain while walking. The x-ray shows total collapse of the left lung (*arrows*). Tension is implied by the fact that the mediastinum and heart shadow are both shifted into the right chest.

Chest pain may originate from the *aorta*. A dissecting aortic aneurysm produces virtually diagnostic pain. The patient, profoundly ill, relates the sudden onset of severe tearing pain which migrates as the dissection progresses. Frequently beginning in the anterior chest, the pain will move to the interscapular area and down the back.

Cardiac pain is common but sufficiently varied to be quite confusing. It usually takes the form of a pressure sensation or a weight on the chest. When describing his discomfort, the patient may place his clenched fist over his sternum—the *Levine sign*. Radiation of the pain into either arm or the neck and jaw is a helpful clue when present. We will discuss this topic in greater length in a later chapter.

Of the *dysfunctional* complaints, dyspnea heads the list. Very simply stated, dyspnea is difficult or labored breathing. Not so simple is the elucidation of the cause, which may be metabolic, hematologic, pulmonary, or cardiac.

Metabolic dyspnea results when the lungs must correct for a metabolic acidosis. The loss of bicarbonate demands that the lungs blow off carbon dioxide in order to restitute normal arterial pH. Diabetic ketoacidosis, renal failure, and lactic acidosis are associated with a large concentration of acidic anions which consume buffers like bicarbonate. The mechanism for compensation is rapid deep breathing. The associated symptoms of these metabolic diseases and the absence of a history of pulmonary or cardiac disease facilitate

the diagnosis. To be precise, this form of dyspnea is best called hyperpnea. The patient may be unaware of the rapid deep breathing and may not have the sensation of true air hunger.

Anemia, when profound, deprives tissues of oxygen, and hyperventilation results in an attempt to saturate all available hemoglobin. In a like fashion, if the hemoglobin is incapable of accepting oxygen, dyspnea results. *Carbon monoxide* poisoning and *methemoglobinemia* are examples.

Any pulmonary disease which interferes with gas exchange will eventuate in dyspnea. Acute illnesses like *pneumonia* with exudation of fluid into alveoli and chronic diseases like *emphysema* with loss of diffusing surface are examples. Lung diseases which increase the work of breathing cause dyspnea. *Interstitial fibrosis* producing "stiff lungs" and *asthma* with partial bronchial obstruction by spasm and mucous plugs are examples. Pulmonary dyspnea worsens with exertion and gets better with rest, and it is often associated with cough, wheezing, or a past history of lung disease.

Cardiac dyspnea has a more subtle origin. The common denominator is an elevation of the pulmonary venous pressure. Congestive heart failure transmits the elevated left ventricular and diastolic pressure to the left atrium and, thence, to the pulmonary veins. The elevated venous pressure soon overwhelms the tissue and oncotic pressure within the pulmonary bed and allows transudation of fluid into alveoli. Once filled with fluid, these air sacs are unable to exchange gas, and dyspnea results. As in pulmonary dyspnea, cardiac dyspnea gets worse with exercise. Two other characteristics, however, are distinct.

Orthopnea refers to dyspnea brought on in the recumbent position. *Paroxysmal nocturnal dyspnea* also occurs in the recumbent position but is a more sudden, profound, frightening dyspnea. Both result from expansion of blood volume. The cardiac patient, when erect, may put several liters of fluid into the dependent tissues. This fluid returns to the intravascular compartment when he is recumbent and can overwhelm the borderline-compensated heart. The patient with orthopnea may prop his head and shoulders on several pillows or may even sleep sitting in a chair. When a paroxysm of nocturnal dyspnea occurs, the patient almost always quickly gets up, walks around, and may throw open a window gasping for air. In an hour or so the dyspnea will subside.

Wheezing affects all age groups and always means airways narrowing. The patient hears the wheeze and may complain of a tightness in the chest with dyspnea. Virtually any disease of airways can be at fault, but the history will narrow the possibilities. *Asthma* is the prototype. A family or personal history of atopy can frequently be elicited in this disorder. Thus, hay fever, seasonal rhinitis, exzema, and drug allergies might be recounted. In asthma, various stimuli cause the bronchial walls to constrict, narrowing the lumen and causing wheezing. *Chronic bronchitis* with excess mucus production also narrows airways and causes wheezing as may a partially obstructing *endobronchial tumor*.

Coughing is a physiologic response to irritation of bronchi. The enormous number of proprietary suppressants attests to the frequency of cough as a patient complaint. Cough originates anywhere from the larynx to the distal bronchioles. Simple chemical irritation from aspirated food or fluid causes immediate coughing. Associated complaints of fever and sputum suggest an infectious cause. Whenever the bronchi become wet, cough results. The fluid might be mucus from bronchitis, pus from infection, or transuded plasma fluid in congestive heart failure. With the last, the patient commonly states that the cough occurs at night while recumbent. Ask the patient if the cough produces sputum. A nonproductive cough means either that the cause is purely irritative or that the cough is ineffective in clearing the airways.

Characterize the *sputum*. Pure mucus is white to clear. Purulence adds color, usually yellow to green. Saliva is not sputum nor is the material cleared from the throat which drains down from the posterior nasopharynx. Plain mucus has no taste, but with infection the patient might complain of a metallic bitterness. Foul smelling sputum usually means that a lung abscess is present. Under such circumstances, the history of alcoholism or recent dental work might point to aspiration as the causative event.

Have the patient quantitate the sputum in terms of both duration and amount. The diagnosis of *chronic bronchitis* is made by the history and is defined as daily sputum production for at least 2 months in at least 2 successive years. Quantity may be hard to get at, but such estimates as a tablespoonful or a cupful can be helpful. *Bronchiectasis,* for instance, usually causes copious amounts of sputum—occasionally as much as a quart per day.

Rarely, a patient may cough out material other than liquid sputum. Stony hard calcific pellets are a complaint with *broncholithiasis.* Calcified mediastinal lymph nodes may erode into bronchi and also be expectorated as stones. Similarly, foreign bodies aspirated from the pharynx may be coughed out—food, broken teeth, etc.

The expectoration of blood or bloody stained material is called *hemoptysis*. The patient's description of the hemoptysis and his associated complaints may go a long way toward establishing the cause. The etiologies of hemoptysis ran the gamut from the very benign to the very malignant. If the breech of the vascular integrity is quick and localized, the volume of blood may be great, the color bright red, and the content pure. An isolated *bronchial telangiectatic vessel* is such a condition. A history of weight loss and cough, on the other hand, would point to the erosion of a blood vessel by a *neoplasm*. A curious observation is that many patients with hemoptysis can tell you from which lung it originates even in the absence of pain.

A more diffuse vascular insult usually causes hemoptysis which is of less volume and more prolonged. Older blood is darker. The mixture of blood and mucus or purulence means chronic inflammation or infection. Necrosis of lung tissue from infection might be associated with fever and cough antedating the hemoptysis. *Infarction* of lung from a pulmonary embolus is preceded by dyspnea, cough, and pleurisy.

Further consideration of the history in diseases of the chest will be given in the chapter on the cardiovascular system.

Anatomic Considerations

Physical diagnosis will be facilitated if you understand a few of the surface projections of the thoracic organs.

The designation of upper and lower lobes of the lung is not altogether correct. The lower lobe is indeed the dependent lobe, but the actual location might suggest that anterior and posterior lobes would be more descriptive. The lower lobes are the posterior lobes and project from the midscapula down to the diaphragm posteriorly and occupy only the very lowest portion of the anterior thorax. The upper lobes, on the other hand, are predominantly anterior. The right middle lobe is wedged between the upper and lower lobes anteriorly, is in intimate contact with the heart, and does not project at all onto the posterior chest. Because of the left chest position on the heart, the left lung does not come as close to the sternum as does the right lung (Figs. 11.2 and 11.3).

The ribs sweep far downward as they go from posterior to anterior. The child's chest is more or less cylindrical but, as maturation proceeds, the lateral dimension comes to exceed the anteroposterior dimension until old age when childhood proportions again appear. Expansion of the chest cavity volume is accomplished by raising the ribs anteriorly and flattening the diaphragm (Fig. 11.4).

Several specific anatomic points and imaginary lines become useful points of reference. The sternum and spine are obvious. The function of the manubrium, sternum, and second ribs anteriorly protrudes prominently as the *sternal angle*. Rib interspaces are conveniently counted from here. The nipples are unreliable for reference, especially in women. A line dropped perpendicularly from the middle of the clavical defines the *midclavic-*

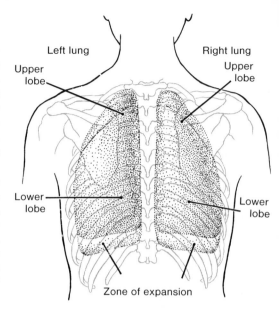

Fig. 11.3. Posterior view of the lungs. The posterior position of the lower lobes is well seen.

ular line. A similar perpendicular line from the fold of the pectoralis and deltoid is the *anterior axillary line.* The *midaxillary line* falls from the center of the axilla and the *posterior axillary line* from the posterior fold (Fig. 11.5).

When accuracy is important, measurements should be taken in centimeters from defined points on the sternum or spine.

Inspection of Chest

Assuming that you have already counted the rate and noted the rhythm of breathing and looked for the presence of cyanosis and clubbing, we now focus on inspection of the chest.

First, determine the static dimensions of the chest, i.e. the lateral and anteroposterior diameters. Next check for symmetry. The asymmetrical chest, like the asymmetrical head, can be the result of abnormal bony structures or abnormal thoracic contents. Do the intercostal spaces appear equal, one side against the other? A chronically collapsed or fibrosed lung on one side frequently retracts the chest wall and closes the intercostal spaces (Fig. 11.6). A parasternal pulsating mass may be an aneurysm of the ascending aorta caused by syphilis (Fig. 11.7). Likewise, check the supraclavicular spaces for asymmetry. A mass may produce unilateral bulging while

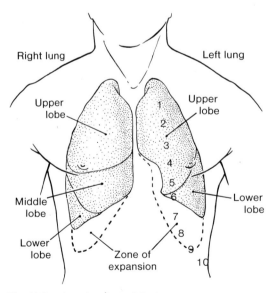

Fig. 11.2. Anterior view of the lungs. Note the relative position of the lobes. The upper lobes are really the anterior lobes, while the lower lobes are posterior. The right middle lobe is a wedge-shaped lobe with its base against the heart.

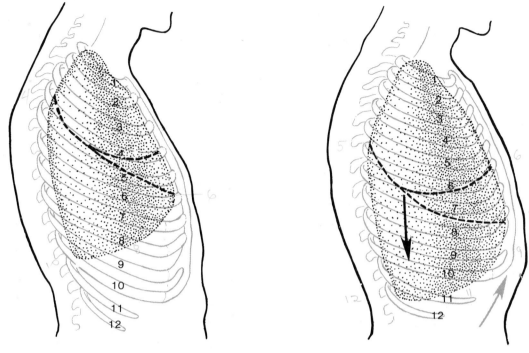

Fig. 11.4.   Changes in lung volume with respiration. During inspiration, the diaphragms flatten and the anteroposterior diameter increases.

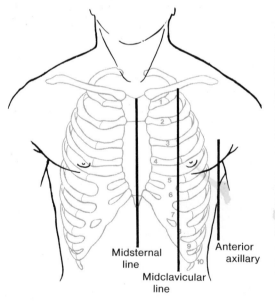

Fig. 11.5.   Reference lines on chest wall. Of the three lines shown, the midsternal line is the most reliable for measurement purposes.

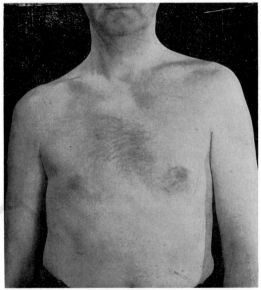

Fig. 11.6.   Asymmetric chest. The right side of the chest is flattened. The reason is extensive chronic fibrous tuberculosis of the underlying right upper lobe and pleura.

fibrosis of the underlying pleura and lung may produce unilateral retraction.

Inspect the bony thoracic cage. Posteriorly, the spine is prominent at the seventh cervical vertebra and then describes a gentle outward curvature (kyphosisl), to meet the normal lumbar inward curvature (lordosis). Abnormal curvatures will be alluded to in greater detail later in the text.

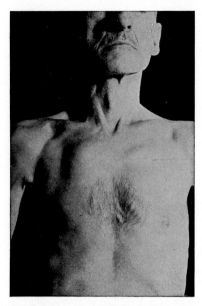

Fig. 11.7. Bulging chest wall due to aneurysm of the ascending aorta. The prominent bulge shown to the right of the sternal angle was found to be pulsatile on palpation. X-rays demonstrated an enormously enlarged aneurysm of the ascending aorta which had eroded through the anterior ribs. The serologic test for syphilis was strongly positive.

Anteriorly, the sternum is relatively straight below the sternal angle. A marked protrusion of the sternum is called *pectus carinatum*—pigeon breast. *Pectus excavatum* is a marked inward displacement of the sternum.

The costochondral junctions may enlarge markedly as a result of rickets or scurvy. The vitamin D deficiency of *rickets* prohibits normal mineralization of the osteoid matrix. The exuberant matrix jams the cartilage plates and becomes prominent as a *rachitic rosary*. The *scorbutic rosary* results from disordered maturation of the cartilage, which then splays out as a bulge.

Now, inspect the dynamics of respiration. Inspiration is normally an active event while expiration is passive. In normal quiet respiration, the flattening of the diaphragm and elevation of the anterior ribs increases the volume of the thoracic cage. The increase causes greater negative intrapleural pressure and allows air to enter down the pressure gradient. Maximum inspiration in the normal individual calls into play the accessory muscles in the neck which elevate the first and second ribs and, somewhat, the clavicle.

Certain diseases of the lungs present with hyperinflation secondary to air trapping. In such conditions as emphysema, the chest is constantly expanded with unusable air. In order to exchange additional air, the accessory muscles are used even in quiet respiration. These muscles stand out to casual observation. Such patients frequently assume a typical posture. He will sit in a chair slightly inclined forward with elbows or hands on his knees to help brace for a maximum effort (Fig. 11.8).

Expiration (E), normally passive, becomes active when obstruction exists. The *abdominal muscles* can be used to increase intra-abdominal pressure, thus forcing the diaphragm upward. If the obstruction results from diseased and weakened bronchi, the patient may also be noted to breathe through pursed lips. By so doing, he maintains a high intraluminal pressure and avoids airways collapse. He, in effect, moves the maximum obstruction out to his lips (Fig. 11.9).

Note the relative times of inspiration and expiration. Inspiration (I) consumes the same time span as expiration in the normal subject breathing quietly. Obstruction, as in asthma, for instance, affects expiration more than inspiration and results in a smaller I:E ratio. In other lung diseases where no obstruction plays a role, the relative times remain normal although both may be significantly reduced—interstitial fibrosis, for example.

Retraction of the intercostal spaces during inspiration always indicates *reduced compliance* of the lung. Many diverse diseases of the lungs have reduced compliance as a common denominator—diseases ranging from the *hya-*

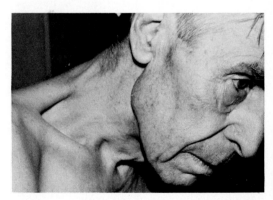

Fig. 11.8. Chronic lung disease. The hypertrophied strap muscles are evident. In addition, note that the patient appears chronically ill.

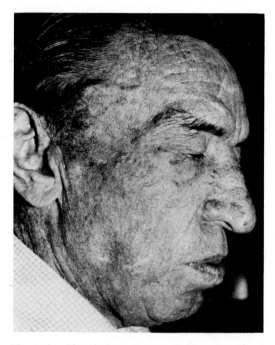

Fig. 11.9. Chronic lung disease. The face is suffused and the patient is expiring through pursed lips.

line membrane disease of newborns to diffuse *interstitial fibrosis* in the adult. In order to expand these stiff lungs, the patient must generate a large negative intrapleural pressure. This high negative pressure, in addition to expanding the lungs, will pull in the intercostal spaces—retraction.

Listen to the patient speak. We have already commented on the breathy voice of cord paralysis (Chapter 7). The patient with a markedly reduced vital capacity may pause several times during a single sentence to take a breath. The patient with severe obstruction may actually sound to be squeezing out his words.

If the patient is producing sputum, inspect it. Note the color, consistency, amount, and odor. Microscopic examination contributes greatly. Wet smears of unstained sputum under a coverslip allow you to recognize eosinophils of asthma, bronchiolar mucous plugs of bronchitis, and hemosiderin-laden macrophages of chronic heart failure. Gram-stained material shows neutrophils of acute inflammation and the predominant bacterial organisms. An acid-fast stain is much faster than the weeks it takes to grow tubercle bacilli in culture. Remember, however, that there must be at least 100,000 organisms/ml before you can expect to see them on a smear.

Palpation of Chest

Cardiac palpation will be discussed later. In general, palpation of the chest provides confirmation of the findings of inspection.

Feel asymmetric and abnormal contours to assess their precise contours and consistencies. Lumps on the chest wall can be of most any origin—tumors of the skin, muscle, or bone; fluctuant abscesses from within the chest; metastatic deposits of malignancy; or dilated vascular structures.

Now, palpate the dynamic events of respiration. Place each hand on corresponding areas of the chest. Have the patient breath deeply. Are the motions of each side synchronous in time and extent? Start by placing the tips of both thumbs on the xiphoid and the hands splayed out over both anterior lower lung fields. Repeat this posteriorly and superiorly.

The neck has already been examined for static abnormalities, but it is worth returning to it for dynamic events. Palpate the trachea during deep inspiration. It descends both in a real and apparent sense. The normal elastic trachea stretches downward as the diaphragm descends and the lungs expand. The sternal notch also rises as the anterior ribs move up and out. The trachea may deviate constantly or only during inspiration. In the latter, a *pendulum* motion from side to side gives an important clue to bronchial obstruction. The lung supplied by an obstructed bronchus cannot inflate. When air enters, then it fills up the remaining lung and lobes and they swing toward the obstruction, pulling the trachea along. A constant deviation results from *traction* from one side to *pulsion* from the other, as in a tension pneumothorax.

You can feel low frequency sounds better than you can hear them, and this can be used to advantage in the lung examination. Low-pitched adventitious sounds are better felt than heard. The normal spoken voice produces a palpable resonance—*tactile fremitus*. Use either the ulnar sides of your fifth fingers or the flat of your hands on the same place over each lung and have the patient say "ninety-nine" (Fig. 11.10). Aberrations of these sounds follow the same principles outlined below in the section on auscultation. Fremitus is more marked in men than women since the low-pitched voice is closer to the natural resonance of the chest. Fremitus in children is prominent, also, since the younger

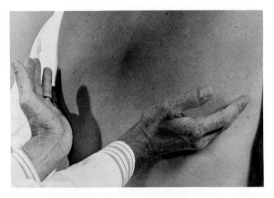

Fig. 11.10. Testing vocal fremitus. The ulnar surfaces of both hands are applied to similar areas of the back. The patient is asked to speak and the vibrations are noted. Consolidation enhances fremitus.

chest has a higher natural frequency, approximating the higher frequency voices of children.

The measurement of chest expansion provides an index of efficiency. If the difference between the chest circumference from full expiration to full inspiration is small, you know that rib excursion has been less than adequate. Air trapping results in this abnormality since in this condition the lung is more or less constantly expanded and can expand very little in addition. Decreased lung compliance, as in pulmonary fibrosis, and arthritis of the spine with bony fixation of the ribs, as in anklyosing spondylitis, cause similar reductions in chest expansion.

Percussion of Chest

Review Chapter 2 for the general techniques of percussion. Recall that the percussion note penetrates only 3–4 cm and that the resonance of the note is a function of the density of the tissue, air-containing tissue being more resonant than solid tissue. Repeat the percussion over your thigh for an illustration of flatness, over the liver for dullness, over the lung for resonance, and over the gastric bubble for tympany.

The resonant note of normal lung tissue can be learned only with practice (Fig. 11.11). The note may vary from individual to individual, but it should be the same over similar areas of the same patient. A thick chest wall, for instance, gives a duller note than the wall of a child but, in both, the note on one side of the chest is the same as on the other. If the pleural space fills with fluid, *pleural effusion,* the percussion note becomes dull. If the effusion is unilateral, the percussion note will differ from side to side. Similarly, a *consolidated lung* filled with fluid and containing no air sounds dull.

If there is an increase in air, there is an increase in resonance—*hyperresonance.* Air in the pleural space, *pneumothorax,* causes hyperresonance (Fig. 11.12). Likewise, the emphysematous lung contains more air per gram of lung than normal because of the destruction of alveolar walls and the hyperinflation. Such a lung will also sound hyperresonant. As we shall see shortly, the combination of bilateral hyperresonance and decreased breath sounds is strong evidence for emphysema.

Two unusual findings of air leaks might be mentioned here. If escaped air dissects into the subcutaneous tissue, it can be felt as *subcutaneous emphysema*—palpable bubbles of air which ripple in the fat as you press on them. If the air dissects along a bronchus to the mediastinum, an audible and sometimes palpable *mediastinal crunch* can be detected. It sounds a bit like a boot in dry snow.

A variety of percussion notes can be elicited over normal lungs. Anteriorly, on the lower left, you may hear a tympanitic note caused by the gastric bubble—*Traube's space.* A bit lateral to this an area of *splenic dullness* may be appreciated. The heart in the left

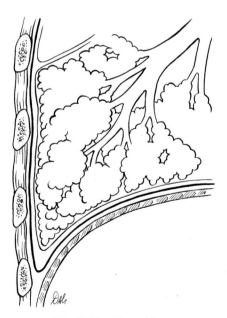

Fig. 11.11. Normal lung.

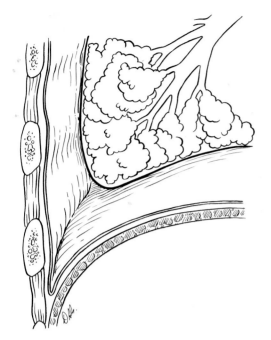

Fig. 11.12. Pneumothorax. The breath sounds are distant and the percussion note hyperresonant.

chest produces increasing dullness as you move toward the sternum. *Liver dullness* appears at about the sixth intercostal space on the right. It is important to determine the upper border of the liver by this means so that you can measure its overall span—normally about 10 cm. Posteriorly, each lung has a strip of resonance over each shoulder, *Krönig's isthmus*, reflecting the apex of the lung. The width of this isthmus should be the same over both lungs. Unilateral narrowing or obliteration of this resonance might indicate apical fibrosis or consolidation as in tuberculosis or an apical tumor.

The diaphragm defines the lowest border of the lung as a change from resonance to dullness posteriorly. If the diaphragm moves normally, this border will change during respiration. Check for this difference by percussing the border at end inspiration and again at end expiration and measure the difference as the *diaphragmatic excursion*. With hyperinflation, the diaphragms are constantly depressed and will move very little. *Paralysis* of one diaphragm can be diagnosed by percussion. The cause might be a *central lesion*, *phrenic nerve injury*, or a sympathetic accompaniment of a *subdiaphragmatic abscess*, or *empyema*.

Auscultation of Lungs

Sound emanates from the chest wall with a variety of origins. The motion of air through bronchi and alveoli produces sound vibrations. The spoken or whispered voice transmits sound waves through the lungs. The presence of abnormalities within the lungs alters either the naturally produced sounds or the transmission of vocal sounds.

We need to review first a few principles of sound transmission before describing pathologic alterations of sounds. The *distance* between the origin of the sound and the stethoscope obviously affects the intensity of the sound. Another factor to consider is the *number of interfaces* of different densities through which sound must travel. The alveoli act as a series of baffles so that bronchial sounds are broken and muted much in the same manner as acoustical tiling. The *nature of the transmitting medium* will affect the intensity and may selectively favor certain frequencies or pitches. The *more dense the medium, the better the transmission*; e.g. fluid transmits better than air. The basic frequency of the chest is about 110 cycles/second, and sounds close to this frequency are selectively transmitted.

Let us review first normal breath sounds. *Vesicular breath* sounds arise from air swirling into alveoli. This sound is low pitched, fine, and heard well at the periphery of the lung since it is produced nearby. Because it comes from air *entering* alveoli, it is an *inspiratory sound*. Listen at the base of the lung posteriorly for an example of this type of respiratory sound.

Bronchial breath sounds stem from air turbulence in cartilaginous bronchi. They are harsher and of higher pitch than vesicular sounds. Bronchial breath sounds melt almost entirely by the time they pass through the alveolar baffles and consequently cannot be heard at the periphery of the normal lung. Auscult over the trachea for an example of bronchial breath sounds.

Bronchovesicular breath sounds are a mixture of the two elements. These can be heard where there is a thin layer of alveoli over large bronchioles. Listen over the right infraclavicular area close to the sternum for an example (Fig. 11.13).

Bronchial breathing and the bronchial component of bronchovesicular breathing en-

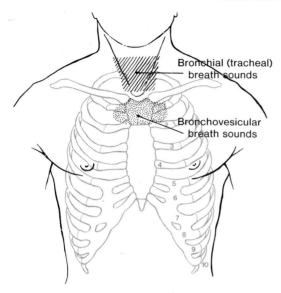

Fig. 11.13. Representative areas for breath sounds. Bronchial breath sounds are not heard over normal lung tissue. They may be simulated by listening over the trachea. Bronchovesicular breath sounds are normally heard in the area designated. Elsewhere over the anterior chest vesicular breath sounds are the rule.

compass the entire respiratory cycle. Since vesicular breathing is an inspiratory sound, vesicular breathing *appears* as though inspiration is much longer than expiration. As more and more bronchial component is added to what you hear, expiration *appears* to become longer and the two phases approximate each other in duration.

We listen over the lung for the spoken and whispered voice as a test of the transmitting medium—the lung. The spoken voice becomes garbled and muted by the time it reaches the periphery of the lung. The chest will favor low frequencies close to its natural resonance. The whispered voice cannot be heard at the periphery. As you listen over areas of bronchovesicular breathing, the spoken voice sounds louder and the whispered voice becomes audible. When you reach areas of bronchial breathing, the spoken voice is very loud and may be decipherable and the whispered voice is distinguishable at least in terms of the number of syllables being enunciated.

The single greatest factor producing abnormal sounds in the lungs is *fluid*—fluid as edema transudate, purulent exudate, or abnormal quantities of mucus.

Rales (pronounced rahls) are the sounds produced by air and fluid in alveoli. Their presence always indicates fluid. (You may hear the term "dry rales," which is clearly a misnomer.) Rales may be heard throughout the respiratory cycle or during either phase alone. A variety of descriptive terms may be used to characterize these adventitious sounds—crepitant or very fine, bubbling, intermittent, musical, etc. Very small amounts of fluid can be detected as *posttussive* rales. To hear these, have the patient expire fully and at the very end of expiration, cough. Now, when he next inspires you will hear a shower of fine crackles. The cough at end expiration collapses some of the wet alveoli so that they crackle when opened again during inspiration. This sign, when present over the apex of a lung, may be the only physical finding of tuberculosis and, elsewhere, it is the very earliest sign of fluid accumulation.

As more fluid accumulates in alveoli, rales get more prominent and last throughout the cycle until they are obliterated because of complete filling.

Rhonchi come from air turbulence around mucus or other fluid debris within large airways. The sound is harsh, continuous, and can vary from breath to breath as the material shifts position. Rhonchi may also originate from a solid, partially obstructing object in the bronchus—a tumor or foreign body, for instance.

Wheezes imply narrowing of small airways. Wheezes are high-pitched whistling sounds from air acceleration through pathologically narrowed lumens. Asthma is the prototype. Wheezes appear throughout the cycle or during inspiration or expiration alone. Because increased resistance accompanies wheezes, the phase of respiration during which wheezes are heard is frequently prolonged. Localized wheezes mean localized partial obstruction. Occasionally, you may hear a continuous wheeze at the end of expiration as air continues to whistle out past a partial obstruction—the *bagpipe sign*.

The respiratory sounds change as more fluid accumulates. With a small amount of fluid, rales become audible. The addition of some bronchial fluid bridges the conducting medium so that bronchovesicular breathing appears over areas which normally have only vesicular breathing. Rhonchi may be noted at this stage. When the area becomes completely filled or *consolidated* with fluid, the baffles are essentially obliterated and a single interface of bronchus to fluid results. The fluid will transmit bronchial breath sounds very well to the periphery. *Bronchial breathing heard in areas normally yielding only vesicular breathing always means consolidation.* The principle to remember is that fluid acts as a sound cable for the noises striking its interface even if that interface is remote from your stethoscope (Figs. 11.14 and 11.15).

You might now be able to predict the changes which occur to the spoken and whispered voice under similar circumstances. *Bronchophony* is the sound of the spoken voice over a bronchus and when heard elsewhere indicates consolidation. The spoken voice, as we noted, sounds very loud over a bronchus and may almost be recognizable; this same sound will appear at the periphery if the bridge of consolidation exists. Likewise, *whispered pectoriloquy* is the reproduction of a bronchial whisper in an area which normally does not register any sound with a whisper. The exact whispered word may not be heard, but the syllabic representation of the word will be clear.

Consolidation of the *right middle lobe* presents an interesting corollary to the above. Since this lobe is in intimate contact with the heart, it will transmit heart tones very well when it is filled with fluid. Audible heart sounds, then, over the right anterior lower

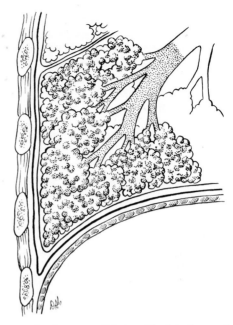

Fig. 11.15. Late consolidation. The lobe is completely airless. The bronchus and alveoli are filled with fluid. There is now only one interface between the bronchus and the chest wall.

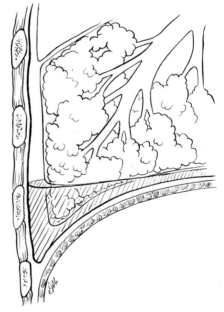

Fig. 11.16. Pleural effusion. The fluid moves the normal lung tissue away from the chest wall so that breath sounds are distant. The area is dull to percussion.

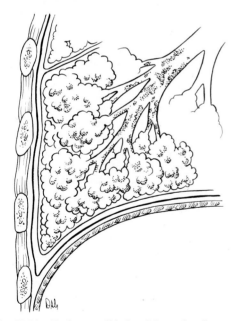

Fig. 11.14. Early consolidation. Many alveoli are partially filled with fluid and there is mucoid material in the bronchus.

chest strongly suggest consolidation of this lobe.

Fluid favors higher pitches and will select these from a varied frequency pattern. This can be used to advantage by having the patient say "e." The lower frequency is damped by consolidated tissue and the sound heard at the periphery approximates "a." In a similar fashion, an area of near consolidation may transmit the spoken voice with a bleating quality—*egophony.*

The degree to which you hear these pathologic changes depends on the degree of consolidation. The less the consolidation (the more interfaces), the less prominent are bronchial breathing, bronchophony, whispered pectoriloquy, and "e"-to-"a" change.

The pleura and pleural space further complicate the findings. A thin layer of fluid normally allows the visceral and parietal pleural surfaces to glide freely over each other during respiration. The pleura calls forth an inflammatory exudate of fibrin and white cells when it is irritated. The raw surfaces now grate and may produce an audible *friction rub.* This sound resembles creaking leather and appears close to your ear. There is a characteristic to-and-fro quality synchronous with respiration.

The accumulation of fluid in the pleural space (effusion) separates the lung from the chest wall. If the underlying lung is normal, breath sounds will disappear. The soft vesicular sounds are moved away from the stethoscope and are not transmitted well. If, on the other hand, the underlying lung is consolidated, weak bronchial breathing may be apparent as the pleural fluid acts as an extension of the fluid of consolidation. The percussion note is full with either consolidation or pleural effusion. They are differentiated on the basis of breath sounds—bronchial with consolidation and absent with pleural effusion (Fig. 11.16).

PRACTICE SESSION

Some of the following can be observed on your own chest and some will require the help of your partner. Begin by identifying significant landmarks. Find the sternal angle where the second rib joins the sternum and count ribs and interspaces downward. Next, follow the spine from the vertebra prominens downward to confirm the thoracic kyphosis which blends into the lumbar lordosis. Take a washable marker pen and outline the positions of the lobes of both lungs on the chest.

Now, watch the respiratory cycle closely. Have your colleague take a slow very deep breath. You will note that the expansion begins at the bottom of the chest and moves up. The neck muscles will stand out as the lungs are near to being completely filled. Now, watch as he exhales completely. At the end, the abdominal muscles contract to force out the last of the air. Remember this when faced with patients with lung disease. If the problem is getting sufficient air in, the neck muscles will be prominent. If the problem is obstruction to outflow, the abdominal muscles will be called into excessive use.

Now, palpate the chest during inspiration and expiration. Begin at the front with both thumbs on the xyphoid and the hands spread out over the lower anterior chest. With inspiration, your hands will flare outwards. Repeat this on the back. Palpate the trachea during deep inspiration and confirm that it descends.

Now, compare tactile fremitus, using the ulnar side of your hands against similar areas of both lungs. Have your partner repeat "ninety-nine." The feeling you get through your hands may be different in different areas but should be the same over similar areas of both lungs.

Now, percuss the posterior aspects of the chest and find the level of the diaphragm. Percuss softer than you think you should and you will be about right. Once you have determined the level of the diaphragm, mark it in full inspiration and then in full expiration and measure the excursions.

Anteriorly, you should be able to find the upper level of the liver and the tympanitic area of the stomach bubble.

Now, auscultate over the lower posterior lungs to hear vesicular breathing. Make sure your colleague does not make breathing noises in his throat. The fine noises of vesicular breathing will be heard throughout inspiration but fade quickly in expiration. It will appear as though the time of inspiration is much longer than expiration. Now, listen just under the right clavicle close to the sternum and there will be an element of bronchial breathing added to the vesicular sounds. Expiration will now sound longer and the sound will be a bit harsher. Listen over the trachea to hear a facsimile of bronchial breathing. It is close but not quite the real thing. True bronchial breathing is always of pathologic significance.

Now, listen over the posterior chest as your partner says "ninety-nine," then whispers, then says "eee." The sound may be audible, but it will be muffled and not clear in the normal. The millions of baffles break it up, effectively acting as a sound dampener.

Clinical Correlations

A few common specific disease entities will serve to combine the physical findings just discussed.

Lobar Pneumonia

This disease, less prevalent than in years past, still strikes previously healthy people as well as the debilitated. The history commonly includes the sudden onset of a shaking chill. The patient bundles in blankets to get warm but to no avail. An hour or so later, when the temperature has reached its peak, he feels relatively comfortable but feverish. A hacking cough increasingly productive of thick sputum ensues. A drenching sweat signals lysis of the fever. The cycle may repeat several times. As the consolidation reaches the pleural surface, chest pain may appear (Figs. 11.17 and 11.18).

On examination, the patient appears ill. Fever, tachycardia, and tachypnea are noted. Signs of dehydration might be expected because of the anorexia, drenching sweats, high fever, and hyperventilation. Pleural pain may cause *splinting* of the affected hemithorax.

Percussion reveals dullness over the af-fected lobe. Tactile fremitus is increased and bronchial breathing, bronchophony, whis-pered pectoriloquy, and "e"-to-"a" change are noted with auscultation.

A sputum smear shows sheets of neutro-phils and a predominant organism. The Gram stain tells you the major category of organism. A pale green sputum with lancet-shaped Gram-positive diplococci suggests pneumococcus. A brick-red sputum with Gram-negative rods suggest a klebsiella or-ganism. A sputum with Gram-positive cocci and an examination disclosing early pleural effusion are most typical of streptococcus. Staphylococcus rarely causes lobar consoli-dation.

Bronchopneumonia will give you more dif-ficulty on physical examination. The disease process is scattered, nonconsolidating, and deeper in lung tissue. The history might be similar, but the findings consist only of scat-tered rhonchi and wheezes with a purulent sputum.

Emphysema

The patient, usually a man in his fifties, complains of dyspnea on exertion. Cough, if present, is dry and nonproductive. Careful

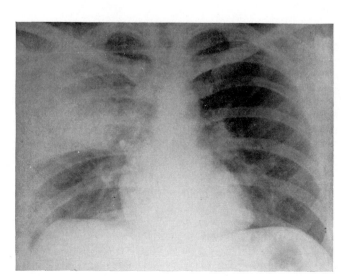

Fig. 11.17. Lobar pneumonia, right upper lobe. The dense infil-trate abuts on the major fissure and extends upward to involve all but the apical segment of the right up-per lobe.

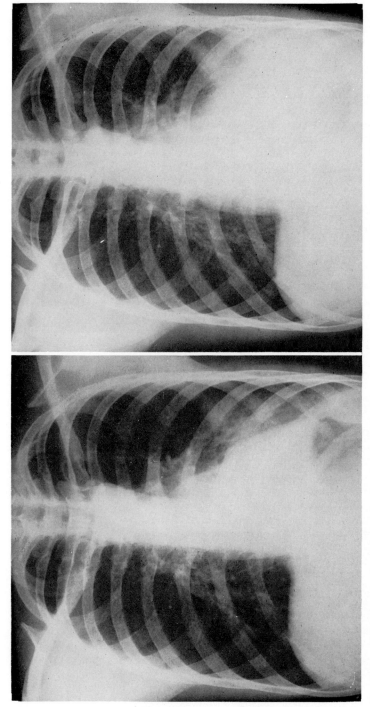

Fig. 11.18.   Lobar pneumonia. *Left,* shortly after the onset of symptoms a patchy infiltrate is noted in the left midlung field. *Right,* a homogenous dense shadow involving the lower third of the left lung field and obliterating the left border of the heart. The physical findings were those of consolidation with bronchial breath sounds, bronchophony, and whispered pectoriloquy.

questioning reveals a slowly progressive illness. If carbon dioxide retention occurs, you might find a slow decline in mental acuity, sleep disturbance, and periodic confusion.

Inspection shows signs of air trapping and inefficient respiration. The anteroposterior diameter of the chest is increased, and hypertrophy of the accessory muscles of respiration is apparent. The patient is usually thin and obviously dyspneic.

The percussion note sounds hyperresonant throughout, and the diaphragms are low and relatively immobile. Breath sounds are distant, and no adventitious sounds are heard unless an element of chronic bronchitis exists as well.

Chronic Bronchitis

The complaints resemble those of emphysema except that sputum production is always a feature. A smoking history is the rule. Symptoms of congestive heart failure such as orthopnea and ankle edema suggest *cor pulmonale* which occurs earlier in chronic bronchitis than in emphysema. Carbon dioxide retention is also more prevalent in chronic bronchitis.

He is more likely to be obese and cyanotic—the *blue bloater* bronchitic as opposed to the *pink puffer* emphysematous. Obvious obstruction to expiration prolongs that phase, involves abdominal muscles, and causes pursed lip breathing. Wheezes, rhonchi, and rales appear in varying combinations. Air trapping leads to the same increased chest diameter and flat diaphragms as in the patient with emphysema.

Interstitial Lung Disease

The patient with interstitial lung disease suffers from inadequate volumes—*restrictive lung disease*—as opposed to bronchitis and emphysema where the problem is reduced flow rates—*obstructive lung disease.* His complaint is profound dyspnea. Symptoms of congestive heart failure or carbon dioxide retention come very late. Look for a history of exposure to dusts, mineral particles, or gases as a clue to a possible *pneumoconiosis.*

Inspection suggests the diagnosis. The chest is small. The lack of lung compliance shows as intercostal retraction during inspiration. The accessory muscles of respiration hypertrophy. The percussion note may be normal or decreased. Very fine rales are common and sound like hair being rubbed together close to your ear. A tremendous disparity may exist between the auscultatory findings and the chest film—either can be strikingly abnormal while the other appears quite normal.

Tuberculosis

This patient has usually had his share of worldy troubles—social and economic. An exposure history is helpful if present, but don't count on it. He has lost weight, complains of anorexia and malaise, and may suffer with drenching night sweats. Cough may be dry or productive and, if productive,

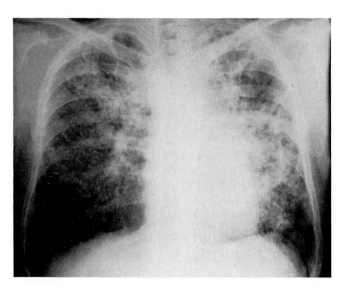

Fig. 11.19. Tuberculous pneumonia. The densities are diffuse, bilateral, and predominantly in the upper lobes. The multiple cavitations are not well seen on this x-ray.

may include hemoptysis. Predisposing factors of alcoholism, previous gastrectomy, previous lung disease, or inanition from any cause are common.

Examination shows a cachectic patient appearing chronically ill. Nonspecific evidence of chronic lung disease is common. Tuberculosis favors the upper lobes, especially the apices. The physical findings, of course, depend on the extent of the disease. Apical scarring retracts the supraclavicular fossa and narrows the normal band of shoulder resonance. Listen for posttussive rales over the apices. Large cavities rarely can be detected by an area of increased resonance, *amphoric breathing* (air swirling in a large cave), and the *coin sign*—listen on one side of the chest and tap a coin laid on the other side of the chest with another coin; the transmitted sound resembles a bell being struck. An associated pleural effusion will give its typical findings of dullness and diminished breath sounds (Fig. 11.19).

Carcinoma of Lung

The patient is usually an adult male with a long smoking history. Cough, anorexia, and weight loss are the usual symptoms. More often than not, he suspects the disease him-

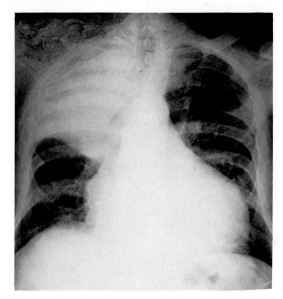

Fig. 11.21. Bronchogenic carcinoma of the right upper lobe bronchus. The malignancy has completely obstructed the right upper lobe bronchus with atelectasis and consolidation of the right upper lobe. The atelectasis is inferred by loss of volume as noted with the elevated major fissure. The heart shadow is also noted to be enlarged. The patient additionally suffered from coronary artery disease.

Fig. 11.20. Consolidation distal to a tumor (*arrow*). Although the lobe is consolidated, the typical findings are not present since no breath sounds get through the tumor.

self. This is a very common disease with some very uncommon manifestations, however. The presenting complaint may direct your attention to areas quite remote from the lungs, e.g. osteoarthropathy, neurologic disease, or skin disease, to mention a few. Pain, if present in the chest, implies involvement of adjacent structures. Hemoptysis indicates an endobronchial lesion.

Look for clubbing, cachexia, Horner's syndrome, vocal cord paralysis, supraclavicular adenopathy, or an enlarged liver as nonthoracic manifestations.

If an endobronchial lesion obstructs a lobe, that lobe will collapse, dragging the trachea toward it. The involved area becomes dull and breathless. Partial obstruction can give unilateral wheezing and the bagpipe sign. Rhonchi are common because of the associated mucous debris, and signs of chronic lung disease commonly accompany the findings (Figs. 11.20–11.22).

Pulmonary Embolus

Predisposing causes to pulmonary embolism include recent surgery or trauma, con-

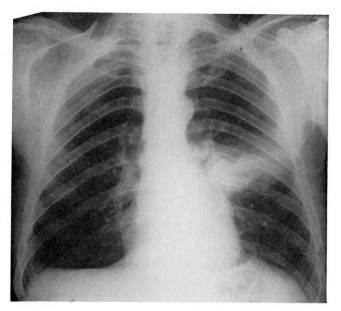

Fig. 11.22. Bronchogenic carcinoma, lingular segment left upper lobe. The tumor mass is the density close to the heart border. The density extending out to the chest wall is caused by secondary collapse and inflammatory infiltration of the lung parenchyma.

gestive heart failure, venous insufficiency of the legs, pregnancy or recent delivery, and birth control pills, all of which can be elicited historically.

The embolus initiates a series of events. Dyspnea and fear appear first. Confusion may be the only clue in the elderly. Examination at this point will show only tachypnea, mild cyanosis, and occasionally diffuse wheezing—presumably a reflex bronchospasm.

The parenchyma, now deprived of blood, quickly exudes fluid, and rales progressing to bronchial breathing and other signs of consolidation appear over 24 hours or so. When infarction results, fever, hemoptysis, pleural pain, and a pleural rub appear following which the dullness and decreased breath sounds of an effusion are commonly heard. The examination of such a patient must always include evaluation of the legs for a possible source of the embolus. The examination of the legs and the cardiac findings of a pulmonary embolus are covered later.

chapter 12

HEART

We come now to what most students consider the most challenging and exciting field of physical diagnosis. These feelings generate from the magnitude of the diseases affecting this organ, from the frequency of those illnesses, and mostly from the presumed difficulty of mastering the techniques involved. Recent investigations have unveiled a wealth of information regarding the physical diagnosis of heart disease. We owe our surgical colleagues a great debt for this, since their skill has made it imperative that we be precise in our interpretation of the data. There was, for instance, very little interest in the physical findings of specific congenital heart defects years ago, as there was nothing that would affect the outcome. Similarly, the diagnosis "coronary heart disease" is imprecise—we need to know if the obstructing lesions are proximal or distal in the artery and whether more than one artery is involved.

The procedure remains the same, however; a thoughtful history, not by rote; a well patterned examination; and then a pathophysiologic approach to the findings. Don't be in a hurry to listen to the heart. If your ears are stuffed with ear pieces, you won't hear the most important sounds, the history.

History

Each of the various structural elements of the heart is prone to specific diseases which can produce pain, dysfunction, or a change from the steady state, and we can consider each one in turn.

Chest pain of cardiac origin results from disease of the coronary arteries or pericardium. Disorders of heart valves yield pain through abnormalities in coronary blood flow.

Angina pectoris is diagnosed from the history, although as we shall see, there are accompaniments to be detected in the physical examination. The oppressive chest discomfort of angina pectoris was well described in 1818 by Dr. William Heberden (Commentaries on the History and Cure of Diseases, Wells and Lilly, Boston 1818, p. 292), and little can be added to his account.

" ... there is a disorder of the breast marked with strong and peculiar symptoms, considerable for the kind of danger belonging to it, and not extremely rare... the seat of it, and sense of strangling, and anxiety with which it is attended, may make it not improperly be called angina pectoris.

"They who are afflicted with it, are seized while they are walking, (more especially if it be up a hill, and soon after eating) with a painful and most disagreeable sensation in the breast, which seems as if it would extinguish life if it were to increase or continue; but the moment they stand still, all this uneasiness vanishes.

"In all other respects, the patients are, at the beginning of this disorder, perfectly well, and in particular have no shortness of breath, from which it is totally different. The pain is sometimes situated in the upper part, sometimes in the middle, sometimes at the bottom of the os sterni, and more often inclined to the left than to the right side. It likewise very frequently extends from the breast to the middle of the left arm. ... males are most liable to that disease, especially such as have passed their fiftieth year... ."

Many patients who relate this illness liken it to a profound weight on the chest and may make a fist over the sternum during their description, the "Levine sign." The compromise of the coronary circulation producing the pain most always occurs during exertion and abates with rest. Severe cases interrupt sleep and very profound degrees may confine patients to bed and chair. Heart pain should always be suspected when the discomfort radiates into the arms.

When coronary circulation fails to the point of *myocardial necrosis*, the pain can be similar, although more profound and prolonged. The inferior and diaphragmatic surfaces of the myocardium have rich vagal innervation. Acute infarction of this area, which is produced by occlusion of the right coronary artery, adds nausea and sometimes vomiting to the clinical spectrum. The anterior myocardium, supplied by the anterior descending branch of the left coronary artery, when infarcted, gives extreme pain and diaphoresis.

Some unusual pain patterns include jaw or neck pain, "gas pain," and arm or hand pain and numbness, which may be the only manifestations of ischemia.

The absence of pain in a myocardial infarction is not rare. The patient relates the following; "I had a routine checkup and my doctor told me that I have had a heart attack, but I have never been sick." This story should always suggest the presence of diabetes mellitus. The presumption is that the neuropathy of diabetes affects the pain fibers of the myocardium.

Angina pectoris rarely occurs in menstruating women or children, but do not discard the possibility out of hand. Premature coronary disease shows remarkable familial tendencies. A history of hypertension, hypercholesterolemia, or hyperuricemia may also come out with questioning.

The pain of *coronary insufficiency* results from some diseases of heart valves, especially the aortic valves. Severe aortic stenosis and, less commonly, aortic insufficiency can so alter the hemodynamics about the coronary ostia as to produce angina pectoris.

The pericardium, as with all serous membranes, is capable of producing pain as a response to inflammation. Described as a sharp stabbing discomfort, *pericardial pain* is many times position related. You may be told that this position or that position aggravates or alleviates the discomfort. It is a disease of all age groups, and a viral etiology is the rule in painful pericarditis.

A potpourri of symptoms comes about with *heart dysfunction.* Fatigue is the earliest such symptom and the most common. There is nothing specific about it, the patient explains that he just tires easily. Previously simple chores now exhaust him, he feels as though he has put in a full day by noon, or that vacuuming or bed making take it all out of her. This is a particularly noteworthy complaint of patients with mitral valve disease.

Congestive heart failure is profound dysfunction, and all of the symptoms can be explained by the compensatory mechanisms which no longer compensate but rather aggravate. The failing myocardium demands increasing diastolic stretch of its myofibrils. To accomplish this, the intravascular volume expands through the renal retention of salt and water. Once the increased volume overwhelms the heart, failure ensues. With failure of the left ventricle, pulmonary venous hypertension occurs. The patient complains of dyspnea on *exertion.* Further, when he is recumbent, the lungs become more dependent in position and accumulate more fluid, and *orthopnea* is the complaint. He might tell you that he sleeps better with two or three pillows beneath his head and shoulders. Sudden episodes of a smothering sensation awake him at night—*paroxysmal nocturnal dyspnea*—and force him to stalk around the bedroom gasping for air. A nighttime cough is another manifestation of these wet lungs. Nonetheless, the pump is moderately more efficient at night while resting, and as renal perfusion increases so does urine flow and *nocturia* may be the complaint.

When the pulmonary venous hypertension is very great, *acute pulmonary edema* results. Akin to drowning, the patient becomes intensely dyspneic and frightened. He may bring forth copius amounts of pink frothy sputum as the alveoli fill with a bloody effusion. Chronic elevation of pulmonary venous pressure from left ventricular failure or from obstruction to the left ventricle as in mitral stenosis sometimes becomes manifest as hemoptysis.

Right ventricular failure causes systemic venous hypertension as the increased volume engorges these vessels. When sudden in onset, this backup engorges the liver, stretches its capsule, and causes pain in the right upper quadrant of the abdomen. When upright, this venous hypertension is added to the pressure of the column of blood from the right atrium down to the ankles. This pressure overwhelms the intravascular oncotic pressure, and fluid seeps into the soft tissues as *ankle edema.* This is uncomfortable but painless and when profound can extend up to the chest wall as *anasarca.* The usual story is, "my ankles are swollen by the end of the work day, but normal when I get up in the morning."

Predispositions to congestive heart failure are hypertension, coronary artery disease, valvular heart disease, and primary myocardial diseases (cardiomyopathies).

The historical hallmarks of congestive failure, then, are fatigue, dyspnea on exertion, orthopnea, paroxysmal nocturnal dyspnea, nocturia, and ankle edema.

Dysfunction of the heart can transiently compromise cerebral blood flow. Neurologic complaints may be a result of this phenomenon. A sudden fall in cardiac output because of a heart arrhythmia might cause dizziness, blurred vision, seizures, or syncope as in the *Stokes-Adams syndrome* of complete heart block with a subsequent ventricular tachycardia or asystole. This form of syncope must be distinguished from the benign vagal faint which carries a far less ominous significance. With a Stokes-Adams attack, patients relate a sensation of spreading warmth, lightheadness, and then unconsciousness. A vagal faint characteristically imparts a sense of impending unconsciousness, produces pallor, nausea, and, if carefully observed, a bradycardia.

All of us have experienced *palpitations*. Variously described as a fluttering, lurching sensation, pounding or sudden catch in the chest, palpitations may signify anything from too much coffee to serious organic heart disease. Careful questioing is required to identify predisposing factors of coronary artery disease, hyperthyroidism, excess catecholamines or other ionotropes, or valvular heart disease. In general, the more serious the arrhythmia the more profound the symptoms. *Paroxysmal atrial tachycardia*, for instance, commonly affects young people and is not very serious. It produces a notable tachycardia, a faint feeling, starts and stops suddenly, and is usually recurrent. *Ventricular tachycardia*, on the other hand, produces a marked fall in blood pressure, syncope or near faint, diaphoresis, and a sensation of imminent death.

An important historical clue to the seriousness of the isolated palpitations is the setting in which they occur. Extrasystoles which "escape" when heart rate is slow at rest are far less worrisome than those that come on with, or just after, vigorous activity.

The past medical history becomes extremely important in the review of heart disease. When was the last physical examination and did your doctor tell you that your blood pressure was normal? That you had a recent murmur? That your chest x-ray and electrocardiogram were normal? Have you ever been rated for life insurance?

The diagnosis of *acute rheumatic fever* can be difficult in retrospect. Many patients with this disease at 55 were never told of rheumatic fever at age 12. Often, the best you can elicit is a history of bad "growing pains" which kept the patient from school, or a long illness when the doctor confined him to bed. Remember, too, that, Sydenham's chorea, or *St. Vitus' dance*, is a manifestation of acute rheumatic fever and question the patient about a period of abnormal, uncontrollable muscular activity in childhood.

Congenital heart disease still eludes detection until adult life in some patients. Was the patient a "blue baby," a poor feeder, or did he have trouble keeping up with his peers in physical activities?

The family history might reveal that all males "dropped dead" before 55 or that hypertension or diabetes were common. A few myocardial diseases, certain arrhythmia syndromes, and Marfan's disease are examples of familial cardiac illnesses.

Anatomic Considerations

Although a rehash of what you have covered in anatomy, we can profitably review a few anatomic relationships which are important to remember in the physical diagnosis of the heart.

The right ventricle is more accurately the *anterior* ventricle, while the left ventricle is *posterior*. The *apex* of the heart is predominantly left ventricle and the interventricular septum runs in a plane which is on a diagonal with the chest. If you face a patient who is turned 45° to his right (your left), you are looking end on at his septum. Likewise, the pulmonary valves and pulmonary arteries are anterior, and the aortic valves and pulmonary arteries are anterior, and the aortic valve and aorta are posterior. These relationships affect what you feel and hear and what you see on a chest x-ray (Figs. 12.1).

The representation of the chambers on a chest x-ray is important to understand. From top to bottom, on the right side of the heart, the bulges are the superior vena cava, the ascending aorta, and the right atrium. On the left side, the bulges represent the left subclavian vein, the aortic knob, the pulmonary artery, the left atrial appendage, and the left ventricle.

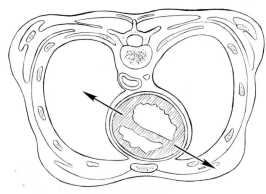

Fig. 12.1.   Anatomic relationships. The septum is on an angle (*arrows*) to the chest wall. The right ventricle is anterior and the left ventricle posterior.

Inspection

We begin with an inspection of the cervical veins. The column of blood in the internal and external jugular veins is not interrupted by venous valves and so can be viewed as a pressure manometer to the right atrium. The fluctuations in the level of blood within these veins then represent a valuable yardstick for the pressure changes within the atrial cavity. The higher the column, the greater the pressure in the right atrium. The top of his column can be demonstrated in every normal individual if you look carefully.

Position your patient by having him on a bed or examining table which can be tilted upward at the hips. Start with the spine straight and elevated about 30° from the horizontal. Light the right side of the neck on a tangent with a movable lamp or your penlight and observe the jugular veins carefully. When the external jugular vein distends, it appears as a discrete vessel over the sterno-cleidomastoid. Frequently, this vein cannot be seen and you must rely on the internal jugular system. These veins do not appear as discrete vessels; rather, the pressure waves are seen as bulgings or lateral distensions on the surrounding soft tissues. If the patient has a thin neck, the underlying carotid pulsation can be mistaken for venous pulsations. A simple test is to compress the root of the neck just above the clavicle, which will obliterate the venous pulses but not the carotid pulses. Remember, too, that *venous pulsations cannot be felt* (Fig. 12.2).

If with the patient at 30° you cannot see pulsations, lower the angle slowly until the patient is flat. If the venous pressure is very

low, as in dehydration, the pulsations should then become apparent. If they still cannot be seen, it might be because the pressure is very high. Therefore, raise the bed again and go up slowly from 30° and 90°.

Note that the top of the column fluctuates during the cardiac cycle. The maximum sustained height defines the *venous pressure*. Measure this by running a horizontal marker to a position over the anterior chest and drop a perpendicular to the level of the right atrium (Fig. 12.3). (Some examiners prefer using the angle of Louis and then adding 5 cm). The height of the perpendicular line is the venous pressure measured in centimeters of blood. This form of measurement avoids the ambiguities of saying "the jugular veins were distended to the angle of the jaw with the patient at 45°." There is too much room for error with this method of description. When you see the horizontal and vertical line, it doesn't matter at what angle the patient is

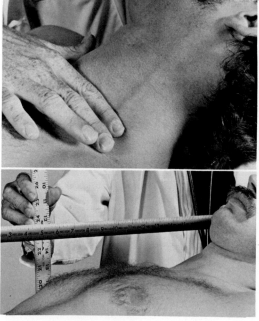

Fig. 12.2. (*Top*).   Locating the cervical vein. Light compression at the base of the neck will distend the cervical vein so that you will know where to look for cervical venous waves.

Fig. 12.3. (*Bottom*).   Measuring the cervical venous pressure. A *horizontal ruler* is run from the top of the cervical vein to a *vertical ruler* applied close to the chest. In using this method, the absolute venous pressure can be determined regardless of the angle of repose of the patient.

positioned. The absolute number will always be the same.

The venous pressure reflects the filling pressure of the right atrium and indirectly the right ventricle. *Obstruction* anywhere in the right heart circuit will elevate this pressure. Thus, a pulmonary embolism, pulmonary hypertension, pulmonic valve stenosis, tricuspid valve stenosis, congestive heart failure, right ventricular hypertrophy, and constrictive pericarditis may all elevate the venous pressure (Fig. 12.4). A low venous pressure results from a reduced intravascular volume.

Now that you have established the venous pressure, observe the fluctuations in the level. Gross changes occur with respiration. With inspiration, the level falls and, with expiration, it rises. When the patient inspires, the intrathoracic pressure falls, and the volume of blood which can be taken in the pulmonary bed increases; hence, the venous pressure drops. The converse happens with expiration. In certain diseases, the pressure paradoxically *rises* with inspiration. With *constrictive pericarditis* or *pericardial tamponade* the descent of the diaphragm with inspiration tenses the

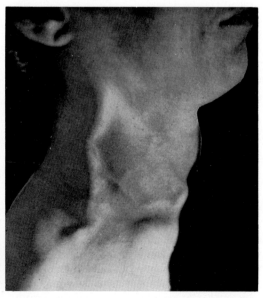

Fig. 12.4. Distended cervical vein. This patient suffered with tricuspid insufficiency. The enormously distended cervical veins were evident even in the upright position. Prominent "v" waves and a systolic murmur which increased with inspiration were the physical findings. When extremely profound, the earlobes will be noticed to move with each "v" wave, and pulsatile proptosis might also be seen.

already diseased pericardium and actually impedes right ventricular filling. The venous pressure then rises, and *Kussmaul's sign* is present (see Fig. 5.4 in Chapter 5).

Now focus on the very subtle fluctuations in the pressure head during the cardiac cycle. Two episodes of rising and falling accompany each normal cardiac contraction. The "a" wave is the first of these waves. It occurs just before the carotid impulse. Watch the right neck veins and palpate the left carotid artery to help time the waves. The "a" wave occurs with atrial systole and is the final kick to fill the ventricle. Because there is no valve between the right atrium and superior vena cava, the pressure rise of atrial systole is reflected up the neck veins as well as down into the ventricle. After the atrium ejects its blood, ventricular systole closes the tricuspid valve and the atrium relaxes to receive more blood and the "a" wave recedes as the "x" descent. The neck veins then again rise as the "v" wave during the end of ventricular systole. This is related to the filling of the atrium and slight upward ballooning of the tricuspid valve and can be seen just after or synchronous with the carotid pulse. The "v" wave recedes as the "y" descent when the tricuspid valve opens and passive filling of the ventricle occurs.

With direct right atrial pressure recordings the "a" and "v" waves are separated by a "c" wave coincident with tricuspid valve closure, but this wave cannot be seen in the neck (Fig. 12.5).

Distorted venous waves occur with a variety of heart abnormalities. The "a" wave becomes exaggerated if there is obstruction to filling of the right ventricle. Such *cannon "a" waves* occur in pulmonary hypertension, severe pulmonary stenosis, tricuspid stenosis, or when the atria and ventricle happen to contract simultaneously, as in complete heart block (vide infra). In all of these conditions, the right atrium finds it difficult to eject blood into the ventricle, and consequently most of its contractile force is reflected backward up the jugular venous system.

Very large "v" waves accompany *tricuspid insufficiency*. The pressure of ventricular systole passes through the incompetent valve directly up the neck vein.

Complete absence of venous pulsations implies obstruction between the atrium and the neck veins, and this most commonly occurs

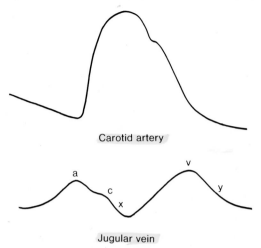

Carotid artery

Jugular vein

Fig. 12.5. Jugular venous pressure. The "a" wave occurs with or just before the carotid upstroke and the "v" wave just after. The "c" wave is not formally visible in the neck. The "a" wave recedes as the "x" descent. The "v" wave recedes as the "y" descent.

with *superior vena caval obstruction*, usually secondary to malignant disease.

Now inspect the precordium. Is the chest symmetrical? Congenital heart disease with ventricular enlargement may deform the chest so that the left anterior precordium bulges forward. If the patient has pendulous breasts, lift the left breast to observe the *apex beat* (the most lateral pulsation). Sight tangentially to the chest. If the apex cannot be seen, have the patient roll slightly to the left to bring the apex into contact with the left chest wall. Ventricular systole begins with an outward pulsation followed by a retraction normally seen in the forth or fifth intercostal space at about the midclavicular line.

Pressure overloading of the left ventricle exaggerates the apex beat. *Volume overloading* of the left ventricle also accentuates the apex beat. The two can be distinguished, however, by the area encompassed by the apex beat. With pressure overloading, the apex is exaggerated but circumscribes a very small area, whereas volume overloading produces a much larger apex beat. A *rocking* apex beat may be observed with a ventricular aneurysm which paradoxically bulges as the ventricle contracts, or by *right ventricular hypertrophy* which overrides the left ventricle.

Palpation—Arterial Pulse

Begin with an assessment of the carotid pulse. The contour and volume of the pulsa-

tions in the carotid artery best reflect the cardiac events. Pulsations are damped and altered by the time they reach more peripheral vessels. We discussed rhythms in the chapter on vital signs, and now we focus on the characteristics of each pulse beat.

Normally, the carotid upstroke occurs about 0.04 second after the first heart sound. Place your first three fingers on the carotid artery and note that the pulsation rises steadily in intensity and falls off briskly. In many normals, a slight hesitation of the falloff may be felt, the *dicrotic notch*. Two forms of abnormalities occur. First the rate of the impulse may be accelerated or delayed, and second the *volume* may be increased or decreased.

A slowly rising carotid pulse results from any partial obstruction between the left ventricle and the artery you are palpating. *Aortic stenosis*, from rheumatic heart disease, calcification of the valves, or congenital stenosis, produces this slowly rising pulse—*pulsus tardus*.

A very rapid upstroke and falloff is called a *water hammer pulse*. A common denominator of the diseases causing the water hammer pulse is a low resistance to the runoff of blood. Incompetence of the aortic valve allows a portion of the ejected blood to fall back into the left ventricle. The total volume ejected with each systole becomes increased to compensate for the amount "wasted." Hence, the upstroke is brisk and of increased volume. The falloff is also quite rapid for two reasons. First, a significant volume drops back into the left ventricle and, second, the peripheral vascular resistance is reduced in order to allow a maximum flow of blood out of the aortic tree.

Reduction of the peripheral vascular resistance occurs with other diseases as well. A large *arteriovenous fistula* which bypasses the resistance of small anterioles "wastes" a great volume of blood. Fistulas can occur anywhere. A particularly dramatic example results when the ductus arteriosus fails to close after birth. Persistence of this channel permits blood from the high-pressure aorta to pour into the low pressure pulmonary artery.

Another low pressure egress for ventricular blood exists when the mitral valve is incompetent. With each systolic ejection, blood squirts both into the aorta and into the left atrium. Just as in aortic insufficiency, the volume of each ejection must be increased, and the peripheral arterial pulses reflect this phenomenon.

The volume characteristics usually follow with the rate of upstroke; that is, a slow upstroke is accompanied by a small volume. The *pulsus tardus* of aortic stenosis is also a *pulsus parvus*. The bounding pulse of aortic insufficiency, arteriovenous fistulas, and mitral insufficiency is also a large volume pulse.

A normal upstroke with a small volume

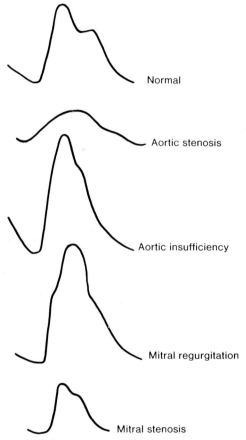

Fig. 12.6.   Carotid artery pulse tracings.

means something else—a low blood volume, for instance, as in shock. A small ejected volume can also result from obstruction to filling of the left ventricle. Mitral stenosis does just that, and the carotid upstroke produced is very weak (Fig. 12.6 and Table 12.1).

Diagnosis of Arrhythmias

The examination of the jugular venous pressure and the carotid pulsation permits the accurate diagnosis of many arrhythmias.

With *atrial fibrillation*, the atrial myofibrils contract more or less at random, and no concerted action results. Stimulation of the ventricle from above, then, also occurs at random, and the pulse generated is irregularly irregular. Without an atrial contraction no "a" wave can be generated. The physical findings are then predictable. The carotid pulse is irregular in timing and in volume. If one beat follows closely on another, the volume will be low, since the ventricle has had little time to fill. The neck veins show only "v" waves, which also vary with the duration of the previous diastole—large "v" waves with a long filling time and small "v" waves with a short filling time (Fig. 12.7).

Atrial flutter is another common supraven-

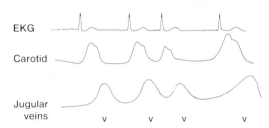

Fig. 12.7.   Atrial fibrillation. The electrocardiogram (*EKG*) shows the random irregularly irregular rhythm. The carotid artery and jugular venous pressure demonstrate the variable magnitude of pressure which depends on the length of diastole. Observe also the absence of "a" waves in the venous pressure (v).

Table 12.1
Upstroke and Volume Characteristics

Characteristic	Denominator	Common Examples
Upstroke:		
Slow	Obstruction	Aortic stenosis mitral stenosis
Rapid	Decreased resistance	Aortic insufficiency arteriovenous shunt
Volume:		
Small	Decreased ejection volume	Shock, mitral stenosis, aortic stenosis
Large	Increased ejection volume	Aortic insufficiency, pregnancy, hyperkinetic states

tricular arrhythmia. The atria contract 300 times a minute. The atrioventricular node cannot transmit this many impulses down into the ventricle. Usually, only every other impulse gets through, so that the ventricular rate is 150 (other ratios like 4:1, 3:1, etc. obtain under special circumstances). Again, the findings are predictable. The carotid artery pulsates regularly 150 times a minute. *Two "a" waves* are generated for every arterial pulse, and "fluttering" of the neck veins becomes evident (Fig. 12.8).

Paroxysmal atrial tachycardia is a regular rhythm usually 120–160 beats/minute. With this rhythm, only one atrial systole precedes each ventricular systole so flutter waves in the venous circuit are not seen. Thus, the flutter waves distinguish atrial flutter from paroxysmal atrial tachycardia, both of which may have a similar rate.

Complete heart block means total dissociation of the atrial and ventricular electrical events. The atria contract at their own rate of about 70–90 beats/minute, and the ventricles at their own rate of about 40–60 beats/minute. The carotid pulse, which reflects ventricular systole, is regular with a slow rate and large volume since the diastolic filling time is long between beats. An interesting phenomenon will appear in the jugular pulse. From time to time the atria and ventricles will contract simultaneously. The atria have no hope of opening the tricuspid valve against the pressure of ventricular systole. The entire pressure wave of the atrial contraction is then reflected up the neck veins as a *cannon "a" wave* (Fig. 12.9).

Premature beats may be initiated by the atrium or the ventricle. As the name indicates, these beats appear sooner than expected. The earlier they occur after the preceding beat,

the less volume they generate. If a beat comes very early, no ejection at all results, and you detect a *dropped beat*. An atrial premature beat produces a premature "a" wave which can occasionally be observed in the neck. Obviously, no "a" wave accompanies a ventricular premature beat (Figs. 12.10 and 12.11).

Palpation—Precordium

Start by laying the palm of your hand and fingers over the area where you sited the *apex beat* by inspection (Fig. 12.12). Normally, the apex feels like a pulsation covering about a half-dollar-sized area of your hand. Now place the pads of two or three digits over the precise apex beat and note the brisk tap and withdrawal. Simultaneously, palpate the carotid artery and apex to appreciate the *ejection time*. The apex beat is best felt in one or two interspaces. Record the exact position in terms of interspace and distance from the sternum—e.g. "the apex is felt in the fourth

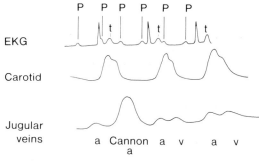

Fig. 12.9. Complete heart block. The atrial contractions (*P*) are independent of the ventricular beats. When they happen to coincide, a cannon "*a*" wave results, which is the combination of a "*v*" wave and an atrial contraction against a closed atrioventricular valve.

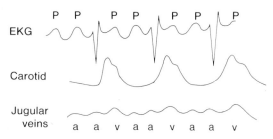

Fig. 12.8. Atrial flutter. The rate is 150 with 300 atrial beats/minute. Only every other atrial beat gets through so that there are two "*a*" waves for every "*v*" wave.

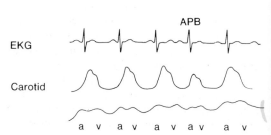

Fig. 12.10. Atrial premature beat (*APB*). The electrographic representation is normal except that the beat is early. The carotid pulse is somewhat reduced. An "*a*" wave although small, is present. *v*, venous pressure.

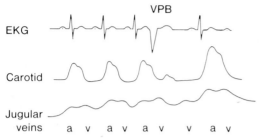

EKG

Carotid

Jugular veins   a   v   a   v   v   a   v

Fig. 12.11. The ventricular premature beat (*VPB*) interrupts the normal cycle. The short diastole accounts for the weak carotid pulse. The subsequent diastole is long, allowing for complete filling and a full carotid pulse. Note that there is no preceding "*a*" wave.

intercostal space, 8 cm from the midsternal line" (Fig. 12.13).

The formation of the apex beat is complex. The right ventricle has a bellows-like action and does not really play a role in what you feel. The left ventricle rotates anteriorly and to the right during systole, thrusting its apex outward against the chest wall, and this is what you feel.

You may have to palpate obese patients leaning far forward or in the left lateral decubitis position. If you cannot feel an apex, try several positions and be sure to check the right precordium so as not to miss a dextrocardia. Also, palpate very carefully at the end of forced expiration to move the air-filled lung from between the heart and chest wall. The absence of an apex beat always means disease. Remember, however, when you roll a patient to one side or the other, the apex will shift toward the more dependent side so that its exact location may be incorrectly determined.

Two kinds of ventricular change cause alterations of the apex beat—hypertrophy and dilatation. The left ventricle hypertrophies as a result of pressure overloading and dilates because of volume overloading. Hypertension and aortic stenosis are examples of pressure overload. The hypertrophied ventricles strike the chest wall vigorously, but in a small area in the expected position. This tap, or lift, is easily felt and seen. The dilated ventricle of aortic or mitral insufficiency enlarges laterally, and the apex may be far from the midsternal line. The area of the impulse becomes very large, and the entire left precordium may heave under your hand.

The apex and the point of maximum impulse are usually the same. The term should not be used interchangeably, however. The apex is the most lateral pulsation of the precordium, and the point of maximum impulse (PMI) is just that and can occur anywhere. A discrepancy between the two frequently results with diseases of the right ventricle. Hypertrophy of this chamber gives a precordial heave just next to the sternum while the apex will still be seen and barely felt lateral to this (Fig. 12.14). The large chest of chronic lung disease can hide the apex beat and the PMI is felt below the xiphoid. The right ventricle is felt under these circumstances by hooking your fingers over the xiphoid and pressing in and up (Fig. 12.15).

Now, palpate the sternal clavicular joints and the suprasternal notch. Normally, very little pulsation can be detected. Diseases of

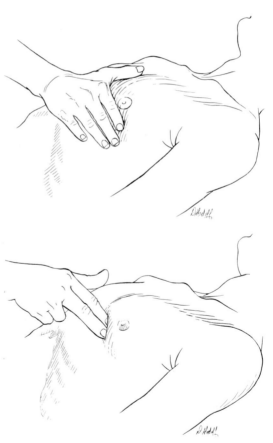

Fig. 12.12 (*Top*). Locating the cardiac apex beat. The flat of the hand is first used to detect the approximate location of the apex.

Fig. 12.13 (*Bottom*). Point localization of the apex beat. After the apex has been crudely localized with the flat of the hand, its specific point is further localized with two fingers.

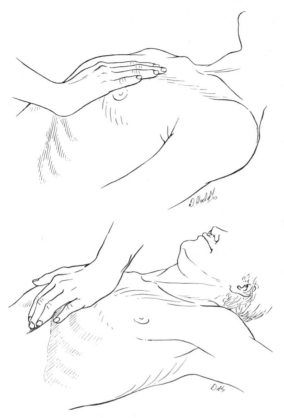

Thrills also occur over the precordium. When close to the left lower sternal border, a systolic thrill suggests a ventricular septal defect, while one at the apex usually means severe mitral regurgitation. Mitral stenosis can produce a diastolic thrill at the apex.

In young patients, or patients with heart failure, a brief outward pulsation can be seen and felt at the apex in early diastole. This is the rapid filling period of the ventricle. Later, we will see that this event is audible as a *ventricular gallop* or *third heart sound*. Similarly, late in diastole, the atria contract and may be seen and palpated as an *atrial gallop* or *fourth heart sound*. The double impulses described above are systolic and diastolic—a systolic contraction impulse and a diastolic filling impulse. *Two systolic* impulses imply a *ventricular aneurysm*. The first impulse is a normal apical beat. As systole progresses, the

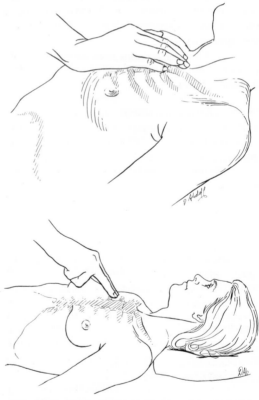

Fig. 12.14 (*Top*). Palpating the parasternal area. Again, the flat of the hand is used to detect a parasternal lift of right ventricular hypertrophy.

Fig. 12.15 (*Bottom*). Palpating the subxiphoid space. The right ventricle when enlarged or hypertrophied will be felt strongly in this position.

the aorta are the most common cause of pulsations here. Aortic aneurysms—syphilitic, arteriosclerotic, or dissecting—all are associated with marked dilatation of this vessel and may produce pulsations in this area (Fig. 12.16).

The second right intercostal space is called the *aortic area* (Fig. 12.17). It lies over the ascending aorta. The closure of the aortic valve can be felt here if the pressure in the aorta is elevated as in *hypertension*. Palpable vibrations, *thrills*, resulting from aortic stenosis can also be felt here, as well as up the carotid arteries.

The second left intercostal space is the *pulmonic area*, and similar events can be palpated there—closure of the pulmonic valve in pulmonary artery hypertension and a thrill in pulmonic stenosis (Fig. 12.18).

Fig. 12.16 (*Top*). Palpating the suprasternal notch. The middle finger is deeply insinuated in the sternal notch to palpate for thrills and aneurysmal bulgings of the arch of the aorta.

Fig. 12.17 (*Bottom*). Palpating the aortic area. The index and middle fingers are firmly applied in the second right intercostal space.

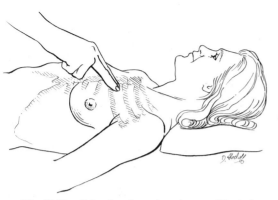

Fig. 12.18. Palpating the pulmonic area. The index and middle fingers are firmly applied over the second or third intercostal space just to the left of the sternum.

thin-walled noncontractile aneurysm bulges outward as the second impulse.

Percussion

Compared to inspection and palpation of the heart, percussion is of limited value. Percussion outlines static borders but gives no information about dynamic events. The x-ray has largely supplanted its value. Still, percussion is cheap and portable (Fig. 12.19).

Begin in each interspace far laterally toward the axilla and percuss toward the sternum and mark the point where the notes change from resonant to dull. Normally, this occurs about 2 cm from the sternum in the third interspace on the right and about 4 cm on the left. The border is detected about 8 cm laterally in the fifth intercostal space the left. Moving from this point to the sternum, the dullness becomes flat about 4 cm from the sternum as the area of absolute cardiac dullness. This is the point at which the heart is in direct contact with the anterior chest wall. This area of absolute dullness enlarges disproportionately when a *pericardial effusion* pushes the anterior heart border against the chest wall (Fig. 12.20).

Auscultation

The stop-go movements of blood, rapid changes in pressure, and openings and closings of the valves combine to produce a wide variety of vibrations, some of which are audible.

Take 10 minutes and study very carefully the composite diagram of pressure waves generated in the heart (Fig. 12.21). Learn it well

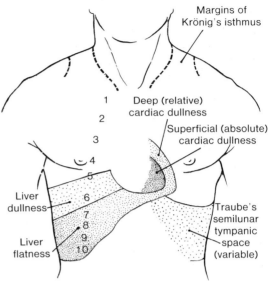

Fig. 12.19. Percussion outlines of normal chest. Krönig's isthmus is a zone of resonance over the apices of the lungs. The area of relative cardiac dullness is the area of heart just under a lip of lung tissue. The area of absolute cardiac dullness is the point at which the heart is in direct contact with the anterior chest wall. This area enlarges markedly with pericardial effusion which pushes lung tissue away from the heart and brings more of the distended pericardium against the chest wall. Traube's semilunar space is the area of tympany registered by the underlying gastric air bubble.

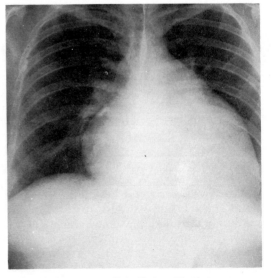

Fig. 12.20. Pericardial effusion. The heart assumes a globular configuration. The apex beat was not palpable, and the area of absolute cardiac dullness was markedly increased. Other physical findings included a marked pulsus paradoxus and Kussmaul sign.

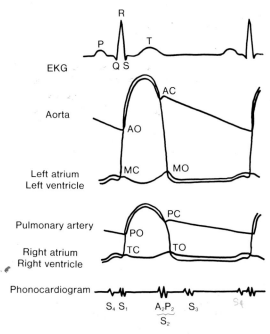

Fig. 12.21. The electric, pressure, and sonic events of the cardiac cycle. AO, aortic opening; AC, aortic closure; MC, mitral closure; MO, mitral opening; PO, pulmonic opening; PC, pulmonic closure; TC, tricuspid closure; TO, tricuspid opening. A vertical line at any point gives the simultaneous events occurring at that time.

enough to reproduce it on paper and auscultation will be much easier for you to master. Let us examine it line by line. The electrocardiogram records the initiating sequence of electrical events which spread the contraction waves down the myocardium. The P wave starts atrial contraction. The Q wave is the passage of current through the interventricular septum, while the R and S indicate waves of current spreading through the ventricles. The straight line between the P and Q-R-S waves is the P-R segment and is the delay in transmission imposed by the atrioventricular node. The T wave is a repolarization wave and produces no contractile action.

Mechanical events lag behind the electrical ones. The next line is a pressure tracing in the aorta, the third line a pressure tracing of the left atrium, and the fourth line the left ventricle. Look at the left atrial pressure line. Shortly after the P wave of atrial depolarization on the electrocardiogram, the atrial pressure rises with atrial systole. Next, the atrial and ventricular lines cross as ventricular systole occurs. At this crossing point, the ventricular pressure rises rapidly above the atrial

pressure, and the mitral valve slams shut, producing the first heart sound shown on the phonocardiogram. Now, follow the left ventricular pressure up to where it intersects the aortic curve. As soon as they cross, the aortic valve opens—normally a soundless event. Follow the curve over the top. The lines again cross as the ventricular pressure falls below the aortic pressure and the aortic valve snaps shut, producing the second heart sound seen on the phonocardiogram. The ventricular pressure plummets rapidly and crosses the left atrial curve allowing the mitral valve to open again—normally soundless. The atrial pressure remains slightly greater throughout diastole, allowing passive ventricular filling until atrial contraction again gives the small kick just before the ventricle again contracts.

The bottom three lines are simultaneous pressure curves from the right heart—pulmonary artery, right ventricle, and right atrium. Comparable events are not synchronous one side with the other. Note that the mitral valve closes before the tricuspid valve and the aortic valve closes just before the pulmonic valve because left ventricular contraction begins and ends before right ventricular contraction. *Ejection*, however, occurs earlier on the right because the pressure rise needs to be less on that side and therefore the pulmonic valve opens before the aortic valve.

We will refer to parts of this sequence to understand the normal and abnormal sounds which appear during auscultation of the heart.

Technique of Auscultation

A relaxed comfortable patient and examiner in a quiet room and a well fitting stethoscope equipped with bell and diaphragm are all that is required.

The entire precordium should be auscultated, but four major areas best reflect sounds from the four valves (Figs. 12.22 and 12.23). The mitral and tricuspid areas generally transmit low frequency sounds from low pressure events. Use the bell here, very lightly applied to the chest making a complete seal with the skin. In the aortic and pulmonic areas, the high frequency sounds are better heard with a diaphragm applied firmly to the chest. An *inching* technique going from the apex through the atrioventricular valve areas up to the base of the heart is an acceptable routine.

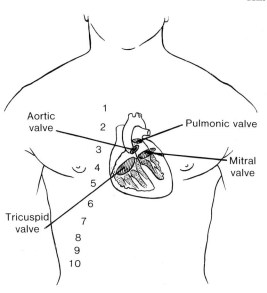

Fig. 12.22.   Anatomic position of the heart valves.

Certain maneuvers change the heart tones. Apical sounds can be accentuated by rolling the patient into a left lateral decubitus position. Sounds from the base—aortic and pulmonic areas—are brought out by having the patient leaning far forward or up on all fours. At end expiration, the heart is closer to the chest wall, and otherwise inaudible sounds might appear. With deep inspiration, the pulmonary circuit pulls more blood and the right ventricle ejects more blood and the right ventricle ejects more blood and takes a little longer to do so. Contrarily, a Valsalva maneuver—forced expiration against the closed glottis—impedes right ventricular filling, and it will eject a smaller volume more rapidly. As we shall see, this maneuver aids in distinguishing right-sided heart sounds from left-sided sounds.

Normal Heart Sounds

Normal heart sounds are two in number, designated S_1 and S_2 and phonetically are mimicked by lubb-tup.

S_1 originates complexly. Its major components stem from the tensing of the chordae of the mitral valve as it is snapped closed during ventricular systole. Both atrioventricular valves participate in the generation of the sound, but the dominant component comes from the mitral valve. S_1 is lower in frequency and slightly longer in duration than S_2. Use the bell over the mitral and tricuspid areas for the maximum intensity of S_1.

S_2 results from the closure of the semilunar valves, is higher in pitch, and has a maximum intensity at the aortic area.

Two features will help you identify S_1 and S_2. First, the carotid and apex pulsations occur at very nearly the same time as S_1. Second, at normal or slow heart rates, systole is shorter than diastole so that the time between S_1 and S_2 is shorter than the time between S_2 and S_1.

When listening to the heart, don't try to take in all the sounds in each cycle. Instead, begin at the apex and identify S_1 and concentrate only on that sound for several cycles. Then move to the other areas, still listening to S_1. Now, go back to the apex and focus your attention only on S_2.

The intensity of S_1 can be altered for a variety of reasons. *S_1 becomes very loud if the valve is wide open when systole occurs.* Presumably the long distance from which they must recoil generates a louder sound. If the atrium is filling the ventricle late in diastole, this situation will occur. Late filling can be the consequence of several abnormalities. Prolonged filling occurs with *mitral stenosis*, and S_1 becomes loud. The large volume of blood required in *high output states* also prolongs filling and intensifies S_1. *Tachycardia* shortens diastole so the normal volume of blood has less time to cross the atrioventric-

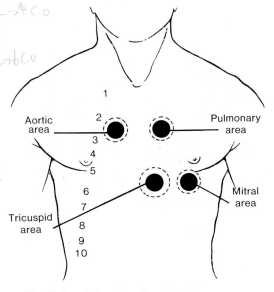

Fig. 12.23.   Valve areas for auscultation. Note that these areas do not conform to the anatomic positions. They are, however, the areas in which sounds emanating from the valves are best heard.

ular valves. Filling occurs then throughout diastole, and the valves are wide open when systole begins. If atrial systole is late, the valves also will be pushed wide open late. Under these circumstances, a loud S_1 tells you that the *P-R interval is very short*.

Conversely, if the valves have had time to float back before systole occurs, S_1 is very soft. *First degree heart block* (a long P-R interval) is a good example, as is bradycardia with a long diastole.

If S_1 varies in intensity from beat to beat, you may surmise that the valve position varies from beat to beat. In atrial fibrillation, the varying positions depend on the length of the previous diastole. In complete heart block (atria and ventricles beating independently), the variation in S_1 depends on when atrial systole happens to come in relation to ventricular systole.

B) Pathologic changes in the valves can also alter the intensity of S_1. Fibrosis of the valves in mitral stenosis contributes to the very loud S_1 characteristic of this lesion. When the valve becomes calcified and immobile, however, S_1 becomes very soft or inaudible.

c) The speed of ventricular ejection probably affects the intensity of S_1. Rapid ejection times of thyrotoxicosis, fever, and exercise cause a loud S_1. Less abrupt closure is seen in myxedema, shock, and congestive myocardial failure.

Refer to the pressure diagram (Fig. 12.21) and note that mitral valve closure slightly precedes tricuspid closure. This *splitting* of S_1 can sometimes be appreciated, especially in young normal hearts. Try to hear this in normals because some other sounds which can occur very close to S_1 are not normal and must be distinguished from physiologic splitting.

S_2 emanates from the sudden closure of the semilunar valves. Listen for this sound at the base of the heart. It has two components, aortic closure, A_2, and pulmonic closure, P_2. A_2 is louder and at the apex it is the only component of S_2 which can be heard. Listen over the pulmonic area to maximize P_2, and both components will become audible and separable (Fig. 12.24).

Note in the pressure diagram (Fig. 12.21) that the aortic valve closes just before the pulmonic valve. Actually, the closures vary with respiration. At the end of expiration, both ventricles complete ejection at about the

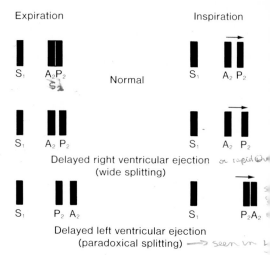

Fig. 12.24. Splitting of the second heart sound.

same time, and S_2 sounds single. As you inspire, more blood is drawn into the right ventricle, the ejection of which takes more time. P_2, therefore, moves away from A_2, and *physiologic splitting* occurs. The ejection time of the left ventricle is relatively constant from beat to beat, and so the variation of the A_2-P_2 interval in normals is a function of right ventricular ejection time. In children and young adults, the phenomenon of splitting can usually be heard. In some adults, S_2 sounds single and no splitting is detected.

Very wide splitting is usually seen in delayed right ventricular ejection or excessively rapid left ventricular ejection. Here, the two components are separable at expiration and become even wider with inspiration. Rapid left ventricular emptying will occur in *mitral regurgitation* and a *ventricular septal defect*, since both these allow the ventricle to vent off in two directions. Slow right ventricular ejection will occur if right ventricular electrical activation is late, as in *right bundle branch block*. Obviously, *pulmonic valve stenosis* will also delay emptying. An *atrial septal defect* shunts blood from the left atrium into the right atrium. This additional blood surges into the right ventricle, prolongs ejection, and widens the splitting. Wide splitting with *no* respiratory variation has proved a very useful sign of atrial septal defect.

Paradoxical splitting of S_2 is more common than wide splitting. Anything which selectively delays left ventricular ejection causes paradoxical splitting. Aortic valve closure now comes *after* pulmonic closure. With inspiration, P_2 still moves out but now, instead of moving out away from A_2, it moves out closer to A_2 and *splitting narrows*. Similar circumstances on the left delay ejection as they do on the right. *Left bundle branch block* with delayed left ventricular activation, severe *aortic stenosis*, increased left ventricular volume as in a *patent ductus arteriosus*, and *systemic hypertension* are common examples.

The recognition of splitting tells you that both valves are functional. It excludes a *truncus defect* and tells you that if stenosis is present at least the valve leaflets still have some movement remaining and are not fixed. Normal splitting also provides a great aid in deciding the functional significance of a murmur. *Very few hemodynamically significant murmurs fail to disturb S_2.*

The intensity of A_2 and P_2 is directly related to the pressure in the artery into which the blood flows. Thus, the higher the aortic pressure the louder A_2, and the higher pressure in the pulmonary artery the louder will be P_2. A_2 is loud in systemic hypertension and soft in hypotension. The intensity of A_2 in aortic stenosis depends less on the degree of stenosis than on the associated aortic pressure. Anything producing pulmonary hypertension will accentuate P_2. Pulmonic stenosis almost always has a low pulmonary artery pressure producing a faint P_2.

S_3 is a low-pitched soft vibrating noise of short duration occurring in diastole shortly after S_2. The origin of this sound is unclear, but it probably arises from the abrupt limitation of the ventricle at the end of the rapid-filling phase. Phonetically, the heart sounds now become lubb-tup-da, hence the term *ventricular* or *protodiastolic gallop*. A common finding in children, S_3 sounds are *not* heard in normal adults. An S_3 must be explained. Its causes include rapid ventricular filling, an alteration of myocardial tone, and an elevated residual ventricular volume at the end of systole. Rapid ventricular filling is a requirement of tachycardia with a short diastole. Rapid filling is also required when the volume which must flow into the ventricle is large. In mitral regurgitation and patent ductus arteriosus, a large volume of blood enters the left atrium (the sum of the distal aortic volume, plus the amount shunted or regurgitated). An S_3 is almost always heard when this extra volume is significant. Myocardial tone changes with diseases directly affecting the muscle. *Congestive heart failure* is the prototype, but such entities as *amyloid* of the heart, *hemochromotosis*, *myocarditis*, and the *idiopathic cardiomyopathies* all will include an S_3 gallop as part of the clinical picture. Congestive heart failure, in addition to altering myocardial tone, is also characterized by less than complete ejection with resultant residual volume at the end of systole giving another reason for the S_3. An S_3 in adults is an ominous finding and indicates serious disease. It may come and go as congestive heart failure worsens or improves.

S_4 is another gallop sound. This one occurs late in systole and coincides with atrial contraction. It is called an *atrial* or *presystolic gallop sound* and results from the forceful distention of ventricles which are already filled with blood. It, too, is common in children, but not so in normal adults. S_4 tends to accompany situations of pressure overload of the ventricles and, thus, frequently is heard in hypertension, either systemic or pulmonic, and aortic or pulmonic stenosis. The stressed ventricle receives this extra presystolic punch to meet its commitment. It does not mean that the ventricle is failing, but rather it is calling on atrial reserve to remain compensated. S_4 can occur very close to S_1 and can, therefore, be confused with a split S_1. Listen for the duller of these two sounds. If the dull sound occurs first, and a crisp one second, it is probably an S_4-S_1 complex. If the first sound is crisper, it is probably a split S_1. If both sounds are crisp, it is probably an S_1-ejection click (vide infra).

Both gallop sounds may be heard in the same patient, producing a lubb-tup-da-da. When tachycardia shortens diastole, the S_3 and S_4 may merge as a *summation gallop*.

When valves open, no audible sound is normally generated. A diseased valve, however, may produce noise which you can hear. The best example is the fibrotic rheumatic mitral valve which causes an audible *opening snap (OS)* early in diastole. An opening snap can be confused with a widely split S_2. Three features help to differentiate a split S_2 from an S_2-OS. First, note that the opening snap is louder during expiration at a time when splitting of S_2 should be minimal. Second, the S_2-OS interval does not vary with respiration and, third, it is heard best at the left lower sternal border, rather than at the base. The time between S_2 and the OS is a measure of the severity of mitral stenosis. If the orifice is very narrow, the atrium must generate a high pressure to force blood into the ventricle. This moves the atrial pressure curve up the ventricular curve so that the OS occurs closer and closer to A_2. Refer to the pressure diagram to appreciate how this phenomenon occurs.

Another sound confused with a widely split S_2 or with an opening snap is a *pericardial knock*. A thickened unyielding pericardium can suddenly halt ventricular filling, and this sudden deceleration of blood produces a diastolic knocking or cocktail shaker sound.

Diseased semilunar aortic or pulmonic valves also can produce an opening sound called an *ejection click*. The genesis seems to be a dilated aorta or main pulmonary artery from a variety of causes, e.g., aortic stenosis, pulmonic stenosis, systemic hypertension, idiopathic dilatation of the pulmonary artery, coarctation of the aorta, and severe tetralogy of Fallot. While the opening snap occurs shortly after S_2 in *diastole*, it will be noted from the pressure diagram that an ejection click occurs early in *systole*. A pulmonic ejection click appears loudest at the pulmonic area and is greatly diminished during inspiration. The aortic ejection clicks do not vary with respiration, are well transmitted to the apex, and may be confused with the split S_1.

We have recently come to appreciate yet another systolic click which may be multiple in number and which occurs later in systole than ejection clicks. Angiographic studies suggest that this click(s) is produced by a snapping of chordae tendinae of the mitral valve which, for one reason or another, is lax. Frequently, this click is followed by midsystolic murmur, as the mitral valve prolapses and permits regurgitation to occur. These midsystolic clicks can be differentiated from ejection clicks by having the patient stand up or perform a Valsalva maneuver. This decreases left ventricular filling, and the click will move closer to S_1. The *midsystolic click syndrome* is important to identify because of the bad company it can keep—sudden death.

Murmurs

The American College of Cardiology defines a murmur as a relatively prolonged series of auditory vibrations of varying intensity, frequency, quality, configuration, and duration.

Intensity of a murmur relates to its loudness. The standard scale is 1–6, where a grade 1 murmur is barely audible and a grade 6 can be heard with a stethoscope just off the chest wall. This, clearly, is a subjective scale. The intensity does not necessarily reflect the severity of the lesion, although a grade 6 is more likely to be serious than is a grade 1 murmur.

The *frequency* is the pitch of the murmur. High-pitched murmurs relate generally to high velocity flows caused by great pressure differences. For example, the murmur of aortic insufficiency is high-pitched since blood rushes from an aortic mean pressure of perhaps 100 mm Hg into the left ventricle with a pressure of 5 mm Hg during diastole. Conversely, the murmur of mitral stenosis is low-pitched as the atrium generates a pressure of perhaps 20 mm Hg compared with the left ventricular diastolic pressure of 5 mm Hg.

The *quality* of the murmur relates to the description of its tonal purity. A pure-toned murmur has a musical quality, whereas one with many frequencies and overtones appears harsh. Many other descriptive terms can be applied.

The *configuration* or shape of a murmur is caused by its change in intensity. A *decrescendo* murmur loses intensity throughout its duration, while a *crescendo* murmur gets louder. A crescendo-decrescendo or diamond-shaped murmur builds to a peak intensity and then fades away. A *plateau* murmur maintains constant intensity.

The *duration* is first categorized by the cycle in which it appears—systolic or diastolic. A long murmur occupies most all of either cycle. A *continuous* murmur is any murmur which runs through a heart sound into the next cycle.

Murmurs are further characterized by their *radiation* or *transmission*. The greatest factor influencing radiation is intensity. A loud murmur transmits farther than a soft one. In general, murmurs tend to be heard best close to their origins and transmit best in the direction of flow of turbulent stream. A murmur changes with transmission. High frequencies become damped by tissue, and low frequencies are selectively transmitted. It becomes important, then, to *track* a murmur in order not to be fooled into thinking that two murmurs are present.

Systolic murmurs result from stenosis of the aortic or pulmonic valves, from incompetence of the mitral or tricuspid valves, or from a ventricular septal defect. Keeping in mind the discussion above, that the frequency depends on the pressure difference and the duration depends on the duration of that pressure difference, we can predict what these murmurs should sound like.

Aortic stenosis can be congenital, acquired, or both. The acquired lesions are rheumatic or calcific, with the latter having a predilection to occur on a congenitally bicuspid valve. Rheumatic aortic stenosis almost always is accompanied by mitral valve disease.

The murmur of aortic stenosis is diamond-shaped because the pressure gradient builds to a peak and then subsides. Since the mitral valve closes before the ventricle begins to eject blood, the murmur begins after S_1. The longer the soundless interval between S_1 and the murmur, the less severe the stenosis. It continues until the aortic valve closes. Aortic stenosis murmurs are loudest at the aortic area and radiate into the neck and carotid artery. Because they usually are intense, they can also frequently be heard at the apex (Fig. 12.25).

The pressure generated by the right ventricle is less than the left, and so the murmur of *pulmonic stenosis* is lower pitched and less intense. It, too, begins after S_1 and continues through to pulmonic valve closure in a diamond-shaped configuration. Best heard at the pulmonic area, the murmur of pulmonic stenosis may radiate to the base of the neck, but not into the carotids.

Stenosis of either the aortic or pulmonary valves can be relative, rather than real. An abnormally large quantity of blood pushed through the fixed orifice leads to turbulence and the murmur. The patient with aortic insufficiency may eject twice the normal volume of blood, producing a murmur of *relative aortic stenosis*. The systolic murmur of an atrial septal defect is caused *not* by blood passing from the left atrium into the right atrium, but by the ejection of that additional volume out of a *relatively stenotic pulmonary valve*.

Murmurs from mitral and tricuspid insufficiency differ from the systolic ejection murmurs in timing, duration, quality, and location. The murmur of *mitral insufficiency* sounds high-pitched since the pressure difference between the left ventricle and left atrium during systole is quite large. Also, since the pressure gradient continues after the aortic valve closes, the murmur runs into S_2. Mitral regurgitation sounds loudest at the mitral area and is transmitted well toward the axilla, but not toward the base. If a large volume of blood is pushed back into the atrium, a diastolic filling rumble may also be heard (Fig. 12.26).

Many disease processes involve the mitral valve. Rheumatic valvulitis, subacute bacterial endocarditis, calcification of the mitral annulus, rupture of the chordae tendineae, and papillary muscle dysfunction constitute the majority. Papillary muscle dysfunction may produce transient murmurs which come and go with various states of myocardial compensation. A ruptured papillary muscle most commonly follows a myocardial infarction and produces a very loud murmur and a very sick patient. The calcified mitral annulus is the most common cause of systolic murmurs in old women. In this entity, the protruding rim of calcification forces the posterior leaflet

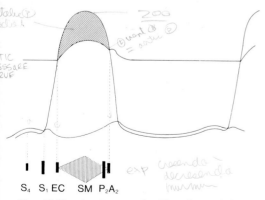

S₄ S₁ EC   SM P₂A₂

Fig. 12.25. Aortic stenosis. The diamond-shaped murmur starts with an ejection click (not always present) and extends to the aortic component of the second sound.

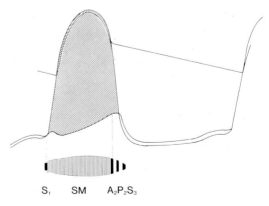

S₁   SM   A₂P₂S₃

Fig. 12.26. Mitral regurgitation. The systolic murmur extends into the second heart sound.

to drape up and over the annulus, preventing normal coaptation of the valves. Both acute rheumatic valvulitis and chronic rheumatic fibrosis of the mitral valve may present as mitral regurgitation. This lesion appears to be more common in men than in women.

Tricuspid regurgitation manifests in more subtle ways. The pressure difference is not nearly as great as on the left side, and so the murmur is of lower pitch and intensity. Consequently, it does not radiate well from the tricuspid area. Frequently, the only sign of tricuspid regurgitation is a prominent "v" wave in the jugular veins and a pulsatile liver. The murmur of *tricuspid insufficiency becomes louder with inspiration* as the right ventricular volume increases.

The last systolic murmur to be considered results from *ventricular septal defect*. The gradient is moderately large during systole and low to nonexistent during diastole when both ventricles are filling. The murmur appears diamond-shaped and extends into the second sound, usually stopping with P_2. Frequently very loud, the intensity does not reflect the volume being shunted from the left to the right. For two reasons, in fact, the worse the lesion the less intense may be the murmur. First, if the defect is very large, there is little resistance, or "stenosis," to the flow. Second, with a severe shunt, the pressure in the pulmonary circuit may approach or equal that in the left ventricle, obliterating the gradient and thus the murmur (Fig. 12.27).

Diastolic murmurs originate from aortic or pulmonic insufficiency and from mitral or tricuspid stenosis.

A large gradient exists when the *aortic valve*

fails. The murmur produced is then high-pitched and decrescendo, falling off as the gradient falls off. It begins with A_2, which may be inaudible and is best heard by having the patient sit up and lean forward. Apply *very firm pressure* with the diaphragm at Erb's point (third left intercostal space). Transmission, if it can be heard, occurs down toward the apex and left lower sternal border. This is the most easily missed murmur. Aortic insufficiency has many etiologies. Aortitis from *syphilis* or *ankylosing spondylitis* may so embarass the valves as to cause incompetence. A *dissection* of the aorta can dilate the aortic ring, rendering the valve functionless. Severe *atherosclerosis* of the aorta in the elderly may cause annulo-aortic-ectasia with resultant insufficiency. *Subacute bacterial endocarditis, rheumatic valvulitis,* and *calcification of the valves* are other causes. *Congenital fenestrations* of the valve allow blood to sift through the sinuses, and an *aneurysm of the sinus* can prolapse a cusp into the ventricle (Fig. 12.28).

Pulmonic insufficiency is uncommonly diagnosed. It is heard best at the pulmonic area as an early diastolic decrescendo blowing sound, especially at end expiration. The major cause is probably severe pulmonary hypertension. When this is the case in mitral stenosis with pulmonary hypertension, the pulmonary insufficiency sound is termed a Graham Steell murmur.

Mitral stenosis causes a diastolic murmur which is low-pitched since the gradient is small (atrial systole with stenosis might generate 25 mm Hg with a ventricular pressure

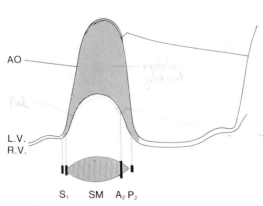

Fig. 12.27. Ventricular septal defect. The pressure tracings of the right and left ventricle have been superimposed. The harsh murmur produced usually is accompanied by a thrill.

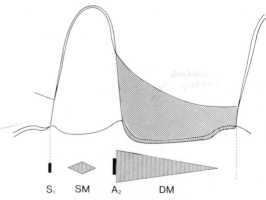

Fig. 12.28. Aortic insufficiency. This diastolic blowing murmur begins with the aortic component of the second sound. The systolic murmur is that of relative aortic stenosis produced by the large volume ejected.

of about 5 mm Hg). Heard best at the apex, the rumbling sound is best appreciated with the patient in the left lateral decubitus position. Review the pressure curves of the left atrium and left ventricle. During diastole, there are two periods of maximum gradient: the first just after the mitral valve opens, and the second late in diastole when the atrium contracts. The murmur of mitral stenosis reflects these events. Initiated with an opening snap, the rumble rapidly falls off and then increases again up to S_1 as a *presystolic accentuation*. The presystolic accentuation obviously will disappear in atrial fibrillation, a common accompaniment of mitral stenosis. The murmur of mitral stenosis can be caused by aortic insufficiency! With profound reflux of blood through an incompetent aortic valve, the anterior leaflet of the mitral valve is caught between two streams into the ventricle, one from the atrium and one from the aorta. The resultant vibrations sound like mitral stenosis, and this is called an *Austin Flint murmur* (Fig. 12.29).

Tricuspid stenosis is usually heard best at the xiphoid and is also a low-pitched rumbling murmur. The diagnosis is facilitated by observing very large "a" waves in the jugular veins. This is an uncommon lesion.

Continuous murmurs begin in systole and continue without interruption through the second sound into diastole. Produced by a communication between two chambers with a continual gradient, they have a number of origins. Continuous murmurs require differ-

entiation from to-and-fro murmurs such as may be heard with aortic stenosis and insufficiency. *Patent ductus arteriosus* is the most common cause of a continuous murmur—sometimes called a machinery murmur. This shunt from the aorta to the pulmonary artery allows flow throughout the cardiac cycle (Fig. 12.30). A similar shunt exists with an *aortico-pulmonary fenestration* (incomplete partitioning of the fetal truncus arteriosus). An aortic *sinus of Valsalva may rupture* into the right ventricle, establishing a continuous gradient shunt. An *anomalous* origin of the left *coronary artery* from the pulmonary artery acts like an arteriovenous shunt with a continual pressure gradient from the aorta through the right coronary into the left coronary thence into the pulmonary artery. A severe *coarctation of the aorta* occasionally manifests as continuous murmur over the back. A recently described cause of continuous murmurs is *pulmonary artery webs*. These residuals of old pulmonary emboli stretch like violin strings across the lumen and vibrate throughout systole and diastole.

Pericardial friction rubs may be confused with valvular murmurs. Rubs occur whenever irritated visceral and parietal pericardial surfaces grate against each other. There are three times in the cardiac cycle when this can occur: first, atrial systole the same time as an S_4; second, during the apex beat of ventricular contraction; and, third, during the diastolic wave of rapid filling or the same time as an S_3. One, two, or three sounds may be heard as scratchy noises, close to your ear. Having the patient lean forward and exhale will facilitate your ability to hear pericardial friction rubs.

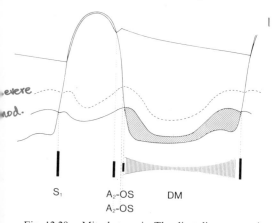

Fig. 12.29. Mitral stenosis. The diastolic murmur is loudest just after the opening snap and again during atrial systole just before S_1. The dotted line shows the difference between moderate and severe stenosis. When severe, the atrial pressure must be higher. The atrial and ventricular pressures cross sooner, and the opening snap occurs *earlier*. The S_2-OS interval is therefore shorter.

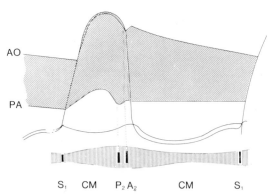

Fig. 12.30. Patent ductus arteriosus. The aortic and pulmonic pressure curves are superimposed to show the persistent pressure gradient which produces the continuous murmur (CM).

PRACTICE SESSION

Begin with a measurement of the venous pressure. Position your partner with the examination table Gatched to about 30°. Do not use a pillow under the head. Now, carefully inspect the neck. Use your penlight to cross-light the neck tissues and watch for pulsatile bulging of the tissues. This will be the internal jugular. If you can see a discrete vein, it is the external jugular. If you can't see any pulsations, slowly lower the head until the pressure column becomes visible. If you still have a problem, you can get an estimate of about what level it should be by using the arm veins as follows. Put the arm in a dependent position until the antecubital vein becomes prominent or at least until the hand veins become visible. Now, slowly raise the arm until the veins just collapse. This will give you an idea of the height of the pressure column which you can now duplicate by positioning the trunk.

Once you have established the top of the column, measure it in terms of centimeters of blood as shown in Fig. 12.3. Confirm that the pulsations which you see are venous and not arterial by applying light pressure to the base of the neck just above the clavicle. The pulsations will cease. If they persist, then you are seeing the underlying arterial pulse.

Palpate the opposite carotid while watching the venous pulsations. You should note the "a" wave just before and the "v" wave just after the carotid pulse.

Now, watch the top of the column as your partner breaths in and out slowly and deeply. With inspiration, the level will fall and at end expiration it will rise quite high.

Now, inspect the precordium with your partner fully supine. Again, use cross-lighting. If he has a thin chest wall, the apex beat should be readily visible. Also, note any pulsations in the subxiphoid region. Now, palpate the apex beat first using the entire hand then just the tips of two or three fingers. Accurately assess the area of the pulsation and measure it carefully in terms of the interspace number and the number of centimeters from the midsternal line.

Simultaneously, palpate the apex and the carotid artery and estimate the ejection time. Also, hook a thumb under the xiphoid to feel the right ventricle. If your partner is thin, you may also see epigastric pulsations from the abdominal aorta.

Now, palpate the aortic and pulmonic areas. It is unlikely that you will feel any pulsations here. Similarly, you will not likely feel any pulsation in the suprasternal notch.

Now that you have established the apex beat, percuss the left chest wall. You will first encounter an area of relative dullness followed by absolute dullness. These are generally 8 and 4 cm, respectively.

Before you begin practicing auscultation, make sure that you have studied and mastered Fig. 12.21. Time spent on that diagram will be repayed many times over as you encounter different heart sounds.

Start by listening lightly with the bell of the stethoscope just medial to the apex which is the mitral area. Simultaneously, palpate the carotid pulse and determine which sound is S_1. Concentrate only on the first heart sound and ignore everything else. Apply increasing pressure with the bell and note how the lower frequency sounds give way to higher frequency sounds as you push harder. Now, inch your way to the right of the sternum at the tricuspid area and then up the right side of the sternum to the aortic area, then across to the pulmonic area, all the time listening only to the first heart sound. You should note that S_1 is loudest at the mitral area and gets softer as you move to the base of the heart. Now, repeat the technique using the diaphragm. Listen just to the second heart sound. S_2 gets louder as you reach the base of the heart.

Now, listen carefully over the pulmonic area for the splitting of the second sound. It is best heart here since the pulmonic component is the softer of the two. Listen very closely as your partner breaths in and out slowly. See if you can confirm that the splitting widens with inspiration and narrows with expiration. Don't be discouraged if you cannot appreciate this the first time. It takes some practice and is better heard in some subjects than others.

Now, repeat the inching technique listening first to the systolic period then the diastolic period. Note any sounds in these periods and carefully characterize them as to timing, frequency, duration, and location of maximum intensity.

Start practicing your description of the cardiac exam. It is best to give all the measurements and descriptions of the expected findings first and then describe the abnormalities.

Clinical Correlations

What follows is not meant to be a compendium of heart diseases, but rather a few common clinical conditions to help you fit together various aspects of the cardiac examination.

Congestive Heart Failure

The patient, usually older than 50, relates that he has been fatigued. Usual amounts of activity make him unusually dyspneic. His weight has increased and toward the end of the day his ankles are swollen. Sleep is fitful. Two or three pillows facilitate breathing at night, but on several occasions he has experienced paroxysmal nocturnal dyspnea. Four or five times each night he must void. The past history might include hypertension, myocardial infarction, or remote rheumatic fever with a persistent murmur.

The pulse is rapid and respiratory rate slightly increased. He may look chronically ill with a grey cast to his complexion and a clammy feel to his skin. The arterial pulse is perhaps a bit thready and rapid. It may also demonstrate *pulsus alternans*, a weak beat alternating with a strong beat. The jugular venous pressure is elevated. Basilar pulmonary rales appear, and pitting edema of the ankles can be detected. The apex is displaced toward the anterior axillary line and may be difficult to feel. A soft first heart sound reflects the slow rate of ejection and the second sounds split paradoxically. A third sound commonly occurs, and a fourth sound is frequent. A soft murmur of mitral regurgitation reflects the ventricular dilatation and malfunction of the papillary muscles.

Acute pulmonary edema is the most profound example of congestive heart failure. The patient, gripped with terror, spews fourth pink frothy edema fluid. Gasping for air, the poor cerebrovascular perfusion may make him irrational. Heart sounds usually cannot be heard through the respiratory noises of rales, rhonchi, and wheezes (Fig. 12.31).

Cor pulmonale is the most common form of heart failure affecting more selectively the right ventricle. Pulmonary hypertension causes cor pulmonale from chronic lung disease, multiple small pulmonary emboli, severe mitral stenosis, pulmonary flooding from a left-to-right shunt, or as a primary disease of the pulmonary arteries.

Dyspnea and edema head the list of com-

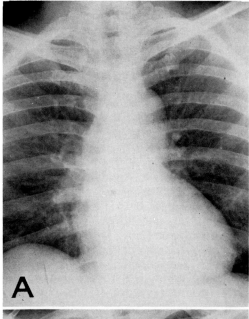

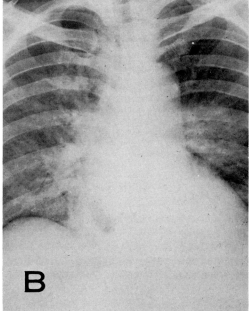

Fig. 12.31. Congestive heart failure. *A*: This x-ray shows the findings of early congestive heart failure. Note that the left ventricle is prominent and boot-shaped. The pulmonary vascular markings are increased, especially close to the heart. Note also that the blood flow has been redistributed so that the pulmonary vessels in the upper lobes are more prominent than in the lower lobes. This is characteristic of pulmonary venous hypertension. *B*: The same patient several weeks later in acute pulmonary edema. The heart has enlarged even more greatly. There is a diffuse density radiating from the hilum on each side—the so-called butterfly wings of acute pulmonary edema.

plaints. A plethoric cyanosis reflects the hypoxia and associated erythrocytosis. The neck veins show extreme distention and elevation with prominent "a" and "v" waves. Edema can be massive anasarca with pitting evident up to the chest wall. A left parasternal lift tells of the right ventricular hypertrophy. The pulmonic component of the second heart sound booms, and a murmur of tricuspid regurgitation is frequently heard.

Angina Pectoris

The patient, usually a man, relates the typical story of squeezing substernal chest discomfort while walking an incline or into the wind. Stopping affords relief and between episodes he may feel perfectly well. As time goes by, these attacks come with increasing frequency and severity. Nocturnal angina or angina decubitus are ominous complaints.

Contrary to previous teachings, there are some physical findings accompanying angina pectoris. These are the signs of transient mild left ventricular failure. Usually, the pulse and blood pressure are *not* altered. S_1 becomes soft and mushy. S_2 splits paradoxically, and a soft mitral regurgitant murmur appears. An S_4 is almost always present, and there may be an S_3. All of these signs revert to normal when the attack is over.

Aortic Stenosis

The patient is usually male but may be of any age. Symptomatically, a characteristic sequence of complaints can be elicited. With early relatively mild stenosis, *exertional syncope* may be the only complaint. During heavy exertion, the patient faints without warning. As time goes by, less and less exertion causes syncope and the story of typical angina pectoris emerges. The tight aortic orifice compromises coronary blood flow. The history of angina in a young man should always alert you to the possibility of aortic stenosis. With late severe stenosis, the left ventricle begins to fail, and you will get the history of congestive heart failure.

The pulse pressure narrows with stenosis, and a blood pressure of 110/90 is not uncommon. The pulse has the typical tardus and parvus wave form. Unless congestive heart failure is present, the neck veins are normal. The apex beat is pronounced but not displaced with a lifting, well localized thrust. The aortic area, suprasternal notch, and carotids may register a thrill. The loud, harsh, diamond-shaped murmur is maximum at the aortic area with wide transmission up the carotids and to the apex. If very loud, it might be picked up by bone and transmitted to the occiput or olecranon of the elbow (Fig. 12.32).

Mitral Stenosis

The patient, usually a woman in her 50s, presents with fatigue and dyspnea. Rheumatic fever occurred as a child, and doctors have told her off and on that she had a murmur. She was fine until her last pregnancy when she was forced to bed during the third trimester. After delivery, she was well again until insidious fatigue began to interfere with her housework. Breathlessness became profound, and she was alarmed by a cough productive of blood.

Examination discloses a haggard woman with a malar flush and palmar erythema. The carotid pulse is rapid and of small volume. The jugular venous pressure may be elevated. The apex is small and difficult to feel. The point of maximum impulse (PMI) represents the hypertrophied right ventricle along the sternum or below the xiphoid. S_1 is very loud. P_2 is louder than A_2. An opening snap is detected along the left sternum and is fol-

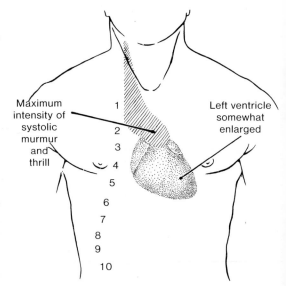

Maximum intensity of systolic murmur and thrill

1
2
3
4
5
6
7
8
9
10

Left ventricle somewhat enlarged

Fig. 12.32. Aortic stenosis. The murmur of aortic stenosis is harsh and radiates upward into the neck as shown by the *hash marks*. A thrill may accompany the murmur. The chest x-ray will show an enlarged left ventricle because of a pressure overload, not because of a volume overload.

lowed by a rumbling diastolic sound which falls off and then increases abruptly to S_1. Basilar pulmonary rales are heard, and ankle edema completes the picture (Fig. 12.33).

Aortic Insufficiency

Complaints are similar to those in aortic stenosis except that syncope is less common and congestive failure occurs earlier.

On examination, you may note a peculiar head bobbing which occurs with each systole. This reflects the enormous pulse pressure which can be generated. The blood pressure might be 180/60. Grasp the forearm with your thumb on the extensor surface and fingers across the flexor surface to best detect the water hammer pulse. Indeed, systolic pulsations may be seen in the capillary beds of the fingernails. Auscultations of the femoral arteries may yield a "pistol shot" sound with each systole.

The apex is diffuse and lifting and may be displaced to the left. The first sound is usually soft since the ventricle is well filled prior to systole. An aortic systolic ejection murmur of relative stenosis reflects the large ejection vol-ume. Following A_2, the blowing diastolic murmur appears. With early insufficiency, the murmur is short. As more blood drops into the ventricle, the murmur gets longer until failure occurs. The elevated diastolic pressure within a failing ventricle reduces again the duration of the gradient, and thus the duration of the murmur. With marked insufficiency, the Austin Flint murmur might be detected. Therefore, a heart affected primarily by aortic insufficiency may have murmurs of aortic insufficiency, mitral stenosis, and aortic stenosis. *The best guide to detecting the major lesion is the quality of the carotid pulse* (Fig. 12.34).

Mitral Insufficiency

When rheumatic in origin, the disease picture is slow in onset. A ruptured papillary muscle, on the other hand, is a sudden event with sudden onset of symptoms. Complaints referable to pulmonary congestion stem from the beat-to-beat pounding backward of blood into the pulmonary veins. The left atrium may enlarge enormously—so much so as to give a tickling cough, dysphagia, or hoarseness from pressure on the recurrent laryngeal nerve. Try to elicit a history of rheumatic fever, a history suggesting coronary artery

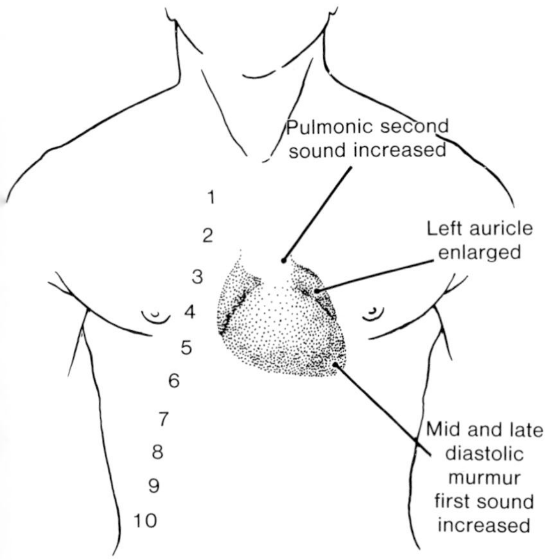

Pulmonic second sound increased

1
2
3
4
5
6
7
8
9
10

Left auricle enlarged

Mid and late diastolic murmur first sound increased

Fig. 12.33. Mitral stenosis. The diagram shows the location of the diastolic murmur at the apex. The first heart sound is invariably increased in intensity. If pulmonary hypertension results from the high pressure in the pulmonary veins, the pulmonic second sound will be increased. The x-ray of such a patient will show marked enlargement of the left auricle. The left ventricle, however, does not enlarge since there is no volume overload on it. *R ventricle does tend to enlarge however.*

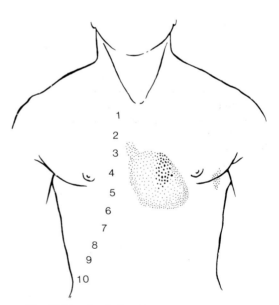

1
2
3
4
5
6
7
8
9
10

Fig. 12.34. Aortic insufficiency. The dots indicate the area of transmission of the murmur. The heavy dots are at Erb's point. The chest x-ray will show enlargement of the left ventricle because of the increased volume and a dilated ascending aorta.

disease with infarction, or a history pointing to subacute bacterial endocarditis.

The neck veins will be elevated with a normal wave form. The carotid has a brisk upstroke and falloff similar to mild aortic insufficiency. The apex is a volume-overloaded impulse. S_1 may be loud or inaudible. A thrill accompanies a very loud murmur which is heard at the apex, lower sternal border, toward the axilla, and sometimes into the back. The diastolic filling rumble results from the extra large volume of atrial blood dropping into the ventricle (Fig. 12.35).

Tricuspid Insufficiency

The chief complaint is edema. Mild right upper quadrant pain results from a congested liver stretching against its capsule. Other complaints refer to the underlying disease—most commonly pulmonary with right ventricular hypertension.

The face appears suffused. Proptosis of the eyes may be seen and may be pulsatile. The neck veins may be so distended that you cannot see the meniscus. In such a case, watch for bobbing earlobes. If the column can be seen, the most important finding is giant "v" waves just following each carotid upstroke. The findings with palpation resemble mitral stenosis with a large right ventricular impulse.

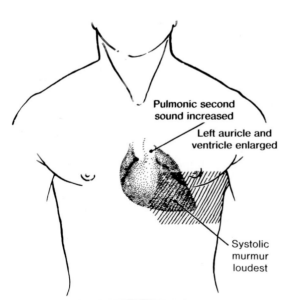

Pulmonic second sound increased

Left auricle and ventricle enlarged

Systolic murmur loudest

Fig. 12.35. Mitral regurgitation. The diagram shows the radiation of the systolic murmur from the apex into the axilla. The heart shadow is as it would appear on a chest x-ray. The left atrium and ventricle are both enlarged because of the volume overload on each chamber.

The holosystolic murmur is rarely loud but accentuates with the enhanced filling during inspiration. The liver enlarges and may be felt to pulsate into your examining fingers. Some amount of edema will always be found.

Tetralogy of Fallot

A common congenital lesion, tetralogy consists of pulmonic stenosis, ventricular septal defect, right ventricular hypertrophy, and a displaced aortic root which overrides the ventricular septum. The right ventricular hypertrophy results from the pulmonic stenosis. With this pressure rise in the right ventricle, venous blood can be ejected up through the septal defect and out the aorta, producing cyanosis.

Early in life, the child is a poor feeder, may lag in growth, and cannot keep up with his peers. With physical exertion he becomes cyanotic and later is cyanotic even at rest. Patients with tetralogy frequently adapt a squatting posture at rest and this history is a valuable clue.

Examination shows plethora, clubbing, and cyanosis, if not at rest, at least with exercise. The neck veins are elevated with a prominent "a" wave. The carotid pulse has a large volume, but a normal configuration. The right ventricular heave with a bulging left precordium is obvious and there may be a systolic thrill. A loud systolic ejection-type murmur maximizes at the left sternal border but is not transmitted well into the carotid. P_2 is soft since the pulmonary artery pressure is low.

Coarction of Aorta

The history suggest hypertension. Usually an athletic young man presents with headaches, epistaxis and, rarely, a cerebrovascular accident. He may or may not have been told that he has had a heart murmur.

The blood pressure in the arms is elevated but not so in the legs distal to the obstruction. The findings with palpation are those of hypertensive heart disease—a pressure-overloaded localized apex pulsation. Thrills may be felt over the posterior ribs accompanied by visible serpentine intercostal vessels which carry blood around the obstruction to a point distal in the aorta. A_2 intensifies, and a loud S_4 is the rule. A systolic ejection murmur at the left sternal border does not radiate well but can be heard over the back to the left of the spine, where it may be continuous.

chapter 13

PERIPHERAL VASCULATURE

We now turn our attention to the physical diagnosis of abnormalities of the peripheral vasculature—arteries, veins, and lymphatics. We have already considered the diagnostic import of the pulses as they reflect cardiac function and the significance of telangiectasia. Raynaud's phenomenon, and cyanosis. Now we focus on primary abnormalities of this system.

History

Pain accurately localizes the disease. The origin may be arterial or venous and results from the vascular compromise of an organ or part capable of registering pain. Cerebrovascular insufficiency, for instance, is painless, whereas acute occlusion of the femoral artery causes exquisite leg pain.

Intermittent claudication permits an historical diagnosis of arterial insufficiency. This complaint, usually rendered by an elderly male, is that of calf pain brought on by exercise and relieved by rest. Moreover, he will relate that walking the same distance always brings on the pain, e.g. one block, and that to continue brings on excruciating discomfort which forces him to halt. Following a brief rest, the pain leaves and he once again can walk the same distance. The compromised vessel, almost always by arteriosclerosis, supplies adequate blood for resting activities but cannot meet the demands of exercise.

If the process involves major pelvic vessels, the *Leriche syndrome* may be present. This consists of buttock claudication and inability to obtain or maintain an erection because of insufficiency of the internal iliacs.

With more profound arterial insufficiency, *rest pain* appears. This becomes most pronounced at night with the leg now relatively elevated and warmed beneath the bedclothes. Such patients may develop a habit of sleeping with the leg hanging down over the edge of the bed or may even be forced to sleep in a chair. This stage presages gangrene of the extremity.

If an artery to an extremity is suddenly occluded, sudden pain results. The historical setting is usually a patient with heart disease. A remote myocardial infarction or rheumatic mitral valve disease both predispose to mural clot formation and the heart arrhythmias which jolt them loose. The patient relates the sudden onset of pain rapidly followed by swelling, weakness, and coldness of the extremity (Figs. 13.1 and 13.2).

Patients with diabetes mellitus have accelerated arteriosclerosis of large vessels and so

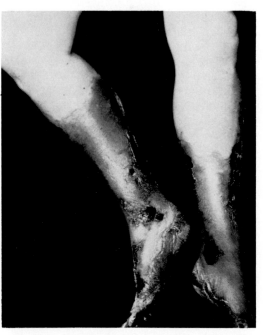

Fig. 13.1.   Gangerene of the skin of the lower extremities secondary to a saddle embolus at the aortic bifurcation. The patient had rheumatic heart disease and atrial fibrillation. A clot was dislodged from the left atrium and occluded the bifurcation of the aorta with infarction of the lower legs. (Courtesy Dr. Robert R. Linton.)

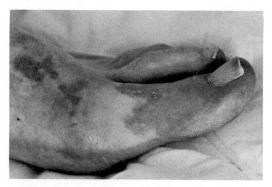

Fig. 13.2. Vascular insufficiency to the foot. This patient suffered from multiple small peripheral emboli which occluded the distal arterioles. In this foot, a well demarcated border between viable and nonviable skin is apparent.

may present with claudication at a younger age. In addition, they suffer disease of smaller arterioles, especially end arteries or those with little hope of collateral supply. These end arteries are more distally located, and the complaint of pain is similarly more distal. Early ulceration from skin infarction and eventual gangrene is common (Fig. 13.3).

Pain of *venous* origin comes from inflammation around an obstructing thrombus. *Thrombophlebitis* most commonly affects the lower extremities. The patient so afflicted may tell of similar previous episodes. A small area of discomfort becomes larger, extremely painful to touch, hard, and red. Progression is linear along the course of the vein, and the discomfort is aggravated by stainding or walking. These are the historical hallmarks of *superficial* thrombophlebitis. Thrombosis of the *deep* system may be without symptoms until this treacherous circumstance eventuates in a pulmonary embolus. A history of recurrent superficial thrombophlebitis, especially of unusual locations like the arm and abdominal wall, should prompt a search for an *occult neoplasm*.

Lymphangitis produces pain through inflammatory reaction around an infected lymphatic. A portal of entry is established through a break in the skin. Some local erythema results, and within 24 hours the patient notes a discrete red streak running up the extremity toward the axillary or inguinal nodes.

Observations related to this system may provide the chief complaint or may appear during the review of systems.

Arterial insufficiency sacrifices the luxury of skin appendages. Consequently, a patient may note the loss of hair over the lower extremities, abnormally slow growing toenails, and the absence of sweating in the feet. He complains that his skin is thin and heals poorly from minor abrasions and that his feet are always cold.

Varicose veins are the most frequently observed disease of the venous system. These dilated tortuous veins appear in one or both lower extremities. They may produce symptoms of heaviness of the legs and easy fatiguability, but they are more often of cosmetic concern. The poor venous return promotes edema of the ankles, stasis hyperpigmentation of the skin, and venous thrombosis from stagnation.

Lymphatic insufficiency causes *edema*. It may be massive, *elephantiasis*, and may be unilateral. The obstruction can be congenital, *Milroy's disease*; malignant, as in *lymphoma* involving pelvic nodes; or infectious, as in *filariasis*.

Inspection

Arterial disease can be suspected following the observation of a tortuous pulsatile vessel. The brachial artery most commonly shows

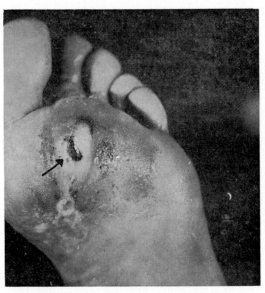

Fig. 13.3. Perforating ulcer of diabetes. The *arrow* points to the crater, and the light-colored zone surrounding the crater is a callous. This lesion is moderately painful and a site for superificial infection. The etiology is probably vascular insufficiency.

this abnormality. When involved with arteriosclerosis, the vessel elongates and runs a serpentine course down the medial aspect of the upper arm. Pulsations are readily apparent. If a normal patient crosses his legs at the knees, the suspended leg will be seen to pulsate, which is a valuable indication of vascular patency.

Varicose veins are usually readily apparent with inspection of the legs while the patient stands. Two major systems can be involved. The *lesser saphenous vein* originates behind the lateral malleolus and perforates the deep fascia at the midthigh level. The *greater saphenous vein* begins just anterior to the medial malleolus, passes just behind the medial femoral condyle, and joins the femoral vein at the groin. The greater saphenous is the longest vein in the body. When varicose, these veins become distended, beaded, and tortuous. Both of these systems communicate with deep veins by way of *perforators* which pass through the superficial fascial layers. The perforators, when normal, are valved to allow blood to pass only from without to within. Similarly, the greater and less saphenous veins have one-way valves permitting only an upward flow. Varicosities may result

from incompetence of valves, either in the perforators or in the saphenous veins, allowing retrograde flow.

When varicose veins appear with standing, a simple maneuver will help you decide which valves are incompetent. Have the patient return to a supine position and raise the affected limb until the veins drain completely. Then apply a tourniquet about the upper thigh and have the patient stand. If the varicosities now fill partially from below, the perforators are incompetent, allowing blood to flow from the deep system to the superficial system. If the veins remain collapsed, remove the tourniquet. Rapid filling from above indicates that the saphenous valves are faulty (Fig. 13.4).

Palpation of varicose veins will serve to confirm this. First palpate the greater saphenous close to the groin where it joins the femoral vein and have the patient cough. If the major valve at this junction is incompetent, you will feel a transmitted impulse. Next, lay your fingers over the mass of *lower* varicosities and tap the *upper* greater saphenous with the fingers of your other hand. Again, a transmitted impulse downward indicates incompetent valves (Fig. 13.5).

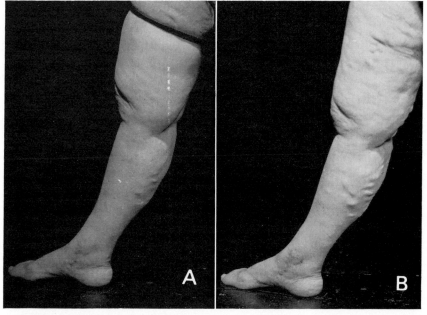

Fig. 13.4. Trendelenburg test for varicose veins. *A:* the patient has a tourniquet in place and has assumed the upright position. Slight filling from below suggests incompetence of communicating veins. *B:* the tournique has been removed and there is rapidly marked filling from above, indicating incompetence of the valve of the long saphenous vein.

Palpation of the arteries should be part of every physical examination, and the results should be carefully recorded for future reference. The lower extremity pulses are most important and seen to give students the greatest difficulties. The *femoral pulse* should be located just as it exits the groin (Fig. 13.6). The *popliteal* artery lies deep and slighly medial to the midline popliteal fossa (Fig. 13.7). Have the patient flex the knee and palpate deeply with one or both hands. The *dorsalis pedis* artery runs just lateral to the extensor *hallucis longis* over the metatarsals (Fig. 13.8). Begin laterally with the tips of several fingers and move medially until the pulse is located. The *posterior tibial artery* courses midway between the medial malleolus and the Achille's tendon (Fig. 13.9). Either of these, *but rarely both*, may be congenitally absent or displaced in the normal.

Next assess the *capillary filling time* of the toes. Compress a toe until it blanches white, quickly release it, and note how long it takes for the pink color to return. Four- to six-seconds is normal.

Now, elevate the leg until the foot becomes

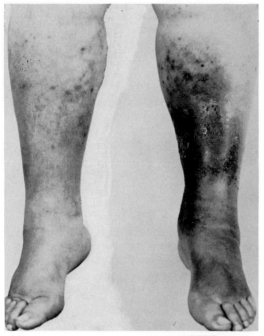

Fig. 13.5.   Stasis dermatitis. This patient had suffered thrombophlebitis of the left leg. The hyperpigmentation, edema, and superficial ulcerations are characteristic.

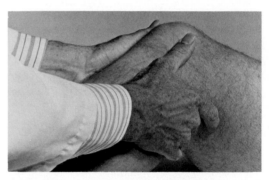

Fig. 13.7.   Palpating the popliteal artery. The knee is slightly flexed, and both hands are palpating the popliteal space.

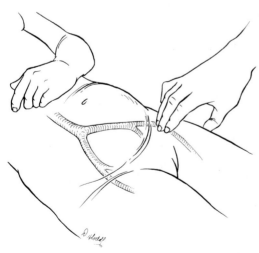

Fig. 13.6.   Palpation of the femoral artery. The femoral artery will be found in the middle-third of the inguinal ligament. The femoral nerve is just lateral to it and the femoral vein just medial.

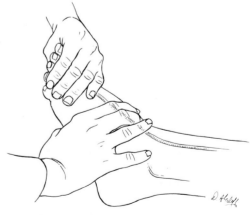

Fig. 13.8.   Palpating the dorsalis pedis artery. The artery is found in the position shown, running just lateral to the extensor tendon of the great toe.

pale. Mild foot exercises hasten this. If arterial insufficiency is present, the foot will assume a cadeveric hue and, when it is again dependent, will turn a deep red-blue in color—*dependent rubor*. At that same time, note the venous filling time. The collapsed veins on the dorsum of the foot should fill within 10–12 seconds of becoming dependent.

Ankle edema may result from lymphatic obstruction, venous disease, or acute arterial occlusion. *Pitting* is detected by applying *very firm pressure* with your fingers over a bony prominence (Fig. 13.10). When the pressure is withdrawn, a visible, palpable depression remains. Venous disease is the most common cause. It may be *local venous hypertension* from varicosities or old thrombophlebitis, *systemic venous hypertension* from congestive heart failure, or *lack of oncotic pressure* from hypoalbuminemia. In the last, the edema has a peculiar spongy quality to the palpating fingers.

Unilateral pitting edema always means local disease, which is frequently arterial or lymphatic. Carefully examine the pulses and regional node-bearing areas for pathology.

When pulses are absent in a limb, move back to the most distal palpable pulse and listen carefully with the diaphragm. An audible bruit may indicate the site of an obstruction.

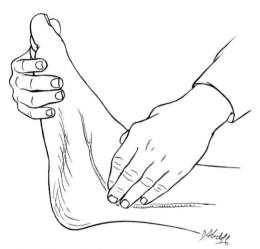

Fig. 13.9. Palpating the posterior tibial artery. This artery runs just behind the medial malleolus.

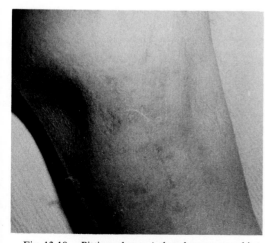

Fig. 13.10. Pitting edema. A thumb was pressed into the soft tissue of the lower leg leaving the indentation.

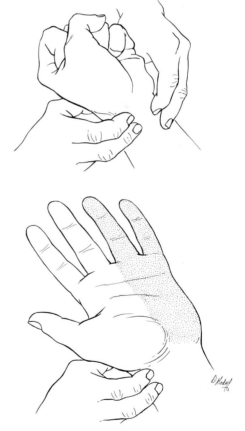

Fig. 13.11. The Allen test. Compress both the radial and ulnar artery and have the patient clench and unclench the fist to drain the hand of blood. Now, let up on one of the arteries and note the pattern of the blush in the hand. In the drawing, there is an absence of palmar arches, hence only half of the hand shows the flush.

PRACTICE SESSION

Locate each of the following arteries on both sides: the brachials, radials, ulnars, femorals, popliteals, dorsalis pedis, and posterior tibials. Next, measure the capillary filling time of the toes and the venous filling time of the feet. Now perform the Allen test (Fig. 13.11).

chapter 14

ABDOMEN

We now consider the physical examination of the abdomen exclusive of the genitourinary tract, which will be covered later. All techniques—inspection, palpation, percussion, and auscultation—are useful, but first let us consider some important clues from the history.

History

It is usually a straightforward matter to determine that abdominal disease exists when the appropriate chief complaint is voiced. It may be far more difficult to localize the offending organ system. As before, rely on an adequately explored history to direct your attention to the appropriate region or organ systems.

The abdomen is conveniently divided into sectors. The easiest is a four-quadrant system with perpendicular lines intersecting at the umbilicus, dividing into right and left upper and right and left lower quadrants. A nine-sector system may also be used with two vertical and two horizontal lines. The two vertical lines run along the lateral borders of the rectus muscles, the upper horizontal line joins the lowest points of the rib cages laterally, and the lower horizontal line joins the anterior superior iliac spines. The segments thus defined from right to left, top to bottom, are the *right hypochondrium*, the *epigastrium*, the *left hypochondrium*, the *right flank*, the *umbilical region*, the *left flank*, the *right iliac region*, the *hypogastrium*, and the *left iliac region*.

Abdominal pain may be simplistically considered as originating from either a *hollow viscus* or a *solid organ*. A hollow viscus concerns itself with storage and/or transport—the latter accomplished by peristaltic waves. Pain originates from distension of a viscus, almost always because of obstruction. The pain of an obstructed hollow viscus is termed *colicky pain* and is familiar to all of us as crampy pain. A base line level of discomfort becomes periodically exaggerated as a wave of contraction transiently increases the intraluminal pressure. Thus, the gall bladder, bile ducts, stomach, and small and large intestines are all subject to crampy abdominal pain.

Biliary colic from inflammation or stones localizes to the right hypochondrium. Most commonly elicited in fat, middle-aged women, this complaint is recurrent. It may be precipitated by overindulgence in food or drink. The patient adds complaints of nausea, vomiting, and fever and may explain that the pain radiates *around* the costal margin to the back.

The *stomach* does not register true colicky pain since it can almost always decompress excessive pressure through vomiting. *Acid peptic disease* of the stomach registers as a gnawing or burning epigastric discomfort usually relieved by, but occasionally exacerbated by, eating. Outlet obstruction produces pernicious vomiting.

A wide variety of lesions may produce *small intestinal pain.* Peristalsis is unidirectional, and so obstruction produces an essentially closed tube proximal to the block. Since peristalsis is frequent in the small bowel, the waves of pain are closely spaced. Small bowel pain tends to localize toward the umbilical region. Acid peptic disease of the duodenum registers as a deep-seated right hypochondrial pain almost always relieved by eating, frequently waking the patient from sleep, and most commonly ensuing in spring and fall.

Colonic pain prefers the right or left iliac regions over the cecum or sigmoid. Frequently related to function, colonic pain may be initiated by defecation or relieved by same. The waves of discomfort are less frequent than in small bowel disease because peristalsis is less frequent.

Acute appendicitis is one form of viscus pain

which, because of its frequency, merits elaboration. Appendicitis is a 5-day disease, and complaints in excess of that probably come from elsewhere. Anorexia is the initiating event, followed by periumbilical crampy pain, occasional nausea, and usually a low grade fever. The pain then migrates to the right iliac region, becomes progressively intense, until it suddenly disappears as the viscus ruptures. Hours later pain recurs as diffuse abdominal peritoneal pain associated with high fever.

With any colicky abdominal pain, you must carefully review functions of all suspected organ systems for clues to the involved structure.

Pain arises from *solid viscera* in a slightly different fashion. The parenchyma itself is insensitive to pain. Each visceral organ is wrapped in a serous membrane, either a capsule or layer of peritoneum. These respond to either inflammation or stretch by producing pain. The two basic mechanisms are swelling or infiltration by leukocytes.

Swelling which gradually progresses is less painful than that which occurs quickly. With either mechanism, pain is constant rather than crampy and may be aggravated by motion.

The *liver* becomes painful in hepatitis or liver abscess through the action of leukocytes. Acute congestion from right heart failure, hepatic vein thrombosis, or rapidly growing metastatic deposits will produce pain through stretch. Usually dull and deep-seated, hepatic pain radiates over the organ in the right hypochondrium and epigastrium. If the dome of the diaphragm is involved by contact, a pleuritic component may be present. This diaphragmatic pain may, in addition, be radiated to the shoulder.

The *pancreas* is a retroperitoneal organ lying on the spine tightly invested with peritoneum. Pain is frequently the only manifestation of pancreatic disease, especially malignant disease. Midepigastric or periumbilical, pancreatic pain frequently goes into the back. The patient will relate a boring constant pain which may force him to assume a fetal position for relief.

The *spleen*, a highly vascular organ, is subject to vascular events such as embolic infarction, subcapsular hemorrhage, rupture, and leukemic infiltration. Infarction produces sudden severe left hypochondrial pain often registering in the left shoulder as well. Acute distention is painful, but chronic distention, as in leukemic infiltration or portal hypertension, at most produces a "dragging heaviness" in the left upper quadrant.

The entire peritoneum is richly innervated and susceptible to pain whenever irritated by blood, intestinal contents, or pus. Peritoneal pain rarely originates *de novo* (primary peritonitis). It may be localized or diffuse. A patient with peritonitis is quite ill, has a short history, and you can frequently elicit symptoms pointing toward a responsible organ.

If pain is not the chief complaint, certain *asymptomatic observations* may bring a patient to see you or may come out with a review of symptoms.

Painless jaundice is more frequently observed by a friend or relative than by the patient himself. A history must include questions to elicit liver disease such as exposure to hepatitis, ingestion of raw shellfish, injections, alcohol, and toxin exposure. Profound hemolysis will also produce painless jaundice. A helpful clue to biliary obstruction is the complaint of pruritus, secondary to retained bile salts. Also question the color of the stools since the absence of bile in the gastrointestinal tract yields clay-colored *acholic stools*.

An important observation to elicit is *melena*. Black stools may indicate the presence of digested blood although iron preparations and bismuth (Pepto Bismol) will do likewise. True melena will be described as tarry, sticky stools and, if a lot of blood is passed, malodorous. *Bright red blood* per rectum means either that the bleeding originates distal to the cecum or, if coming from above, it is so brisk and profound as to be passed unaltered. With colonic bleeding, blood will be noted mixed in the stool. Rectal bleeding, on the other hand, merely coats the stool and stains the toilet paper.

Hematemesis may be painless but is always frightening. Any lesion from the nasopharnyx to the duodenum can eventuate in hematemesis. Volume estimates are difficult. A small amount of bloody emesis in the toilet bowl assumes enormous proportions in the patient's eyes. Carefully question for clues to the source. A recent dental extraction, epistaxis, esophageal varices from cirrhosis, gastritis from alcohol or drugs, gastric malignancy, and acid peptic ulcerations are the major causes (Fig. 14.1). *Coffee ground emesis*

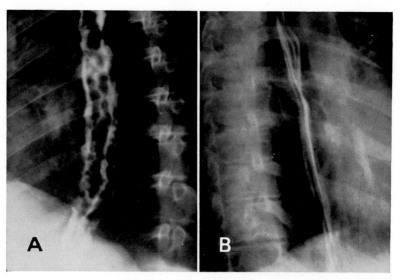

Fig. 14.1. Esophageal varices. When portal hypertension is severe, collateral circulation develops between the portal and systemic venous circuit. One of these includes the blood vessels between the spleen and stomach. When the pressure is extremely high, these venous channels dilate enormously, as demonstrated on the left film (*A*). The serpiginous black spot in the barium-filled esophagus represents these veins in relief. On the right (*B*) is a normal esophagus.

is that which has the appearance of perked grounds—particulate and brown-black. This tells you that gastric acid is present, since coffee ground hematemesis gains its color from acid hematin. This excludes the diagnosis of gastric cancer, which occurs in patients who are achlorhydric.

The patient may complain of *increasing abdominal girth*. Ascites and slowly growing neoplasms give constant distention, whereas gaseous expansion of the bowel is variable. Belt notches are usually an inch apart, and the number out over time is a crude index of the rapidity of distention. Dress sizes and vanity allow no such estimates in women.

Two changes from a *steady state* can implicate gastrointestinal disease. *Anorexia* and *weight loss*, while nonspecific, command a careful review of abdominal symptomology. Fever from abdominal origin usually has localizing pain to help with the diagnosis.

Nausea, vomiting, constipation, and *diarrhea* head the list of dysfunctional complaints. Characterize these complaints in terms of normal function, temporal sequence, and associated complaints. Obtain an accurate description of the emesis or stool. Emesis containing green or golden bile would exclude the diagnosis of pyloric or common duct obstruction. Diarrhea may mean to the patient either loose or frequent stooling. Stools of malabsorption tend to be particularly malodorous and bulky. Constipation for one patient may be normal for another, so that the change in frequency is the important determinant.

Also frequent among dysfunctional complaints is food intolerance. The etiologies are as diverse as food allergies, pancreatic, biliary, and intestinal mucosal disease. Associated symptoms and past history provide helpful diagnostic clues.

Examination

Inspection

The patient should be exposed from the nipples to the symphysis for adequate inspection. Tangential lighting is again helpful.

Distention may be obvious. One must first exclude fat. First determine if the swelling encompasses the entire abdomen or is localized. If localized, where is it noted? The suprapubic region might be a distended bladder (Fig. 14.2) or enlarged uterus; the iliacs—ovaries, cecum, or sigmoid; the flanks—kidney, spleen, or liver; the epigastrium—aortic aneurysms, stomach, or pancreas; the hypochondrial regions—liver or spleen (Fig. 14.3).

Ascites, free abdominal fluid, causes the flanks to bulge when the patient is supine and, when it is profound, everts the naval (Fig. 14.4).

The *scaphoid abdomen* is one in which the contour resembles a boat keel sinking sharply away from the xiphoid. Emaciation causes this, and in such cases pulsation of the abdominal aorta may appear quite prominent.

The skin of the abdomen deserves inspection as does the rest of the skin, but it is also the site of some peculiar abnormalities. Abdominal, pelvic, or retroperitoneal hemorrhage may manifest as localized ecchymosis in the flank or around the umbilicus (Fig. 14.5). *Striae* from recent weight change and *pigmented striae* of Cushing's syndrome show prominently on the abdominal wall. A *linea nigra*, hyperpigmented line from umbilicus to symphysis, testifies to a previous pregnancy. *Venous distention* accompanies deep venous obstruction, especially inferior vena caval blockage (Fig. 14.6). Here, the abdominal veins distend to provide collateral drainage. A peculiar example is the *caput medusae*, which is a circle of dilated veins radiating from the umbilicus. This is a collateral flow around the liver in cases of portal hypertension. Rarely, blood will drain downward over the abdomen to bypass a superior vena caval obstruction.

Note carefully any and all scars. Previous surgical invasions alter the diagnostic possibilities of abdominal disease.

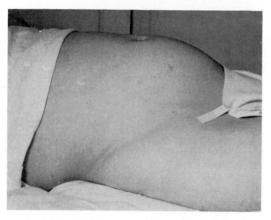

Fig. 14.2.   Bladder distension. The rounded prominence of the lower abdomen was confirmed by percussion, which demonstrated dullness in a circular pattern just above the symphysis pubis. The patient had not voided in 24 hours.

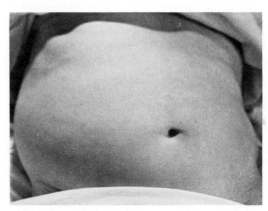

Fig. 14.4.   Bulging flank of ascites. The patient had cirrhosis of the liver with marked fluid accumulation in the peritoneal cavity.

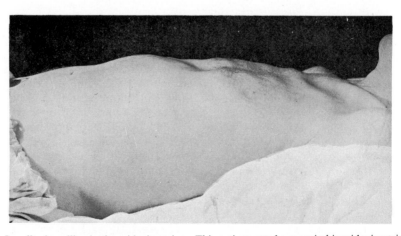

Fig. 14.3.   Localized swelling in the midepigastrium. This patient noted a mass in his midepigastrium 10 months after resection of his sigmoid colon for carcinoma. Biopsy proved metastatic disease to the liver.

Abdominal hernias can sometimes be diagnosed by inspection alone. An abnormal umbilical protrusion or bulge around a scar may be the clue. *Diastasis recti* resembles a hernia but is not truly such. Weak abdominal musculature permits the rectus muscles to pull away from the midline. Have the patient lift his head and a strip of midline-protruding abdomen will appear.

Peristaltic waves appear in normal patients with a thin abdomen and may be accentuated in patients with partial bowel obstruction. The suggestion must be pursued by palpation and auscultation.

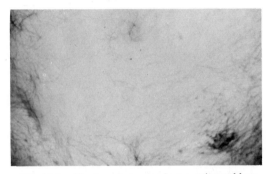

Fig. 14.5.   Flank ecchymosis of retroperitoneal hemorrhage. The small ecchymotic lesion can be seen below and to the left of the umbilicus. This patient had recently suffered a ruptured aortic aneurysm.

Finally, recall that abdominal muscles aid in the expiratory phase of respiration. With any obstruction to expiration, the effort becomes exaggerated and easily appreciated with inspection.

Palpation

The information obtained varies directly with care used. Approach the abdomen cautiously, gently, and without haste. Abrupt pokes, cold hands, or a quick once-over will thwart all efforts.

If there is reason to suspect pathology or pain in any area, *palpate that area last.* Using the pads of the fingers with the hand parallel to the abdomen, first palpate slowly and gently. A tense patient may relax more with the hips and knees flexed and just the heels on the examining table. In each area, slowly increase the pressure of the examining hand until you are satisfied with the results. If additional depth is required, put your other hand on the palpating hand to add additional pressure. A helpful way of getting deep into the abdomen is to have the patient respire slowly and deeply. With each expiration the abdomen falls in, and the position is maintained until the next expiration, when the hands are advanced even deeper.

The experienced examiner will distract the patient during the examination with idle dis-

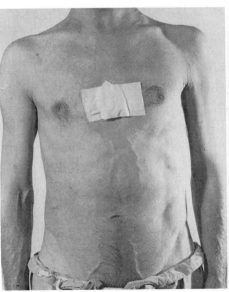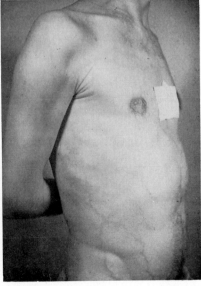

Fig. 14.6.   Distended thoracic and abdominal wall veins. This patient suffered superior vena caval obstruction from bronchogenic carcinoma. The patch covers a recent sternal biopsy incision.

cussion, all the while watching the patient's face for signs of discomfort during palpation.

True *bimanual palpation* is useful to outline organs, vascular structures, or masses by getting on opposite sides of the structure. This is accomplished either front to back through the flanks or by indenting both sides of the structure anteriorly (Fig. 14.7).

Ballottement is useful, especially when the abdomen is large or full of fluid. This technique may reveal a mass or organ enlargement not appreciated on ordinary palpation. The fingertips are pressed with a quick stabbing motion into the abdomen, momentarily displacing overlying fluid (Fig. 14.8).

Tenderness, either local or generalized, should be characterized. The patient who is tender may *guard* voluntarily or involuntarily by flexing the abdominal musculature. Voluntary guarding can be overcome by distracting a patient, whereas *true rigidity* is not under such control. The truly rigid abdomen indicates inflammation of the underlying peritoneum which reflexly establishes spasm of the abdominal muscles. Under such circumstances the involved area will feel wooden or *board-like*. Tugging on the umbilicus may help to confirm peritonitis since this is one area where the peritoneum is attached to the skin without intervening muscles (Fig. 14.9).

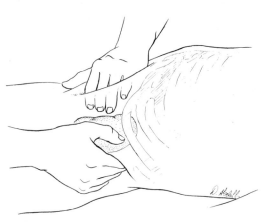

Fig. 14.7. Palpating the kidney. Deep pressure is applied by both the anterior and posterior hands.

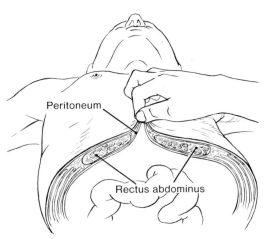

Fig. 14.9. Tugging on the umbilicus. This is an occasionally helpful way of demonstrating peritoneal inflammation.

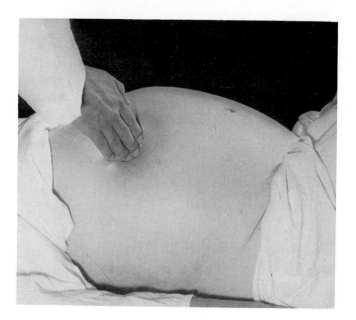

Fig. 14.8. Ballottement. Correct position of examiner's hand and fingers.

Palpation of the distended abdomen aids in the differentiation of dilated bowel, fluid, and fat. Of these three, only fluid will transmit a pressure wave. Have an assistant fix the midline with the edge of his hand. Lay one of your hands against the flank and tap the opposite flank with your fingertips. If free fluid exists, the palpating hand will feel a *fluid wave* strike it (Fig. 14.10).

Rebound tenderness is an important sign of peritoneal irritation. The palpating hand is slowly and gently pressed deep into the abdomen and then quickly released. As the peritoneum snaps back, the patient may wince or cry out. If so, immediately ask him where he felt the pain and he will point to the site of maximum irritation—sometimes quite remote from your palpating fingers.

Peristaltic waves can be felt anywhere in the abdomen and generally indicate mechanical obstruction, either partial or nearly complete. Infant *pyloric stenosis* is characterized by an *olive-shaped mass* in the right upper quadrant which can be detected if the examiner is patient and the child quiet. To localize waves elsewhere, remember that the radix of small bowel mesentery runs from the left upper to right lower quadrant and that the umbilicus crudely divides this line with jejunum above and ileum below.

The right hypochondrium may reveal an enlarged liver or gall bladder to palpation. The liver edge may be felt in normal subjects. When palpable, the *span* must be ascertained by percussion. To palpate the liver, place your left hand on the lower right rib cage and point the fingers of your right hand toward the right shoulder, applying pressure. Have the patient take a deep breath. The edge will be felt to flip over your fingertips as the organ descends with inspiration. Palpate progressively lower until you reach the iliac crest. *A massively enlarged liver is more often missed than a marginally enlarged one* (Fig. 14.11).

An alternate technique is to lay the heel of your left hand on the costal margin and curl your fingers in over the edge while having the patient breathe. Enlargement of the left lobe of the liver can extend well across the midline to the left hypochondrium. When the liver is palpable, note whether the edge is painless, tender, sharp, or blunted. The last is a sign of diffuse swelling.

An enlarged gall bladder usually presents on the imaginary line separating hypochondrium from epigastrium. Its contours will be

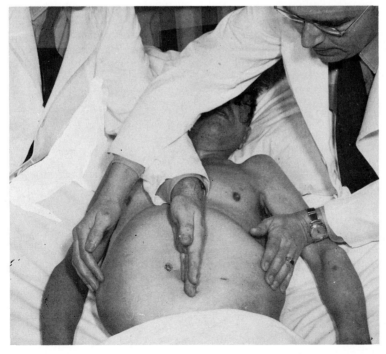

Fig. 14.10. Testing for a fluid wave. An assistant exerts pressure along the midline of the abdomen. The examiner's right hand taps on the abdomen while his left hand feels for the wave of fluid striking it.

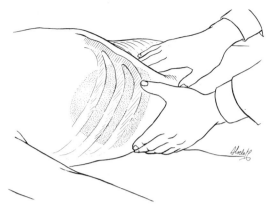

Fig. 14.11. Palpating the liver. The left hand provides support to the flank, and the right hand palpates deeply below the right costal margin while the patient inspires.

smooth and discrete. Primary gall bladder disease rarely causes enlargement. More often the cause of enlargement is distal obstruction. *Courvoisier's law* states that a jaundiced patient with a palpable gall bladder suffers from a carcinoma of the head of the pancreas. Primary gall bladder disease is more painful than palpable. The supposition is that the chronically inflamed gall bladder from infection or stones has developed a thick wall which is incapable of distention. A helpful maneuver to detect a tender or enlarged gall bladder is to hook your thumb over the edge of the costal margin in the region of the organ. A chronically inflamed organ so pinched between the thumb and ribs will cause the patient to "catch" as he breathes (Fig. 14.12). Similarly, deep hepatic tenderness can be elicited by jarring the organ. Lay one hand on the costal margin and with the other fist strike your hand sharply. A wince or grunt is a positive response.

The epigastrium generally registers tenderness to deep palpation in normal patients. The left lobe of the liver may encroach on this region as a palpable mass. The pancreas crosses the spine here and the stomach lies between it and the palpating fingers.

The left hypochondrium contains the fundus of the stomach, the spleen, and the left kidney. Splenic enlargement is best detected by palpating in a similar fashion as to the liver. Support the left coastal margin with your left hand and insinuate the right fingers deep into the subcostal area while the patient respires slowly and deeply (Fig. 14.13). If

uncertain about splenic enlargement, roll the patient toward you into a right lateral decubitus position to force the organ more superiorly and anteriorly (Fig. 14.14). It may be quite difficult to differentiate a large spleen from a large kidney. The latter can usually be noted to fall away from your fingers as you approach the costal margin. Also, the kidney does not move with respiration.

The flanks contain the kidneys. With bimanual palpation, one behind and one in front, the right kidney can often be felt in thin subjects and, rarely, the lower pole of the left kidney.

The umbilical region merits close attention. The umbilicus is a common site of herniation. Press your index finger into the naval. With a hernia, you will feel a sharp ring of fascia around a soft central depression. Intra-abdominal malignancy may deposit here as a *Sister Mary Joseph node*. The abdominal aorta lies beneath and slightly above and to the left of the umbilicus. Palpate bimanually on both sides of the aorta to ascertain its lateral diameter. Lateral pulsations tell you that you are truly on both sides of the vessel and not on top of it. In the elderly, it may be difficult to distinguish ectasia from true aneurysm. Just below the umbilicus, the aorta bifurcates over the sacral promontory. Accentuated lordosis and a thin abdomen can make this promontory strikingly apparent.

The right iliac region contains McBurney's point—the region of maximum tenderness with appendicitis, about halfway between the umbilicus and the anterior superior iliac spine. A dilated cecum is a common finding

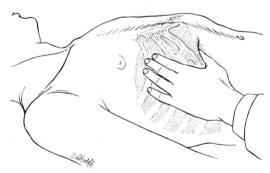

Fig. 14.12. Murphy's sign. The thumb is hooked over the right costal margin at the area of the gall bladder. The patient is then asked to take a deep breath. A positive sign is a catch in inspiration as the inflamed gall bladder is trapped between the thumb and the costal margin.

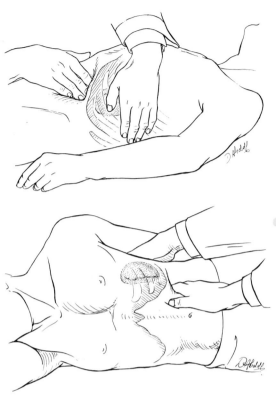

Fig. 14.13 (*top*).   Palpating the spleen. The left hand supports the left costal margin while the right hand presses deeply into the left hypochondrium. The patient is then asked to take a deep breath. A normal spleen cannot be felt.

Fig. 14.14 (*bottom*).   Palpating the spleen. In this illustration, the patient is rolled to his right. The left costal margin is supported with the left hand and the spleen felt for with the fingertips of the right hand.

in the right iliac area. It will feel like a soft sausage-shaped mass and may be tender when rolled under the fingertips. Just medial to this lies the right ovary, palpable with pathologic enlargement, although confirmation rests with the pelvic examination. Similarly, on the left, the sigmoid, or left ovary, might be found in the left iliac fossa.

The hypogastrium contains bladder and uterus, either of which might be outlined by palpation when enlarged.

Percussion

As elsewhere, percussion delineates static borders between tissues of different densities. The technique is the same as in percussion of the thorax. The passive finger applied gently to the abdomen is struck a staccato blow by the plexor finger, and the sound and resistance are recorded.

If the liver edge has been felt in the right hypochondrium, you must determine if it is enlarged or merely displaced downwards. Percuss over resonant anterior lung and move downward until hepatic dullness outlines the upper border. The *span* varies from patient to patient, but a measurement in excess of 12 cm is probably abnormal. The epigastrium and left hypochondrium register *tympany of the gastric bubble.*

Splenic dullness frequently appears between the 9th and 11th intercostal spaces in the anterior axillary line. Slight splenic enlargement manifests as tympany changing to dullness when the lowest interspace is percussed and the patient takes a deep breath—*splenic percussion sign.* The distinction between an enlarged spleen and kidney is further aided by percussion. Recall that the splenic flexure of the colon lies behind the spleen, but in front of the kidney. Frequently filled with air, the presence or absence of tympany, together with a palpable mass, provides a good clue. A mass *without* right upper quadrant tympany suggests spleen enlargement. A mass with tympany points toward renal enlargement.

Tympany anywhere means air. The air might be in distended gut loops or free in the abdomen. Together with the other techniques, the location of tympany can localize obstruction.

Percussion confirms many findings of palpation. *Rebound tenderness* may be elicited by percussion as well as palpation. *Shifting dullness* indicates ascites. Free fluid causes air-containing gut to float up to the most superior position, and this fact is used to advantage with percussion. With the patient supine, begin percussion in the midline and move to both flanks. Mark with your pen the point at which resonance changes to dullness. Now have the patient roll slightly to one side and repeat the procedure on the dependent side. Roll him the other way and repeat. The distance between the supine and rolled lines indicates the amount of fluid since the fluid will seek its level. There are two potential sources of error, however, Sometimes the shift of relatively airless gut will give a false positive result and localized peritoneal fluid will give a false negative result. A peculiar example of the latter is an *ovarian cyst.* These may attain enormous size and appear as marked abdominal distention. The fluid is encompassed, however. Two clues exist: first, the umbilicus does not protrude; and second, the fluid dullness remains anteriorly with the patient supine as resonant bowel is pushed down and laterally—just the reverse of free ascites (Fig. 14.15 and 14.16).

In questionable cases of ascites try the *puddle sign*. Have the patient get on all fours. Fluid will now puddle at the umbilicus where you can detect dullness (Fig. 14.17).

Auscultation

Many diagnosticians prefer to auscult the abdomen prior to palpation or percussion in order not to induce abnormal peristalsis. As elsewhere, auscultation yields information about dynamic, as opposed to static, events. The origins of sounds from the abdomen include peristalsis and noises from vascular structures or serous membranes.

Peristaltic activity emits various gurgling and bubbling sounds as air and fluid interfaces change with contraction waves. They vary with intensity, frequency, and pitch. The intensity relates to the vigor of peristalsis, the frequency to the rapidity of the waves, and the pitch to the tension in the wall of the contracting viscus. Small bowel peristalsis tends to be more frequent and higher pitched than the long low rumbles of the colon.

Partial bowel obstruction requires that the

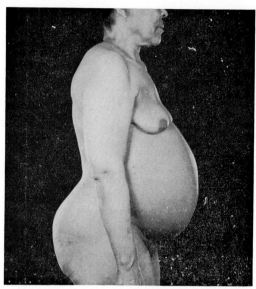

Fig. 14.16.   Ovarian cyst. This huge mass contained loculated fluid. This can be inferred because the photograph does not demonstrate protrusion of the umbilicus. (Courtesy Dr. Thomas H. Green, Jr.)

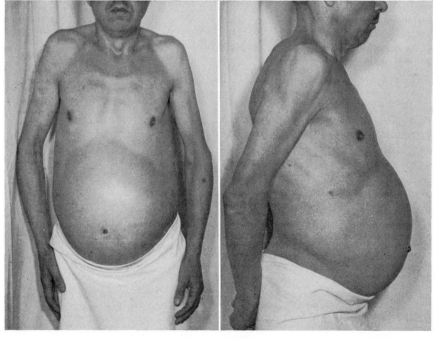

Fig. 14.15.   Ascites. The patient has cirrhosis. The marked abdominal distention is secondary to fluid accumulation. On the lateral view (*right*), the protruded umbilicus is evident. Note also the lack of body hair.

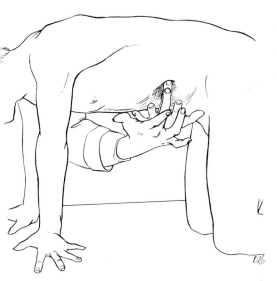

Fig. 14.17. Puddle sign. The patient is up on all fours. Percussion is carried out from one flank to the other. The note of dullness is marked. Small amounts of fluid will puddle in the dependent position and be detected as an area of dullness.

proximal bowel generate high pressure and strong contraction to force material through a compromised lumen. Bowel sounds here are high-pitched, and long, intense *peristaltic rushes* are heard. These sounds occur with any hypermotile state such as partial obstruction, diarrhea, gastroenteritis, and blood in the small bowel. When rushes accompany partial obstruction, they usually produce pain, so it is wise to watch the patient's face while ausculting the abdomen—a grimace plus a rush is cause for concern (Fig. 14.18).

If obstruction becomes complete, the proximal bowel becomes progressively more dilated and weaker, and the sound assumes a *tinkling character*—high-pitched short sounds. In late obstruction or in *paralytic ileus* from peritonitis, postlaparotomy, bowel infarction, electrolyte imbalance, and occasionally pneumonia, bowel sounds cease altogether. A silent abdomen portends a serious problem. *One must listen for at least a full 3 minutes before pronouncing that peristalsis has ceased.* The abdomen transmits peristaltic sounds well, and so it benefits you little to move from place to place. Listen close to the umbilicus.

Vascular sounds, as elsewhere, indicate turbulent flow in a dilated, constricted, or tortuous vessel. Any of the major abdominal arteries may be the source of a bruit. The aorta is heard best right over the vessel in the epigastrium, the hepatic artery in the right hypochondrium, and the splenic artery in the left hypochondrium. Renal artery bruits are important since in patients with hypertension the etiology might be renal vascular stenosis. These bruits are heard in the umbilical region or in the flanks and occasionally in the costovertebral angles posteriorly. The aorta bifurcates into the iliac just below the umbilicus: a common site of atheroma formation, turbulent flow, and thus a bruit.

It may be difficult to separate an abdominal bruit from a transmitted heart murmur. Carefully palpate the cardiac apex or carotid artery while listening over the abdomen. The transmitted murmur will be synchronous with the impulse, whereas a local bruit occurs sometime later.

A low-pitched, faint, continuous *venous hum* is heard in many normals and presumably originates from the vena cava and its branches. A recannulized umbilical vein from portal hypertension also generates a hum, as will large intra-abdominal hemangiomas.

Irritated fibrin-laden peritoneal surfaces will grate with motion, producing a friction

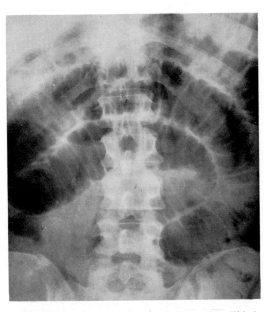

Fig. 14.18. Complete intestinal obstruction. This is the stepladder appearance of small bowel obstruction. A band of adhesions from a previous abdominal surgery obstructed this patient's small intestines. The dilated loops are readily apparent. Note also the absence of gas in the colon and rectum. The patient's bowel sounds were high-pitched and tinkling in quality.

rub much as their counterparts, the pleura and pericardium. The liver and spleen most commonly contribute to these rubs. The liver involved with metastatic disease or abscess, or a spleen recently infarcted, irritate overlying capsule and peritoneum, which rub during respiration.

One stethoscope trick aids in static border definition. Not uncommonly, the liver appears to be enlarged, but you are just unable to delineate the lower border. Place the bell of your stethoscope over the liver just below the costal margin and scratch the abdominal wall with the nail of your little finger. Do this in a semicircle at several points *equidistant* from the stethoscope. When you scratch over the liver, the sound will be very loud, but as soon as you pass over the edge the sound abruptly diminishes—the *scratch test* (Fig. 14.19).

The stethoscope may also be used for palpation. The nervous patient or malingerer recognizes the instrument as something you listen with. He does not anticipate that you might palpate with it. Gently apply the bell and slowly and continuously apply pressure

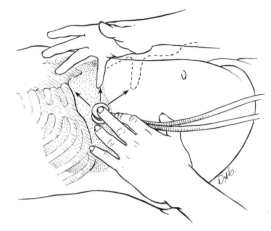

Fig. 14.19.   Scratch test for hepatomegaly. The stethoscope is applied over the enlarged liver while the little finger of the other hand scratches the skin surface at points equidistant from the stethoscope. When the scratching finger crosses over onto the liver, a marked increase in the intensity of the referred sound is detected.

to the abdominal wall. An area quite tender to your fingers might tolerate very deep exploration this way.

PRACTICE SESSION

Make sure the subject is completely supine and exposed from the nipples to the symphysis pubis. Inspect carefully for scars and note their exact location and color. (Recent scars are red, older ones white.) Note the linea alba or nigra. The borders of the rectus muscles should be apparent. Have the patient lift his head and note any bulges of hernias particularly between the rectus muscles.

Gently palpate all four quadrants without using much pressure, so that the subject gets accustomed to your hands. Now, palpate seriously the right upper quadrant. The border of the liver may be just palpable beneath the ribs. If so, note the contour of the edge. Your partner can help you by telling you how hard you can press without giving discomfort. Try the bimanual approach with one hand behind the right lower ribs and the other palpating the liver edge. Repeat this maneuver in the left upper quadrant to find the spleen. It will probably be palpable in one or two of your classmates. Remember to try the decubitus position.

Now, palpate deeply for the kidneys. If your subject is thin and tolerant, you should be able to feel at least one pole of both kidneys. Now, using both hands, palpate the lateral borders of the abdominal aorta. This is best felt above the umbilicus. Start somewhat lateral to the midline on both sides and move toward the midline. Measure the width of the aorta at this point.

Now, palpate both lower quadrants. It is common to feel the cecum in the right lower quadrant and even more common to feel the sigmoid colon in the left lower quadrant.

Now, percuss the liver borders carefully. Measure the span of the liver from top to bottom in the anterior axillary line. It should be no more than 10 cm. Percuss over the spleen. This would be over the lower left ribs. Have your partner breathe in and out and see if you can detect the spleen moving up and down by the change from resonant to dull. Percuss the remainder of the abdomen. Are there areas of air-filled bowel? Is the bladder full?

Now, listen over the four quadrants for bowel sounds and bruits. It is quite likely that one or two of your classmates will have a soft midepigastric bruit (16% of normal young adults). Perhaps you can swap the palpation of your spleen for the auscultation of his bruit.

chapter 15

RECTUM AND UROGENITAL TRACT

Rectum

History

Rectal pain almost always means local disease and is rarely a referred pain. *Rectal tenesmus* is painful straining at stool and clearly links function with symptoms. Any locally irritating phenomenon may be causative. The nature of the stool helps clarify the issue. A rock-hard stool implicates constipation, while a loose bloody stool might point to ulcerative proctitis. Tenesmus with normal stools suggests a rectal fissure. Constant rectal pain suggests a thrombosed hemorrhoid. Previous episodes of the same problem relieved by a sudden discharge of blood are common. Such a patient characteristically sits on one buttock while relating his misery.

Itching and burning are less specific. Local disease is again likely, with hemorrhoids leading the list. Perineal infection with fungus or infestation with pinworms is also possible. Fungus infections are particularly common in the diabetic, while pinworms are common in children. Ask if others in the family have the same complaint when considering pinworms.

The *observations* of bright red rectal bleeding and melena have already been discussed in Chapter 14. The patient may complain of a rectal lump or protrusion. Skin tags of resolved hemorrhoids are painless. A rectal prolapse is dramatic and intermittent, and usually the patient states that he can "push it back in." A rectal polyp on a stalk may also intermittently protrude. The observation of a change in stool caliber must be pursued. If all stools have recently become "pencil-like," a fixed stenosis secondary to stricture or malignancy is suspect. *Intermittent* thin stools usually means spasm.

As mentioned previously, constipation and diarrhea are the most frequent dysfunctional complaints. Remember, however, that obstruction may cause diarrhea instead of constipation. With a significant block, liquid diarrheal stool may be the only stool that can get by. Many fecal impactions are mistreated through inattention to this fact. *Fecal incontinence* attends severe debility, grand mal seizures, and lesions of the cauda equina.

Examination

Any position which flexes the hips will suffice for an appropriate rectal examination. The *Sims' position* calls for the patient to lie on his side and draw the top-side knee up toward his chest. The *knee-chest position* has the patients' knees and chest in contact with the table, jack-knifing his backside. An alternate method has the patient standing, bent at the waist with his chest on the examining table.

This unpleasant examination is made worse by poor technique. Invariably, the patient will tense, pulling his buttocks together and tightening his rectal sphincter. Have him breath slowly and deeply through the mouth and proceed gently.

Spread the buttocks and carefully inspect the anus and perineum. *Hemorrhoidal tags* are loose bits of skin arranged around the anus. Lichenification from pruritis and excoriation can be noted. A *thrombosed hemorrhoid* is a turgid red-purple mass with surrounding erythema (Fig. 15.1 and 15.2). A *rectal fissure* can be hard to see. Have the patient perform a Valsalva maneuver. A thin crack, usually weeping, then appears at the anal verge. A *prolapse* will also appear with this maneuver. Hairy males are prone to have a *pilonidal sinus*. This sinus tract can be found at the tip of the coccyx. It is lined by hairy skin and the accumulated secretions predispose to infection with local tenderness and a foul discharge.

Now, perform a digital examination (Fig. 15.3). Use a well lubricated finger cot or

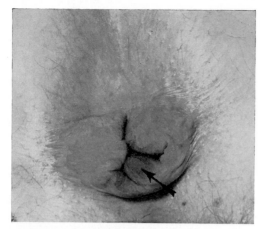

Fig. 15.1. Hemorrhoids. These large blood clot-filled hemorrhoids were the source of rectal pain in this patient. (Courtesy Dr. E. Parker Hayden.)

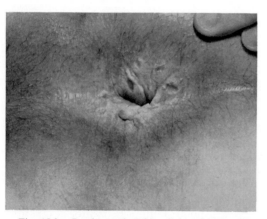

Fig. 15.2. Pruritus ani. White, lichenified perianal skin, hypertrophied folds, and ulcerations presumably secondary to infection created by scratching. (Courtesy Dr. E. Parker Hayden.)

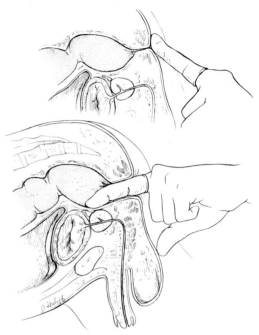

Fig. 15.3. Rectal examination. The pad of the examining digit is pressed on the anal verge (*top*) and rocked into the rectum. The prostate, and usually the seminal vesicals, can be palpated anteriorly (*bottom*). Examine for symmetry, consistency, and size.

glove. Lay the *pad*, not the tip, of your index finger on the anus and have the patient "bear down like you are going to have a bowel movement." Press with the finger and rock it slowly forward through the sphincter. Slowly, but firmly, insert your finger the full length. Posteriorly, feel the sacrum. Anteriorly, in the male is the prostate. In the female, the cervix should be palpable. There is a potential space between the rectum and vagina—the pouch of Douglas—a convenient collecting area for infection and intra-abdominal malignancies. Again, have the patient bear down. More distal pathology can be pushed against your examining finger. A high percentage of colonic neoplasms are within your reach. True polyps are soft, easily movable and difficult

to feel. The *villous adenoma*, precursor to malignancy, is a bit more firm, flat-based, and circumscribed. *Cancer* is hard and relatively easily detected.

The *prostate gland*, small until puberty, enlarges to a greater or lesser extent throughout adult life. Two lateral lobes are distinguishable with a dividing midline sulcus. The normal consistency is that of a pencil eraser. With enlargement, obliteration of the sulcus and asymmetry are common. *Cancer* of the prostate appears as a *stony-hard* nodule. *Prostatitis* causes the gland to become boggy and tender to palpation. Occasionally, the seminal vesicles can be felt as softer ribbons of tissue radiating inward and laterally from the top of the prostate.

Be sure to rotate your finger through the full 360° at different levels. *The examination is not complete until you test the stool on your glove for occult blood.*

Urinary Tract

History

Pain originating from the urinary tract may be of solid parenchyma or hollow viscus character as was described in Chapter 14 on the abdomen. Parenchymal pain from the kidney results from distension or irritation of the capsule. Patients usually describe this as a dull, deep-seated ache in the flank. *Acute pyelonephritis* is the prototype. The pain will be unilateral and the history may reveal previous episodes of cystitis or a recent urologic or gynecologic procedure or manipulation.

Acute glomerulonephritis gives less marked pain, lower fever, and bilateral discomfort. A recent pharyngitis or skin infection strengthens the clinical suspicion. Patients with heart disease predisposed to embolization—mitral stenosis, subacute bacterial endocarditis, cardiomyopathies, or a recent myocardial infarction—may throw clots or infected debris into the renal parenchyma with infarction, sharp flank pain, and later hematuria.

Hollow viscus pain stems from the ureter or bladder. As with this discomfort elsewhere, it indicates increased tension in the wall as a result of obstruction or spasm. *Ureteral colic* produces excruciating pain. Lancinating in quality, it has less periodicity than small bowel pain and is more or less constant. The pain moves as the calculus moves. It radiates from the flank down over the hypogastrium to the groin or testicle. When the stone falls into the bladder, pain ceases. A subsequent micturition passes the stone through the urethra, again producing stabbing pain and perhaps a drop of blood.

The bladder, especially in women, is prone to infection—*cystitis*. The history allows an almost certain diagnosis. The patient complains of stinging and burning with each voiding. Moreover, the irritated bladder is in spasm and holds very little urine, and there is a constant urge to void with the passage of very small volumes.

Hematuria is a startling observation. Carefully question the patient. Smoky or Coca Cola-colored urine usually means glomerulitis. Bright red blood may come from upper or lower tracts. Clots almost always come from a point distal to the pelvis of the kidney. A drop of blood initiating micturition suggests *urethritis*. Other colors may be noted. Myoglobin has a chocolate hue and may follow crush injuries or profound muscular activity. A variety of ingested dyes can appear in the urine. Urine which turns black leaving a stain on the underclothing may be your only clue to ochronosis, a rare metabolic dis-

ease. Urine containing high quantities of protein will foam in the toilet bowl. Jaundice darkens the urine. Urine of low specific gravity appears pale to clear.

The patient may note an asymptomatic mass. Many cases of *polycystic kidney disease* are discovered this way. The patient complains of a mild flank discomfort and palpable mass on both sides. A positive family history, if present, substantiates the diagnosis (Fig. 15.4).

Change from the steady state presents the diagnostic challenge. Such symptoms as weight loss, malaise, fatigue, and loss of cognitive functions are routine in failure of any major organ system. A renal etiology might be suggested by the notation of *uremic frost*—flaky white crystals on the skin or the detection by the family of an *ammoniacal odor* about the patient. A change in voiding habits and a previous history of renal disease are obvious pointers.

Attempt to understand *dysfunctional* complaints in a pathophysiologic framework. *Nocturia* and *polyuria* are common symptoms. The volume path puts you in two different physiologic categories. *Nocturia or polyuria with large volumes means a loss of concentrating powers.* The causes vary, but the urine always has a low specific gravity. An osmotic load as in *diabetes* with glycosuria is one example. *Renal tubular disease* as with chronic pyelonephritis represents another category. *Hypokalemia* and *hypercalcemia* paralyze the concentrating mechanism. Nocturia with congestive heart failure reflects the enhanced renal perfusion at night and an attempt to void the excess plasma volume.

Nocturia or polyuria with small volumes means a mechanical difficulty. Obstructive uropathy, as in benign prostatic hypertrophy, or bladder neck constriction demands a considerable voiding pressure. The bladder distends until that pressure is achieved and empties only until the pressure again falls. This is called *overflow voiding*. The postvoiding residual predisposes to infection and the high pressure predisposes to tubular dysfunction. An atonic bladder from neurogenic impairment also overflows and cannot empty completely because of loss of the detrusor tone.

When urine is lost with coughing, sneezing, or laughing, *stress incontinence* is the term applied. This mechanical problem usually obtains in multiparous women who have lost pelvic support. In males, it suggests obstruc-

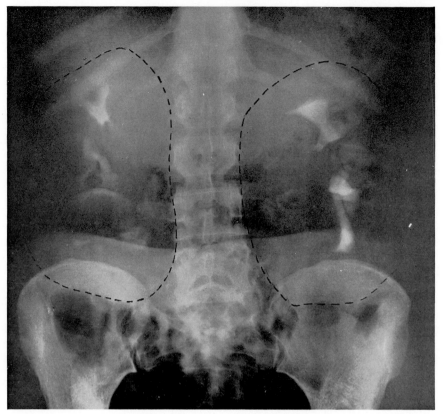

Fig. 15.4.   Polycystic disease of the kidneys. The intravenous pyelogram demonstrates the enormous enlargement of both kidneys which are outlined by black dashes. These kidneys are easily palpated through the abdominal wall. The disease is congenital.

tive uropathy where it is accompanied by dribbling, decreased stream, and a long voiding time. A useful question in regard to the voiding time is to ask "do you stand at the urinal longer than your friends?"

A careful review of renal symptomatology is important in all hypertensive patients. Individuals with hearing problems warrant a similar review, especially those with nerve deafness.

Examination

Hypertension, asterixis, uremic frost, hearing tests, and renal artery bruits have been covered.

Palpate the flanks carefully and deeply since the kidneys are well protected. A jolt to the costovertebral angle can be done only once in blatant pyelonephritis. The patient will not allow a second round. Bimanual palpation is the best technique, using one hand behind pressing in from the posterior flank and the other pressing through the ab-

dominal wall. A ballotting motion of the posterior hand thrusts the organ against the anterior one.

Bladder distension fills out the suprapubic region and can be readily defined by percussion of the supine patient. The dull fluid-filled bladder is surrounded by tympanitic air-containing bowel.

Examine the urethral meatus for inflammation or discharge. Complete the examination by analysis of a freshly voided urine specimen.

Male Genitalia

History

Pain in the groin, penis, or testicles may originate there directly or be referred. The area is encompassed by the 11th and 12th thoracic nerves.

Groin pain immediately suggests hernia. Frequently sudden in onset following straining, it can also onset gradually. If the pain accompanies an intermittent swelling, the diagnosis is almost established without examination.

The urethral meatus when inflamed causes pain exacerbated by micturition. The usual complaint is sharp pain when initiating the stream. Scrotal or testicular pain can be inflammatory or mechanical in origin. In the adult, mumps has a propensity to affect one testicle with swelling and pain. Other viruses are capable of producing *orchitis*. The patient may relate an infectious disease in the immediate family.

Sudden scrotal pain in a young man accompanies *testicular torsion*. Previous episodes which subsided spontaneously are commonly related. In this condition, the testicle twists about its vascular stalk, impeding venous return and arterial inflow.

Priapism, sustained erection, is generally painful and can be seen in spinal cord disease, chronic myelogenous leukemia, thrombosis of the corpora cavernosa for any reason, and, rarely, in prostatitis.

The most distressing *dysfunctional complaint* is *impotence*. Carefully elicit the nature of the difficulty. Inability to attain or maintain an erection *at any time* while the spirit is willing strongly suggests organic disease. If the difficulty varies with the partner, if the patient can masturbate successfully, or if he has morning erections, a psychologic factor is probably operative. Neurologic disease and pelvic obstructive vascular disease must be considered as organic etiologies. The cirrhotic patient becomes impotent but usually is not concerned about the lack of prowess.

Observations prompt several complaints. A penile sore concerns most men enough to seek medical attention promptly. When painless, syphilis is suspect. If the observation includes pain and swollen inguinal nodes, the diagnosis is more likely *chancroid, lymphogranuloma inguinale,* or *lymphogranuloma venereum.* In the older male, epidermal *carcinoma* might be suspected.

A tight foreskin might be noted either over the glans, *phimosis,* or caught behind the glans, *paraphimosis. Atrophy* of the testicle implies local disease, while atrophy of both testicles points to testosterone deficiency or estrogen excess. Parents may observe undescended testicles in the child, *cryptorchism.*

Examination

First, inspect the pubic hair. The male escutcheon points up the midline toward the umbilicus. This point disappears with estrogen excess as in cirrhosis. The pubic hair might harbor crabs, *pediculosis pubis.* These thin pinhead lice attach to the shaft of pubic hairs. The associated irritation frequently induces pruritus and you may note excoriations. The warm moist skin of the groin and perineum favor a variety of inflammatory and infectious processes. *Dermatophytosis* produced marked erythema with sharp raised borders covered with pinpoint vesicles. *Moniliasis* similarly produces erythema and pruritis, although the lesion is usually fairly dry with flaking (Fig. 15.5). *Intertrigo* is a nonspecific inflammation caused by maceration of tissue. It accompanies obesity and poor hygiene.

Palpate the groin for swellings. Lymph nodes can invariably be felt here in the adult, and it takes experience to identify truly pathologic enlargement. Unilateral painful enlarged nodes should prompt a search for a skin rupture on the penis, scrotum, and ipsilateral leg. Unilateral swollen nontender lymph nodes prompts a similar area search for malignancy.

A reducible bulge below the inguinal ligament laterally characterizes a *femoral hernia.* Apply gentle pressure on the bulge while the patient is supine, and the contents should return to the abdominal cavity, allowing you to feel the hernia sac.

An *inguinal hernia* presents more medially. The rupture is called *indirect* when it occurs through the internal ring, traverses medially through the inguinal canal, and presents in the scrotal sac. The *direct hernia* perforates

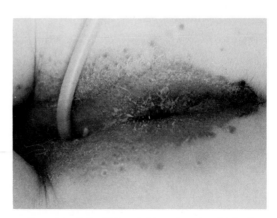

Fig. 15.5. Monilial vaginitis and perineal infection. The patient was admitted to the hospital in diabetic ketoacidosis. The lesion is erythematous and desquamative.

directly forward into the scrotum through the external ring (Fig. 15.6).

To palpate for hernia, have the patient stand facing you. Place the tip of your little finger through the external ring at the top of the scrotal sac. Rotate your finger and push up laterally along the canal to the internal ring. Your finger now will lie parallel to the inguinal ligament running from scrotum toward the anterior superior iliac spine. If no mass is encountered, have the patient cough. A direct hernia will strike your finger at a 90° angle, while an indirect hernia will strike the tip of your finger end on (Fig. 15.7).

A femoral hernia does not enter the scrotal sac. If either femoral or inguinal hernia is thought to be the cause of a lump, *listen to it with your stethoscope.* The presence of bowel sounds confirms the diagnosis. A hernia which cannot be reduced is cause for immediate concern. When associated with abdominal pain, it is cause for alarm.

Now examine the penis. Have a clean glass slide handy for immediate smear of any penile discharge. If the patient is uncircumcised, pull back the foreskin. *Balanitis* is chronic inflammation of the glans and will not be discovered unless the foreskin is moved back. The urethral meatus is checked by compressing the tip of the penis in an anteroposterior direction. Observe for patency, inflammation, or discharge. *Hypospadias* is a low placed meatus. Failure of complete closure of the raphe may place the meatus anywhere along the undersurface of the penis. This common congenital anomaly can accompany a bifid scrotum, undescended testicles, or hermaphroditism. In hypospadias, the penis always curves downward.

Inspect the glans and shaft for ulcers. The primary chancre of syphilis appears as a round ulcer with a hard indurated rim. The lesion is painless (Fig. 15.8).

Condyloma acuminata, veneral warts, are most common in the coronal sulcus in men and in the perineum in women. These are moist, irritated, and foul smelling. They must

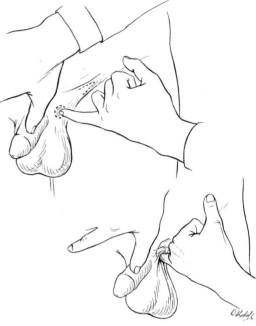

Fig. 15.7. Testing for a hernia. (*Top*), The pad of the little finger is applied to the external inguinal ring. (*Bottom*), The finger is then passed along the inguinal canal to the internal inguinal ring. When the patient coughs, an indirect hernia will strike the tip of the finger while a direct hernia will strike the side of the finger.

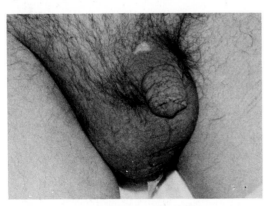

Fig. 15.6. Inguinal hernia. The hernia sac extends well into the scrotum, where bowel sounds can be heard.

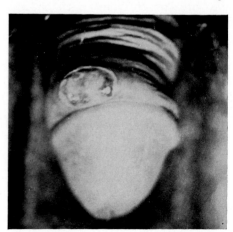

Fig. 15.8. Chancre of penis. This was a painless lesion. The dark field was positive for the spirochetes of syphilis.

be distinguished from *condyloma lata* of syphilis. These are flat, moist, well circumscribed papules. A serous nonmalodorous drainage can be seen (Figs. 15.9 and 15.10).

Next palpate the floor of the urethra from the perineum to the glans. When done in this direction, a drop of discharge may appear for immediate culture and Gram stain. The discharge of *gonococcal urethritis* characteristically has a pale green color, while that of nonspecific urethritis is more likely to be clear.

The scrotum and testes must now be examined.

Scrotal edema will occur when leg edema is massive. When isolated to the scrotum, edema suggests cellulitis or otherlocal pathology. *Chancres* and *carcinoma* may appear on the scrotum as well as the penis.

Epidermal inclusion cysts are very common in the scrotum as small scrotal skin nodules. When infected and draining, they discharge a foul cheesy material.

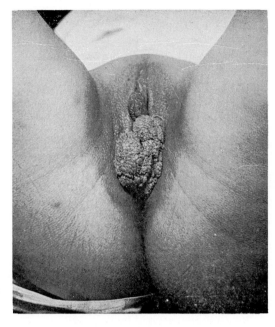

Fig. 15.9.   Condyloma acuminata.

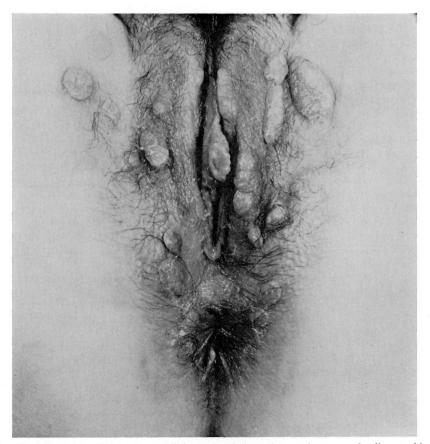

Fig. 15.10.   Condyloma lata. These are syphilitic warts of the vulva, perineum, and adjacent skin. Note the difference between this lesion and condyloma acuminata, the so-called nonveneral warts.

Palpate each testis, attending to the testis proper, the epididymis, and the cord. First ascertain that both testicles are descended. If not in the scrotum, palpate the inguinal canal. If the testis is there, attempt to draw it into the scrotum. A true undescended testis cannot be moved into the scrotum. Are both testicles the same size and are they normal in size? An atrophic testicle presents mute testimony to old inflammatory disease, usually mumps. Bilateral small testicles imply hypoplasia as in Kleinfelter's syndrome, hormonal atrophy as in cirrhosis, or hypogonadotropic states as in pituitary lesions. Atrophy can be distinguished from hypoplasia by examining the epididymis and cord. With atrophy, these structures remain normal in size, whereas in hypoplasia they are reduced to a size commensurate with the testes (Fig. 15.11).

The epididymis caps the testes above and behind. It is more irregular and granular feeling than the testes. The cord contains the vas deferens and neurovascular bundle. It attaches posteriorly. The vas can be distinguished by its hard whipcord-like feel, discrete from the rest of the cord.

The nature of a lump in the scrotum can usually be diagnosed fairly accurately by physical examination. First, find out if it arises in the scrotum or comes down from above. Try to get above the mass at the root of the scrotum. You will be unable to get above a hernia.

Next, note the position of the mass in relation to the testis. A *hydrocele* will be anterior to the testis and will be smooth and fluctuant (Fig. 15.12). *Tumor* of the testis is painless and clearly is in the testis proper. *Orchitis* will swell the entire organ (Fig. 15.13). *Epididymitis* presents as a hard nodule on top and behind the testis—tuberculous epididymitis is the prototype. A *spermatocele* arises from the epididymis and is discrete from it. A *varicocele* involves the cord as a serpentine mass of dilated veins. This latter

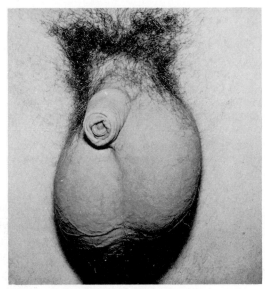

Fig. 15.12. Bilateral hydrocele. This painless enlargement of both sides of the scrotum transilluminated brilliantly.

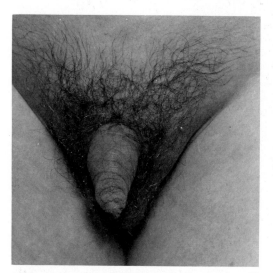

Fig. 15.11. Eunuch. The short phallus, sparse pubic hair, and tiny scrotum are apparent.

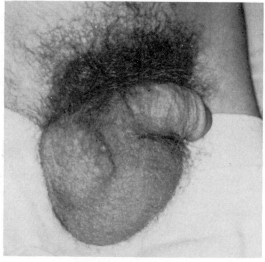

Fig. 15.13. Mumps orchitis. The left testis is enlarged and exquisitely tender. Associated parotitis confirmed the diagnosis of mumps. (Courtesy of Dr. Louis Weinstein.)

anomaly, when recent in onset and on the left, may be the only sign of a renal cell carcinoma. Recall that the left spermatic vein drains into the left renal vein.

Transillumination is very helpful in sorting out scrotal masses. Darken the room and press your penlight against the mass. The cystic hydrocele and spermatoceles are brilliantly illuminated.

Female Genitalia

History

The routine history of any woman should include age of menarche, regularity of periods, duration of periods, and interval between menses. A very crude estimate of amount of flow is garnered by a pad count. Note the presence or absence of cramps—*dysmenorrhea*. The history of parity includes number of pregnancies, miscarriages, stillborns, and normal deliveries, and the sequence. Ascertain the age when periods ceased and the completeness of estrogen deficiency by the presence or absence of vasomotor instability—heat flushes and sweats.

Gynecologic *pain* wears many faces. Characterize it temporally in relation to menses. Cramps with periods, while uncomfortable, usually mean at least that ovulation has occurred. *Mittelschmerz* is pain at the time of ovulation—14 days prior to menses. Pain in both lower quadrants with fever during or just after menstruation is characteristic of gonococcal pelvic inflammatory disease.

Painful intercourse, *dyspareunia* has a host of etiologies. Urethritis commonly leads to this complaint. The inflamed tissue becomes traumatized against the symphysis during intercourse. Similarly, any pelvic inflammation exacerbates with coitus.

Pain with vaginal itching signifies a vaginitis—most commonly *Monilia.* Pursue the possibility of diabetes with this complaint.

Dysfunctional complaints relate mostly to pregnancy or the lack thereof. A careful history is paramount to elucidating the cause of sterility. Previous gynecologic infections, family history of similar difficulty, lack of any cramping with periods, irregular menses, dyspareunia, or habitual abortions are all relevant to the investigation.

The most common *observation* is secondary ammenorrhea—periods having been present previously have now ceased. The most common cause is pregnancy. Consider, too, a history pointing to any gross endocrinopathy—thyroid, adrenal, pituitary, or ovarian. *Primary amenorrhea* means no menses at any time. Consider this in light of the family history. A girl of 18 who has not yet menstruated may not be abnormal if this trait was also present in her mother and grandmother. True primary amenorrhea stems from disease anywhere along the delicate hypothalamic, pituitary, ovarian, uterine axis. Note developmental landmarks of breast development, axillary and pubic hair, and growth curve.

Intermenstrual spotting and *postcoital bleeding* must be taken seriously. Intermenstrual bleeding most commonly arises from breakthrough bleeding while on birth control pills. If pills are not being taken, it generally means uterine pathology—fibroids, endometrial carcinoma, retained products of conception, etc. A cervical lesion is the most common cause of postcoital bleeding. Chronic cervicitis and cervical cancer are the diagnoses to consider first.

A disturbing observation is uterine prolapse (Fig. 15.14). The cervix intermittently presents itself through the vaginal orifice. Usually it can be reduced with ease by the patient. This profound loss of pelvic support also causes stress incontinence.

Examination

It is prudent to have the presence and assistance of a female attendant. Make sure

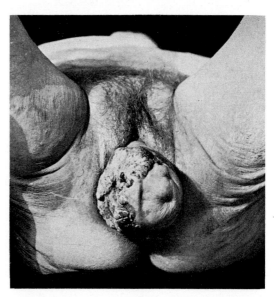

Fig. 15.14. Cervical prolapse with extensive carcinoma. The enormous carcinoma of the cervix had caused it to prolapse through the vaginal orifice.

the patient has recently voided. The lithotomy position affords the best examination and should be used in all but the most unusual circumstances. With her heels in the stirrups angled away 45°, have the patient bring her buttocks to the very end of the table. Drape the abdomen. Have her drop her knees to the side. Sit on a chair or stool between her legs and provide strong lighting over your shoulder.

Inspect the pubic hair, perineum, and external genitalia for the same lesions described in the male. Spread the labia majora. Anteriorly, inspect the hooded clitoris for irritation, ulcers, or enlargement. *Clitoromegaly* indicates excess testosterone and distinguishes *virilization* from hirsutism.

Below the clitoris in the midline lies the urethral meatus. Inflammation here is quite common. Exuberant granulation tissue indicates an *urethral carbuncle*.

Estrogen deficiency, as in *menopause*, causes atrophy of the labia minora and introitus. The mucosa assumes a shiny, dry, thin appearance. The perineum between vagina and rectum will show scars from the episiotomy or the spontaneous rents of delivery (Fig. 15.15).

Now, spread the labia widely and have the patient bear down hard. Loss of pelvic support, usually from multiple pregnancies, may allow the urethra and/or bladder proper to bulge down into and out of the vagina as a cystourethrocele (Fig. 15.16). Suspect this lesion with a history of *stress incontinence*. Posteriorly, this phenomenon allows the rectum to protrude up and out of the vagina as a rectocele.

Continue inspection with a speculum in the nonvirginal woman. Warm the speculum and introduce it with water as the only lubricant. If you use vaseline or a jelly, chances are that you will distort the cytological examination. Introduce the speculum with the

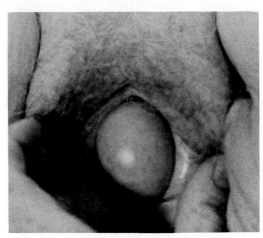

Fig. 15.16.　Cystocele. The patient has been asked to "bear down like you are going to have a bowel movement." The bladder is easily seen in this photograph protruding through the vaginal orifice.

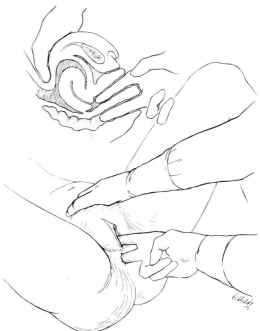

Fig. 15.17.　Rectovaginal examination. The cervix can be felt and, by gentle anterior pressure, the uterus can be brought down on the examining finger. The pouch of Douglas is between the two examining fingers (*top*). By moving laterally to the side of the cervix with both the examining finger and the palpating hand on the abdomen (*bottom*) the ovary and adnexa may be felt.

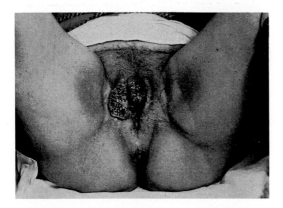

Fig. 15.15.　Carcinoma of the vulva. The lesion was extensive and painless.

blades in a vertical position. Expand the introitus by pushing *posteriorly* on the perineum. Avoid pressure on the anterior urethra and clitoris. Once introduced, rotate the speculum so that the blades are horizontal and spread them until the cervix appears between. The lateral vaginal walls can be seen and should be moist and corrugated. The nulliparous cervix displays a round os while the parous cervix will have a linear "fish mouth" os.

Normally, clear thin cervical discharge from the endocervical glands can be expressed. *Purulent discharge* indicates endometritis or other pelvic inflammatory disease, as in gonococcal infection of the Fallopian tubes. A cervix which bleeds with manipulation is not normal. *Cervicitis* is recognized as inflamed erythematous tissue around the os. An *endocervical polyp* may protrude as a raspberry-colored mass from the os. The *gravid* os assumes a deep blue hue.

Inspect the posterior fornix for shed debris. *Monilia* infection presents as a cheesy white exudate which when stripped from the mucosa leaves a bright erythematous patch. *Trichomonas* exudate is copious, bubbling, and white without such patches.

The *Papanicolaou* smear is an invaluable test for early uterine malignancy. Two smears are usually taken. First a specially cut wooden spatula is fitted into the os and rotated 360° to scrape the junction of cervix and endocervix. This is then smeared on a glass slide and immediately fixed. Next, a specimen of exfoliated cells from the posterior fornix is picked up and spread on another slide. A third smear may be taken from the vaginal mucosa to assess estrogen effect on cell maturation.

Now, carefully withdraw the speculum, keeping the blades partly open so that you do not pinch tissue.

Now, palpate with the gloved lubricated hand. First palpate *Bartholin's glands.* They are located deep in the labia majora close to the perineum and are common sites of infection and retention cysts. *Skene's tubules* are periurethral and exit on both sides and just below the urethra. Milking the urethra from the bladder to the meatus may eject fluid from the urethra or either of Skene's ducts.

The hymenal remains can be seen and palpated as an irregular curtain of small tags of tissue around the introitus.

Next, introduce the lubricated index finger into the vagina with pressure posteriorly against the perineum. Once some relaxation occurs, insert your middle finger as well. With the other hand palpate the abdomen starting at the umbilicus moving downward pressing in and down toward the vaginal fingers. The cervix is readily distinguishable by its shape and relative firmness. The consistency resembles the tip of your nose—during pregnancy more like your earlobe. With the cervix against the vaginal fingers, the abdominal hand will bring the uterus to lie between the two. Normally a pear-shaped organ, it should be freely movable and nontender.

Symmetrical enlargement of the uterus suggests *pregnancy.* Lobulated asymmetrical enlargement is characteristic of uterine *fibroids.* The cervix, too, should be freely movable from side to side and up and down.

Now, slip your fingers to one side of the cervix and the other hand to the same side of the abdomen to palpate the ovary. This small organ is frequently difficult to feel, especially when the woman is corpulent or tense. Normally, Fallopian tubes cannot be felt. Tenderness and/or enlargement should be noted and quantitated insofar as possible.

Complete the examination by performing a rectovaginal palpation (Fig. 15.17). Change gloves and introduce the index finger into the vagina and the middle finger into the rectum. In this manner, the pouch of Douglas comes to lie between your fingers and all of the pathology which frequents this catch basin. This exam can also be augmented by abdominal palpation.

chapter 16

BONES AND JOINTS

The orthopedic examination may be accomplished in toto with specific areas examined in conjunction with other structures in the same area. For convenience, we will consider this section separately since the afflictions of bones and joints and the techniques involved are somewhat similar regardless of the area. This will not be a compendium of all abnormalities in physical findings since the volume of material is considerable. Rather, the general principles and some common diseases are presented.

History

Bones and joints participate in many systemic diseases as well as local pathology. Usually, multiple areas of distress reflect systemic disease, but the converse is not always true. A brown tumor of hyperparathyroidism can give local disease and represent systemic pathology; likewise gout, or a pathologic fracture through a metastasis, can present as unifocal disease.

Also a chief complaint which the patient ascribes to a bone or joint might not originate there, as in the shoulder pain of splenic infarction or the low back pain of a pelvic malignancy.

Exercise extreme care, then, in eliciting all facets of the chief complaint, pursue associated symptoms, and take a careful review of systems.

Orthopedic *pain* originates from the periosteum, joint capsules, or attendant connective tissue. Bone, proper, is insensitive to painful stimuli. A broken bone yields pain through rupture of the periosteum and soft tissue hemorrhage, a malignancy through invasion and stretching of the periosteum, and arthritis through irritation of the joint capsule. Wherever large connective tissue bundles must glide over joints, you can anticipate the presence of a bursa—a fluid-filled thin sack acting as a cushion glide. These, too, register pain when irritated and inflamed.

Pain resulting from injury is straightforward enough. Careful questioning may allow you to predict the nature of the disease. A football knee injury resulting from a lateral tackle most likely pushed the joint medially, tearing the medial collateral ligament and perhaps the medial meniscus as well. The child with elbow pain following an arm jerk by a parent quite likely has dislocated the radial head. Try to determine the exact nature of the injury and mentally plot the axis of stress.

Pain without injury is a bit more difficult. Determine the duration and tempo, i.e. getting worse, constant, intermittent, etc. Are there temporal relationships such as the morning stiffness of *rheumatoid arthritis* which warms up during the day? What aggravates the pain and what relieves it? Aspirin may help rheumatoid arthritis and not *gout*, heat may alleviate *osteoarthritis* and aggravate an *infected joint*. Associated symptoms must be noted. A recent penile discharge in a man complaining of arthritis immediately suggests *gonococcal joint infection*, while the addition of conjunctivitis and a skin rash leads to a suspicion of *Reiter's syndrome*. Weight loss and bone pain might indicate metastatic or *primary bone cancer*.

The *absence* of pain in the presence of obvious joint disease should make you suspect *neuropathy*. Destruction of pain fibers from a joint allows repeated trauma to go undetected and accelerates osteoarthritis. *Diabetes mellitus* is the most common cause of a *neuropathic joint*. *Tabes dorsalis* and *syringomyelia* are other causes (Fig. 16.1).

A common *observation* in older patients is a loss of height. *Senile osteoporosis* reduces the height of each vetebra slightly with a marked total reduction which is measurable.

A *limp* usually stems from pain but may be

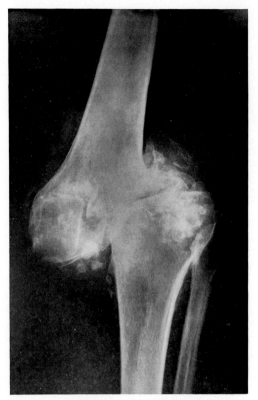

Fig. 16.1. Neuropathic joint. This 59-year-old man had long standing tabes dorsalis. The knee joint has been virtually destroyed. Numerous bone fragments are evident in the joint space.

a painless observation. Muscle disease is usually at fault, but the observation in a child might suggest disparate limb growth, metabolic bone disease such as *Gaucher's*, or *congenital joint dislocation*. Muscle weakness will come out as your history sorts out those specific movements which bring out the limp. A foot drop, for instance, is most noticeable climbing stairs or curbs.

The patient may complain of an asymptomatic lump. Determine its duration and the rate of growth. Carefully review the patient's activity record to learn of repeated unsuspected trauma. A physician recently became intrigued by bony lumps occurring just below the tibial prominence in young adults and found that they were all avid surfboard enthusiasts. The repeated trauma explained the *surfer's knots*.

Similarly, joint swelling can be asymptomatic. An inciting injury frequently is trivial and forgotten until careful questioning calls it to mind.

Dysfunctional complaints can be particularly aggravating to a patient. Slowly acquired dysfunction, as with the rheumatoid hand, can be remarkably compensated for. Sudden dysfunction hurries the patient to your office. The history must be geared to determine if the dysfunction results from bone or joint or from the motor apparatus required to move it. The "locked knee," for instance, can result from a loose body, *joint mouse*, or from instability of the supporting ligaments or tendons.

Examination

Inspection and palpation are the major techniques. Auscultation rarely helps and percussion not at all.

Surface anatomy plays a critical role in the orthopedic examination. Bony landmarks outline the underlying structures and must be learned. Paired joints provide a comparative parameter of great value, and they should be so examined and measured. The range of motion about a joint is the functional test, the results of which are quantitative. For each joint, there is a defined neutral position about which measurements are accomplished and from which arcs can be measured.

The terminology of orthopedics confuses students. So let us begin with some definitions.

A *varus deformity* is that in which an angulation occurs *toward* the midline. It refers to the deviation of the distal structure. Thus, genu varum is the description for bowlegged knees. A *valgus deformity* is the opposite—the distal extremity points away from the midline. *Genu valgus* is the term for knock-kneed. *Abduction* moves a part *away* from the central axis while *adduction* moves it *toward* the axis. These terms obviously cannot be used for spine motion since the spine represents the axis. *Rotation* implies a circular movement and applies only to ball and socket joints. *Active* motion requires an intrinsic motor unit. *Passive* motion does not.

Spine

Inspect the spine with the patient standing (Fig. 16.2). The usual hospital gown with back ties is convenient, providing maximum exposure and modesty. Viewed from the side, note a normal thoracic outward curve commencing at the vertebrae prominens followed

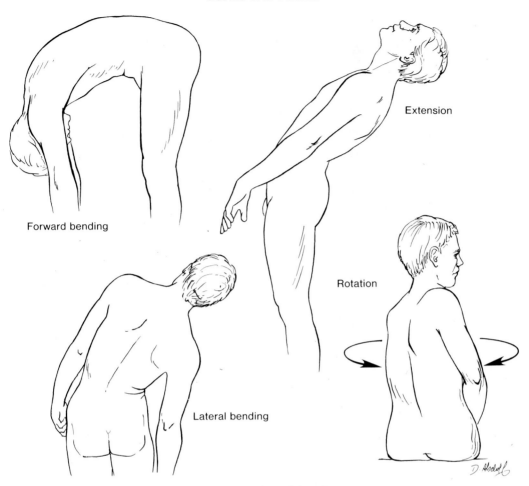

Forward bending

Extension

Rotation

Lateral bending

D. Abdulf

Fig. 16.2. Motions of the spine.

by a gentle inward lumbar curve and then a sacral outward swing.

An abnormal outward hump is termed *kyphosis* (Fig. 16.3). When very sharp angulation is noted, the term *gibbous deformity* is used. A gibbous suggests spinal tuberculosis, fracture, or metastatic disease to the spine. A more gentle kyphosis of the thoracic spine accompanies aging and osteoporosis. An associated finding is folds of loose skin over the back radiating from the spine. Poor posture may also participate in the production of a mild kyphosis and is usually associated with sloping shoulders.

Exaggerated incurving, called *lordosis*, is detected by noting a very deep furrow between the lumbar paraspinous muscles and a pot belly. Whenever a patient must rock backwards to maintain a center of gravity, lordosis results. Pregnancy, flexion contrac-

tures at the hips, and short Achilles tendons are examples.

Loss of all curves gives a poker back spine. Fusion of the vertebral body, as in late ankylosing spondylitis is the cause (Fig. 16.4). Usually the head is bent forward, the back straight and immobile, the belly prominent, and the gait wide based. Such a patient is forced to roll back on his heels in order to look you in the eye.

Lateral curvature of the spine, *scoliosis*, is best detected looking straight at the back (Fig. 16.5). Normally the spine is quite straight top to bottom when viewed posteriorly. While in this position, note the tips of the scapulae, the tops of the hips, the dimples of Venus overlying the sacroiliac joints and the general symmetry of left and right. A mild scoliosis can sometimes be felt more readily than seen by running your thumb

Fig. 16.3.   Kyphosis. The marked protrusion of the thoracic spine is clearly apparent. This lesion was present at birth.

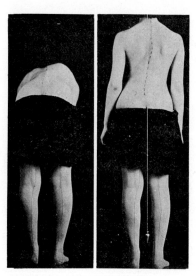

Fig. 16.5.   Scoliosis. In the upright position a scoliosis to the right is barely apparent. When the patient leans forward, however, the right hemithorax is thrown into relief, and the effect is much more obvious.

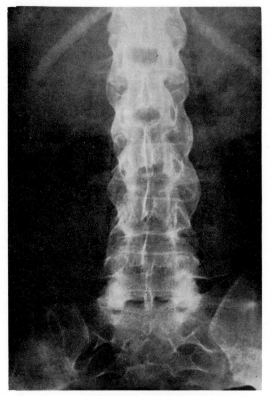

Fig. 16.4.   Ankylosing spondylitis. Each vertebra is fused to its neighbors by bony ridges. This fusion results in relative immobility and a poker back spine. This is a disease of men with an onset usually in the teens to twenties.

over the vertebral spine. The scoliosis should be described fully: where is it and which way does it go; is it single, S-shaped, or more complex; and is it accompanied by rotation? The latter, rotational component, tells you much about the significance of the scoliosis. When the scoliosis is postural, there is no rotation of the spine. When structural, rotation is always present. Demonstrate the presence of rotation by having the patient bend forward at the waist. A postural nonrotated scoliosis disappears. A functionally rotated scoliosis appears more prominent by asymmetry of the back of the chest. The side toward which the scoliosis points humps out higher than the other side.

Questionable spine deformities can be further evaluated by having the patient hang from a trapeze. Structural abnormalities usually persist, and postural ones disappear.

Now, palpate the spinous processes for tenderness. Recall that these structures overlap each other much as shingles on a roof. To get additional information about tenderness, locate each spine with your finger then strike your finger with a percussion hammer.

When the patient with back pain arrives with compensation forms in hand, use the *pointing test.* Have him point to the spot of maximum discomfort. Mark the spot with a pen and later in the exam have him point again. Reproducible results suggest genuine trouble.

A common abnormality of the spine is *spondylolisthesis*—the anterior displacement of the vertebra, usually the fourth or fifth lumbar. Not always symptomatic, this lesion can be detected by running your finger down the spine. It will sink in over the displaced vertebra and you will detect a shelf produced by the next vertebral body. When symptomatic, pain results from anterior bending and will be referred along the fourth or fifth dermatome.

The patient with a *herniated disc* walks carefully into your office. He leans forward, with the hips and knees flexed, standing on the balls of his feet. The normal lumbar lordosis is lost, and marked spasm of the paraspinous muscles can be felt. To detect this sign, have the patient bend laterally to one side and then the other. Normally these muscles soften when the patient bends to that side. With spasm they feel hard and tight even with lateral bending. Further, with an herniated disc, flexion and extension are limited, and pain will radiate along the affected nerve root dermatome. Local tenderness occurs lateral to the spine and sometimes over the affected spinous process.

The sacroiliac joints lie under the dimples of Venus and should be palpated carefully in complaints of back pain. Local tenderness is almost always elicited in early ankylosing spondylitis. Such patients will complain of pain with flexion of the spine while standing. However, when seated, with the pelvis supported, they can bend forward more or less freely. Similarly, rotational movements of the spine cause pain in sacroiliitis. There are a group of lymph nodes over this joint which commonly enlarge with underlying inflammation.

To complete the back examination, have the patient lie supine. The *straight leg raising test* is a most valuable tool (Fig. 16.6, *top*). With one hand behind the ankle and the knee extended, the entire limb is raised slowly until the patient registers pain. The test must be passive with no assistance from the patient. Note the exact angle at which pain results. Repeat the test to the point just short of producing pain and dorsiflex the foot. This applies traction on the sciatic nerve (Fig. 16.6, *bottom*). Pain with this test implies disease of

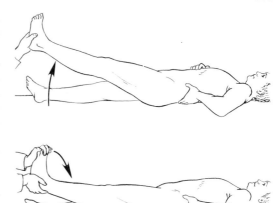

Fig. 16.6 (*Top*) Straight leg-raising sign. When the leg is passively flexed at the hip, the patient complains of pain in the low back. (*Bottom*) Straight leg raising plus dorsiflexion of the foot. The leg is raised to a point just short of that which produces pain. The foot is then forcibly dorsiflexed, which puts traction on the sciatic nerve. The patient with pain of sciatic or sciatic root origin will complain of discomfort with this maneuver.

the sciatic or one of its nerve trunks, as with a herniated disc. Sciatic pain most often occurs from 40 to 60° elevation. Sacroiliac pain, if it results at all, only comes on with full elevation. Sacroiliac pain can be further tested by placing one hand on the patient's shoulder while the other hand pushes the flexed knee toward the contralateral shoulder—painful with sacroiliitis.

By flexing both knees and hips and pushing toward the chest—fetal position—the *lumbosacral articulations* will be stressed, adding further diagnostic information. Lumbosacral pain, usually a disease of women, has a few other clues. The patient so afflicted does not like lying supine and will attempt to obliterate the lumbar lordosis by flexing the knees and hips. Straight leg raising is painless.

Now, roll the patient to the prone position. Hyperextension of the limbs causes pain in herniated disc, sacroiliitis, and lumbosacral sprain. Helpfully, the pain localizes fairly accurately to the affected area.

With any case of back pain, a careful pelvic and rectal examination is mandatory. Too many malignancies are treated symptomatically as low back pain.

Shoulder

Three landmarks identify the shoulder triangle (Fig. 16.7). The tip of the acromion

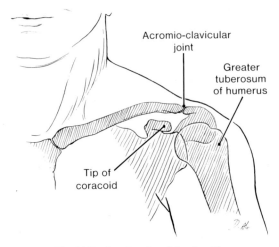

Fig. 16.7.   Landmarks of the shoulder.

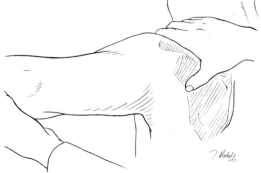

Fig. 16.8.   Fixing the scapula while abducting the shoulder.

presents at the top of the shoulder. This tip of the spinous process of the scapula is the area of tenderness in lesions of the *rotator cuff* of the shoulder. The second point of the triangle is the *greater tuberosity* of the *humerus.* This area is tender with *bicipital tendinitis,* also a common shoulder lesion. The third landmark is the tip of the *coracoid process* which triangulates the other two points. It is felt in the hollow of the anterior shoulder.

Shoulder motion is complex. To observe the range of motion, examine the patient from behind and watch active motions. The reason for looking from behind the patient is so that you can observe the scapula and, if necessary, hold it fixed. Movement of the scapula can compensate for a restricted shoulder (Fig. 16.8).

To test *abduction,* have the patient move the arms out and up—"flapping bird style." The scapula will begin to rotate at about 30°. Fix it with your hand, and the patient should be able to abduct to 90°.

Adduct the shoulder by having the patient place his hand on the top of the opposite shoulder. The elbow should rest on the sternum (Fig. 16.9).

Internal rotation is a back scratcher. Have the patient place his thumb between the shoulder blades. *External rotation* is the back swing for a forearm tennis shot. With the elbow partially flexed the hand is pushed backwards as far as possible.

If difficulty occurs with active motion, then examine with passive motion. Bone disease or fixed muscle contractures will limit all motion, but acute motor injuries will allow full passive motion.

Of these motions, slow abduction gives the most information. If the patient can abduct to 90° and then raise the limb over his head, serious injury or disease is not present. The arc of 30–60° is the *painful arc of the rotator cuff.* This cuff or cap of the shoulder is made up of the supraspinatus tendon and tendons of the infraspinatus, subscapularis, teres minor, and the shoulder joint capsule. Injury or rupture of any of these limits motion through pain. If you assist the patient to 60°, he can then complete the arm lift using the deltoid and rotating the scapula. *Arthritis* of the shoulder gives pain at the very onset of abduction and usually other motions as well.

Bicipital tendinitis can be confused with rotator cuff disease. When the long head of the biceps is inflamed, the *Speed test* is helpful. Have the patient extend the arm straight away palm up and press down on the palm. Resistance requires bicep contraction, and pain results with tendinitis (Fig. 16.10).

Elbow

Again, a triangle of bony landmarks helps orient you to the joint (Fig. 16.11). The medial and lateral humeral condyles are on line, and the tip of the olecranon forms the third point. Compare one triangle with the other. Dislocation dramatically alters the triangle while a supracondylar fracture does not.

Examine the carrying angle. Fully extend the forearms at the sides. Males deviate outward about 10° and females deviate outward about 20°. A wide carrying angle accompanies many congenital anomalies—Turner's syndrome for instance. *Flexion,* "make a muscle," normally leaves about 30° between upper arm and forearm. Full extension should

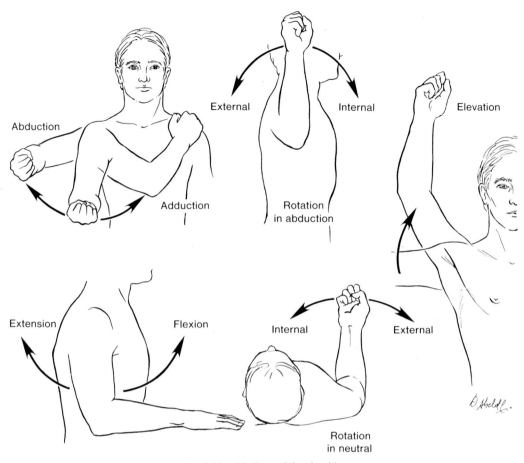

Fig. 16.9. Motions of the shoulder.

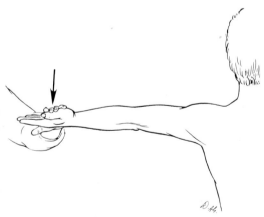

Fig. 16.10. Speed test. The arm is extended and pressure is applied to the palm. If this elicits pain in the shoulder, it is most likely secondary to bicipetal tendinitis.

make a straight line of forearm and upper arm. The total arc from full flexion to extension approaches 160° (Fig. 16.12).

Pronation and *supination* of the forearm can be accomplished at the elbow or at the shoulder by abduction and adduction. To test the elbow, you must fix it at the side of the body and have the patient pronate and supinate the forearm palm out. Pronation is slightly less than 90° and supination a full 90°.

The tip of the olecranon locates a small bursa which can fill with fluid as a fluctuant pouch. Usually painless, it alarms the patient. The elbow participates frequently in the symptoms of *rheumatoid arthritis.* The boggy synovium and joint fluid obliterate the normal bony landmarks and limit motion. The nodules of rheumatoid arthritis are felt along the lateral edge of the ulna just below the elbow (Fig. 16.13). Gouty tophi and tuberous xanthomata also prefer this location for deposition.

The epitrochlear node can be palpated just above the medial humeral condyle. Infectious diseases cause most enlargements of this node.

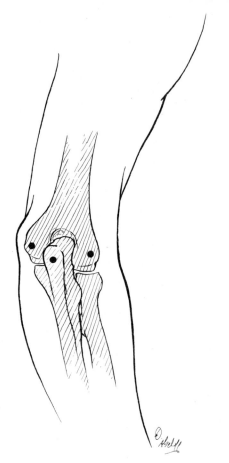

Fig. 16.11. The triangle of the elbow. The lateral condyles of the humerus and the tip of the olecranon.

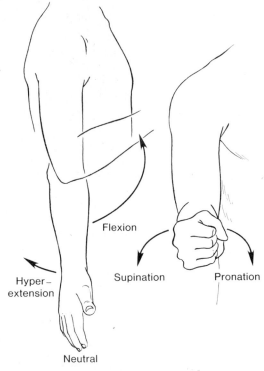

Fig. 16.12. Motions of the elbow.

Hip Joint

Begin the examination with the patient standing. Since this is a weight-bearing joint, standing and walking are functional tests. In the standing position, the patient with hip arthritis will support most of his weight on the noninvolved side. Look for a slight flexion of the knee on the involved side. The gluteus maximus may atrophy on the affected side.

Facing the patient, look for a pelvic tilt. This is best detected by placing your thumbs on the anterior superior iliac spines and middle fingers on the greater trochanters of the femurs. If the iliac spine is elevated on one side *and* the greater trochanter is also elevated, either the opposite limb is shorter or it has a fixed flexion contracture. If the anterior spine is lower but the trochanter is on line with its mate, shortening of the femoral neck is present. Under such circumstances, the glu-

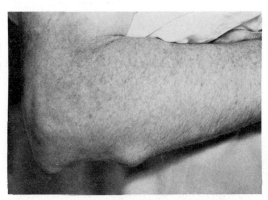

Fig. 16.13. Rheumatoid nodule. This swelling was apparent along the ulnar surface of the forearm in a patient with advanced rheumatoid arthritis.

teus medius on the affected side will be lax, soft, and easily indented.

Now, watch the patient walk. Abnormalities of gait frequently indicate hip disease. A fixed hip forces the patient to *circumduct* by throwing the affected pelvis up and the limb out and around in front. The *Trendelenburg* gait causes the trunk to list to the side of the affected hip in congenital dislocations. When bilateral, this gait is an exaggeration of the

Hollywood buttock attention getter. It is occasioned by relative weakness of the gluteus medius muscles because of shortening of the femoral neck (Fig. 16.14).

Have the patient lie supine and go through the range of motion of the hip (Fig. 16.15). Straight leg raising is a test of *flexion*. If tightness of the hamstrings gives discomfort, have the patient flex the knee and continue. The thigh should rest on the abdominal wall. With the leg straight down, *abduct* it laterally to a normal of about 45° then *adduct* a similar distance for total arc of 90°. In normal adduction, the middle third of the opposite thigh can be crossed. *Rotation* is the first motion lost in hip arthritis. The reason is that the femoral head gets out of round so that the joint will behave as a hinge joint rather than a ball and socket structure. Extension and flexion may be preserved until quite late. Test rotation by placing one hand above and one hand below the knee and roll the extremity so that the patella first points in for *internal rotation* then out for *external rotation*. Alternatively, flex the knee and move the foot out to test internal rotation then move the foot in to test external rotation. With this maneuver be sure to keep the knee in a constant position.

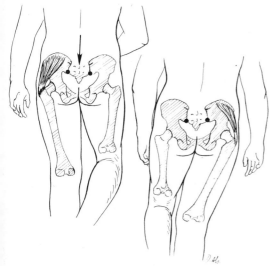

Fig. 16.14.   Trendelenburg's sign. (*Left*), The patient has lifted his right leg. In so doing the sacroiliac notches, marked with *black dots*, remain on a horizontal line. (*Right*), On the right the patient is attempting to stand on his right foot. Note that the left sacroiliac notch is now lower than the right. This indicates that the patient's right iliotrochanteric muscles are not powerful enough to hold the pelvis in alignment.

A flexion contracture represents the most common result of hip disease. The patient compensates by accentuating the lumbar lordosis in order to tilt the pelvis forward allowing the foot to stand flat. *Thomas' sign* readily demonstrates a flexion contracture. Have the patient pull the knee of the uninvolved side up to his chest. This flattens the lumbar lordosis. The affected limb will now rise above the table. The degree to which it rises is the amount of flexion contracture present (Fig. 16.16).

Now, roll the patient to a prone position. Place one hand on the buttock and lift the leg to test extension—normally about 40°. If a pelvic tilt was detected, now check the comparative limb length, measuring from the anterior superior iliac spine to the medial malleolus. To be accurate, the limbs must be in the same position. Palpate up the inguinal ligament until you strike the first bony prominence and stretch the tape measure just inside the patella to the malleolus.

Knee

The knee is a very complex joint in spite of the fact that it is a simple hinge joint. It is the largest joint, bears weight, and must do so while correcting the normal varus tilt from the hips.

The neutral point for the knee is *extension* in line with the femur. Hyperextension may occur to 10°. *Flexion* to 130° is normal (Fig. 16.17).

Landmark anatomy is again helpful. With the patient supine locate the patella. Note that it moves freely side to side but is fixed up and down. There is a deep sulcus below it medially and a more shallow sulcus laterally (Fig. 16.18). Below the patella and somewhat lateral, note the prominent tibial tuberosity. The entire articulation of the knee involves only femur and tibia, the fibula resting on the lateral condyle of the tibia. On both sides, the joint articulation can be felt as the *joint line*. With your fingers on the sides at the joint line between bony prominences, you are directly on top of the collateral ligaments and the meniscus.

The stability of the knee and its free gliding motion are facilitated by two straps of external ligaments, the medial and lateral collateral ligaments, and an internal mechanism consisting of cruciate ligaments and a lateral and medial cartilaginous meniscus. Certain

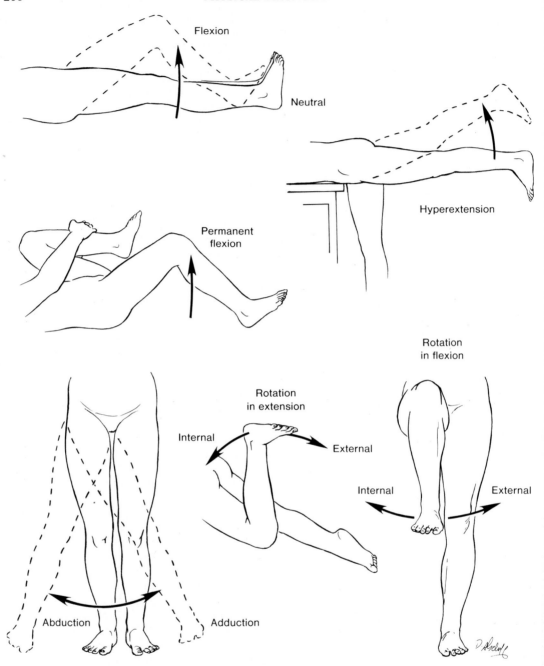

Fig. 16.15.   Motion of the hip.

maneuvers help you to test the integrity of these stabilizing structures (Fig. 16.19).

The medial meniscus suffers more trauma than the lateral one. With a tear of the *medial meniscus*, there will be tenderness *over* the joint line. If the *medial collateral ligament* separates, the pain will be found *above* the joint line over the femoral condyle.

McMurray's test allows you to characterize further meniscus tears. Flex the knee and with one hand palpate the joint line, holding the heel of the foot in your other hand with the ball of the foot on your wrist. *Lateral rotation* of the foot tests the posterior horn of the *medial meniscus*. *Medial rotation* of the foot tests the posterior horn of the *lateral*

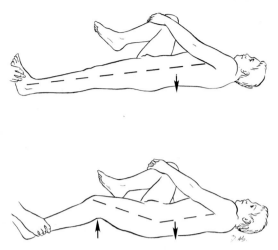

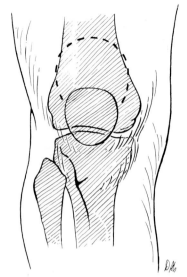

Fig. 16.16. Thomas's sign. (*Top*), A normal response. The patient grasps his knee, which flattens the normal lumbar lordosis. In so doing, he is still able to keep his left leg in flat contact with the examining couch. (*Bottom*), A demonstration of a flexion contracture of the left hip which is brought out by this maneuver. Once the lumbar lordosis is obliterated, the fixed flexion contracture of the hip is thrown in relief.

Fig. 16.18. Landmarks of the knee. Note the suprapatellar pouch, the position of the patella in reference to the joint line and the greater tuberosity of the tibia.

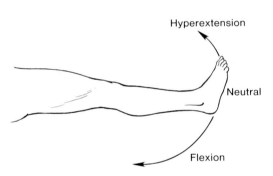

Fig. 16.17. Motions of the knee.

meniscus. A positive test is when a palpable click occurs with this maneuver (Fig. 16.20).

The *cruciate ligaments* are tough and require great force to disrupt. The *drawer sign* allows the diagnosis. The patient is supine with the knee flexed and foot flat on the examination table. Sit on the foot and grasp both sides of the tibia. If the tibia can be pulled far anteriorly, the anterior cruciate ligament is torn. If it can be pushed far posteriorly, the posterior ligament is torn (Fig. 16.21).

Knee effusions are common in injuries and in arthritis. When the effusion is great, inspection alone will demonstrate it. Moderate effusions must be distinguished from synovial thickening. The true knee joint capsule ex-

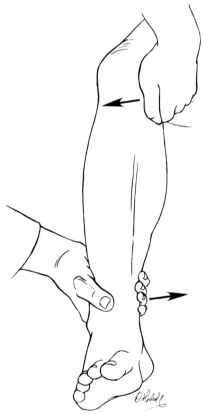

Fig. 16.19. Testing the integrity of the lateral collateral ligaments. The examiner exerts pressure on the medial side of the knee and on the lateral side of the ankle. If the collateral ligaments have separated, abnormal mobility of the knee will be demonstrated and pain elicited.

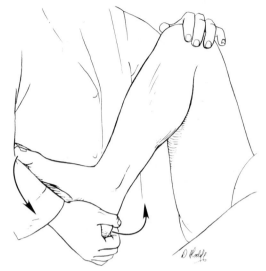

Fig. 16.20. McMurray's test for posterior tears of the meniscus. The examiner has his left fingers on the joint line. The patient's foot rests on the examiner's right forearm. Lateral rotation of the lower leg tests the posterior half of the medical meniscus. The reverse, medial rotation, tests the posterior segment of the lateral meniscus.

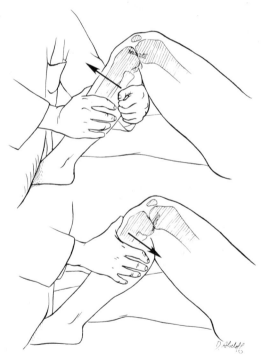

Fig. 16.21. (*Top*), Anterior drawer sign. The examiner is sitting on the patient's foot and with his right hand he pulls forward on the lower leg. If the anterior cruciate has been disrupted, an abnormal forward displacement of the lower leg will become apparent. (*Bottom*), Posterior drawer sign. The examiner exerts pressure posteriorly to test the integrity of the posterior cruciate ligament.

tends above the patella as the suprapatellar pouch (see Fig. 16.22C). Fluid extends freely into this pouch. Compress this area with one hand and ballotte the patella. If fluid is present, the *patella tap sign* will be present as a click when the patella strikes the femur (Fig. 16.22 A).

Palpate on both sides of the patella simultaneously. With an effusion, a true fluid wave is felt. With a thick boggy synovium, the pressure may be transmitted, but it is much different from a fluid wave. Small amounts of fluid may be detected. Light from above the patella so that the medial recess below the patella is in shadow. Carefully and slowly milk the suprapatellar pouch downward and a bulge will ripple into this recess as fluid forces it outward (Fig. 16.22 B).

The joint capsule does not extend into the popliteal space unless diseased. A posterior rupture of the capsule occurs as a *Baker's*

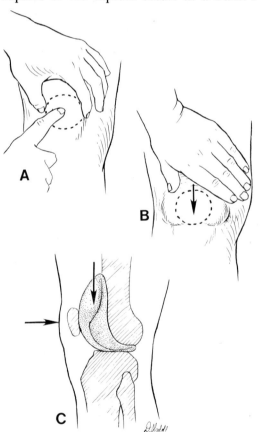

Fig. 16.22. *A*, Ballotting the patella. A quick thrusting pressure will produce a clicking sound if there is interposing fluid. *B*, By milking the suprapatellar pouch downward, a bulge will appear on both sides of the patella if fluid is present. *C*, the bursa extends upward as the supra-patellar pouch.

cyst, a fluctuant mass in the popliteal fossa. Have the patient stand to best demonstrate this lesion. Palpate it carefully to distinguish it from a popliteal artery aneurysm.

Foot

Four motions are tested (Fig. 16.23). Extend the knee and *dorsiflex* the foot toward the patient and *plantar flex* away. *Inversion*

twists the foot sole inward and *eversion* twists the sole outward. These last two maneuvers must be accomplished at the subtalar joint, not by twisting the metatarsals (Fig. 16.24).

The important landmarks are the lateral and medial malleoli of the fibula and tibia, respectively. A bony prominence on the lateral side of the foot is the tip of the fifth metatarsal. Between it and the lateral malleolus, fan several important ligaments joining the fibula with the calcaneum and the talus with the calcaneum. These are the sites of the common *ankle sprain* (Fig. 16.25).

The *Achilles tendon* is a favored site for xanthomata and gouty tophi (Fig. 16.26). It is also frequently tender at its insertion in rheumatoid arthritis. Shortening of this ten-

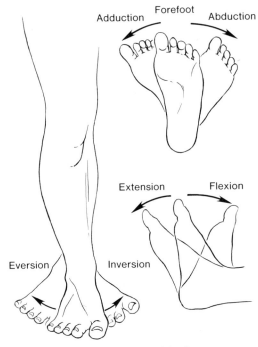

Fig. 16.23   Motions of the foot.

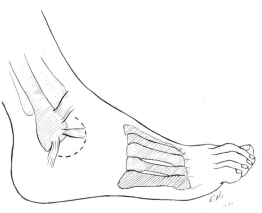

Fig. 16.25.   Landmarks of the ankle.

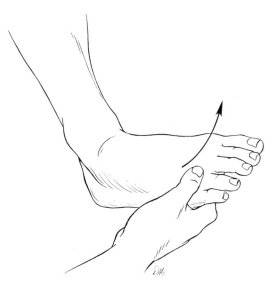

Fig. 16.24.   Inversion of the foot. Note that the entire foot is inverted, not just the digits.

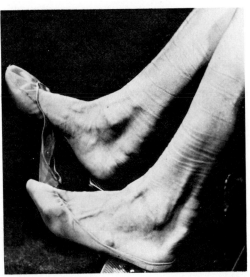

Fig. 16.26.   Achilles xanthoma. This patient demonstrates the xanthomatous nodularity of both Achilles tendons. The horizontal marks are related to Ace bandages applied for ankle edema.

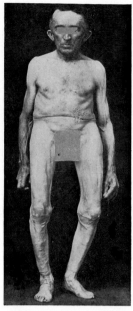

Fig. 16.27. Advanced Paget's disease. The striking features are enlargement of the cranium with a prominent forehead, body shortening, and bowing of the thighs and legs. (Courtesy Dr. John B. Stanbury.)

weakness. A discrepancy in wear, one shoe with the other, indicates unilateral disease.

Bony Abnormalities

The tibia tells tales. *Congenital syphilis* commonly causes a sabre shin deformity in which the leading edge of the tibia bows forward. *Scurvy* causes subperiosteal hemorrhage and exquisite tibial tenderness. This disease is more common than recognized, especially in alcoholics. Confirmatory findings are corkscrew hairs on the legs and perifollicular petechial hemorrhages.

Hypertrophic pulmonary osteoarthropathy is new subperiosteal bone growth associated with pulmonary malignancies. Pain with compression of the tibia or forearm results. This new bone growth can be extremely rapid and produce a red streak over the bone with a zone of increased warmth.

Paget's disease, when blatant, can be diagnosed on the street. The patient has a prominent forehead, stooped gait, a flared pelvis, and bowed leg (Fig. 16.27).

Fat embolization results from fractures of bones, either accidental or surgical. When clinically apparent, the syndrome occurs suddenly 36–48 hours after the insult. The patient appears apprehensive and intensely dyspneic. Fever occurs. Look for a fine petechial rash, especially over the upper chest and neck.

Any draining skin sinus is suspect for *osteomyelitis*. The underlying bone will generally be tender, with erythema and warmth surrounding the major site of infection.

Osteogenesis imperfectum is a rare, probably hereditary, disease of connective tissue characterized by extremely brittle bones that break easily and heal poorly. A peculiar feature is blue sclera caused by thinning which allows the pigment of deeper layers to show through.

don, congenital or acquired, leads to an *equinus deformity* in which the patient walks only on the ball of the foot.

The great toe at its junction of the metatarsal and phalanx is the site of typical gout. In rheumatoid arthritis, this joint appears prominent as a *hallux valgus* because of the lateral drift of the toes. Rheumatoid subluxation in the toes causes a "cock-up" deformity of the toes and the metatarsal heads are palpable in the ball of the foot.

Examine the patient's shoes and note the points of wear. Normally, the lateral heel wears down first. Medial wear usually indicates pes planus—flat feet. Toe wear might indicate *foot drop* from anterior compartment

PRACTICE SESSION

Although you will not do a complete orthopedic examination with all patients, you need to practice just what it entails so that when the need arises you will be prepared to do so. Begin with the *spine*. With your partner standing, first observe the normal spinal curves by looking at the spine from the side. Confirm a thoracic kyphosis, lumbar lordosis, and sacral kyphosis.

With a felt-tipped pen, mark each of the vertebral spines from the vertebra prominens down to the sacrum. Do the marks describe a straight line, or is there a minor scoliosis? A mild curve is very frequently present. If more than a little curvature is noticed, have your partner flex at the waist and look for prominence of one side of the posterior chest over the other. See if the curve disappears when he chins himself on a door sill.

Move the back through the full range of motions as shown on Figure 16.2. Locate the dimples of Venus to confirm the location of the sacroiliac joints. A line drawn across the superior iliac spines across the back will intersect the spinous process of L-2. This usefull landmark will find applications when you perform lumbar punctures.

Now, lying supine, have the subject perform a straight leg raising, first without aid, then with you doing the lifting. At about 60° elevation, forcibly dorsiflex the foot. Have your partner tell you where he feels the greatest strain.

Locate the greater tuberosity of the femur and palpate deeply just behind it. On some subjects, you will be able to feel the sciatic nerve as it traverses this point.

Now stress the *sacroiliac joint*. Put your one hand on his shoulder, flex the knee and hip, and force the knee toward the contralateral shoulder. Now, do the same with both knees at once to put maximum stress on the lumbosacral articulations.

Now, role your partner to the prone position and hyperextend each leg. This maneuver will localize pain in a variety of back conditions.

Locate the major landmarks of the *shoulder*, the coracoid prominence, the acromioclavicular joint and the greater tuberosity of the humerus. Standing behind, put one hand on the elbow and fix the tip of the scapula between the thumb and forefinger of your other hand. Lift the arm out (abduction) and note at what point the scapula begins to rotate. Now, go through active and passive abduction, adduction, internal and external rotation. Try the Speed test for tenderness of the long head of the biceps.

Now, focus your attention on the *elbow*. Identify the medial and lateral condyles of the humerus and the tip of the olecranon. Feel in the groove between the medial condyle and the olecranon for the location of the ulnar nerve. This groove fills with soft tissue in most diseases of the elbow. Measure flexion and extension of the elbow. Now, fix the elbow against the side and pronate and supinate the forearm. Palpate the radial head while doing this and confirm its rotary motion as the radius roles over the ulna. This is the point of maximum tenderness in the "tennis elbow."

Now examine the *hips*. With your partner standing and facing you, put your index fingers on the anterior superior iliac spines and your thumbs on the greater tuberosities of the femurs. The span should be the same if there is no shortening of the femoral neck from fracture or congenital dislocation of the hip. Now, viewing from behind, mark the location of the sacroiliac joints and have him stand on one foot, then the other, and look for weakness of the iliotrochanteric muscles. If such weakness is present the sacroiliac joint on the unsupported side will sink.

Now supine, move the hip through the full range of motions including abduction, adduction, flexion, and rotation. Practice testing the rotation two ways. First, with the leg straight, roll the knee back and forth. Now, flex the knee and alternately abduct and adduct the lower leg.

Practice doing Thomas' maneuver. Put your hand under the small of the back while the subject pulls his knee to his chest. Notice how the lumbar lordosis disappears when he does this. If there were a flexion contracture of the hip, the subject would have exaggerated this lordosis to maintain his leg flat on the table. When you take away this compensating mechanism, the leg will lift up.

Examine the *knee*. Palpate the patella with the leg relaxed. It moves freely laterally but not up and down. Locate the tibial tuberosity and note that it is just a bit lateral from midline. Now, palpate the joint line on both sides. It will be a bit lower than you expected; it is usually just about on line with the lower border of the patella.

Measure the extent of flexion and extension of the knee. Now, stress the medial collateral ligaments, then the lateral ones, and note how much play there is in the joint. Now, go through McMurray's test to check the meniscus and the anterior and posterior drawer signs to check the cruciates of the knee. Look carefully for any effusions in the bursa or a posterior Baker's cyst.

Move now to the *foot*. Dorsiflex and plantar flex the foot at the ankle joint. Now, evert and invert it at the subtalar joint. Find the head of the fifth metatarsal. With your partner standing, slip you fingers under the foot to assess the arch. Also, note the location of foot calluses and note the wear points on the shoes, both of which may tell you of abnormal gaits.

chapter 17

NERVOUS SYSTEM

Neurologic diagnosis is both exciting and frustrating. A careful history and physical examination frequently pinpoint a neurologic lesion to a very accurate anatomic position. On the other hand, the variety of influences of one area of nervous tissue on another may lead to confusing, uninterpretable findings. Also, neurologic abnormalities are frequently evanescent or changing. Many students are chagrined to relate their findings to the attending physician only to have him elicit quite different results.

Neurologists, neurosurgeons, neuro-ophthalmologists, and neuropathologists combine hands in explaining neuropathophysiology. The most valuable tool in their armamentarium remains the carefully performed and recorded physical examination.

History

All *pain*, is, of course, mediated by the nervous system. Some pain, however, originates in nervous tissue. Inflammatory disease of sensory fibers, *neuritis*, results in pain along the course of the infected nerve. A curious phenomenon is that a distal lesion may give pain proximally as well. Thus, the *carpel tunnel syndrome* of entrapment of the median nerve may have elements of upper arm and shoulder pain as well as the typical hand pain and weakness of the thenar muscles. Nerves run a very predictable route with little variation from individual to individual. Therefore, a careful description of the area of pain by the patient is a reliable map to its origin.

Question carefully in regard to inciting factors. Lumbar disc herniation with sciatic pain typically worsens with activity and improves with rest. Tic douloureux, an exquisitely painful affliction of the trigeminal nerve, often has a trigger point. The patient relates that touching a particular area of his face fires off a paroxysm of pain.

Paresthesias, abnormal sensations, are closely allied to pain. The patient complains of tingling numbness or burning, usually of an extremity. When paresthesias are bilateral and symmetrical, direct your questioning to systemic diseases. Toxic and abnormal metabolic states are the usual causes—lead and arsenic intoxication, diabetes mellitus, and nutritional deficiencies, for example. Headache of nervous system origin has already been discussed.

Since the nervous system participates in all voluntary and most involuntary activities, it is not surprising that *dysfunctional* complaints are legion.

Syncope, the sudden loss of consciousness, has been alluded to in the discussion of the cardiovascular system. The diagnosis of neurogenic syncope rests on the history unless you are fortunate enough to witness the event. Neurogenic syncope usually has a prodrome—some warning that a faint is about to occur. Syncope may be precipitated by orthostatic hypotension, a paroxysm of coughing and, rarely, by micturition. Hysterical syncope usually does not cause injury, but true syncope often causes broken glasses, bruises, and cuts.

Epileptic seizures, sometimes called fits by patients, have several patterns. Again the history is all important. An observant relative or friend can lend invaluable aid. *Grand mal seizures* are the most dramatic. They may originate with a aura—a sound, smell, or sensation. The patient loses consciousness and stiffens—tonic phase. Shortly thereafter, a to-and-fro jerking motion of arms and legs and head ensues. Apnea, tongue biting, and urinary incontinence also occur. This clonic phase resolves, and a postictal period of unconsciousness and deep sleep follows. The nature of the aura sometimes indicates the site of origin of the seizures in the cortex.

Petit mal seizures differ in the nature of the

seizures. Generally, a disorder of children and young adults, the attacks are characterized by sudden brief lapses in consciousness. Minor twitching or jerking can occur but falling does not happen. Following the seizure, the patient is immediately awake and alert.

The history of *psychomotor seizures* includes episodes of alteration of behavior, perception, or affect. Some of these indicate temporal lobe pathology.

Focal seizures reliably indicate the anatomic location of the lesion. Such focal seizures may be confined to an area such as arm and shoulder seizure or may begin in an area and progress to a grand mal episode—the *Jacksonian march.*

Weakness is a protean complaint. If the patient complains of focal weakness, a neurogenic cause is likely. Diffuse weakness, especially proximal muscle weakness, is more likely to be metabolic. Sudden unilateral weakness, *hemiparesis,* usually means vascular insufficiency to the contralateral cerebral hemisphere. If transient blindness, *amaurosis fugax,* coincides, carotid insufficiency is highly suspect. If the patient notes that weakness is progressive and associated with *muscle wasting,* a lower motor neuron lesion is at fault.

A *transient ischemic attack* is another example of a neurologic diagnosis made by the history. These episodes frequently presage a major cerebrovascular accident. A brief diminution in blood supplied by a carotid or one of its branches yields transient weakness or numbness, dysarthria, mild confusion, and sometimes syncope. Basilar artery insufficiency affects balance and produces nausea, tinnitus, and a gait disturbance. Carefully elicit the frequency of such attacks, the duration, and a detailed description of them.

Vertigo, a sensation of spinning, is a common and confusing symptom. The cause may be anywhere from the semicircular canals, through the eighth nerve, to the brain stem. The history will help you sort these possibilities. *Labyrinthitis* for any reason results in vertigo which is aggravated by head motion. Many times, the patient volunteers that if he holds his head just so, he is free of symptoms—presumably putting the affected canal in a neutral position. Associated hearing symptomatology, either loss of hearing or *hyperacusis,* indicates involvement of both divisions of the eighth nerve.

Observations by the patient or family may come to the fore in the review of systems. *Fasciculations* and *myoclonic jerks* are involuntary twitches of individual muscle fibers or entire motor units. They are *sine qua non* of lower motor neuron disease. *Dementia* includes a characteristic set of changes which may be observed. The patient suffers recent memory loss. The events of many years gone by are vividly encounted, but the patient can not remember what he had for breakfast. Motor slowing and lack of attention to dress and hygiene also occur.

A change in personality noted by the family is probably the most common presenting complaint with *brain tumors.*

Examination

As elsewhere, the neurologic examination follows a logical sequence. With this system, the examination proceeds from the highest function to the most basic—from the cortex to the brain stem.

The general appearance and behavior of the patient during the interview can present valuable clues to neurologic dysfunction. Are his affect, appearance, and demeanor appropriate? The patient who cleverly *confabulates* can be very distracting. Your first reaction will be "he said something but I'm not on the same wavelength." The patient with Parkinson's syndrome is identifiable at a glance—mask-like facies, akinetic posture, and a typical rest tremor. Similarly, the patient with *forebrain disease* will appear unemotional, mildly disoriented, and inattentive to personal details.

The snarling smile and lisp of *myasthenia gravis* are characteristic as are the long facies, temporal baldness, and nasal voice of the patient with *myotonic dystrophy.* The *scanning speech* of multiple sclerosis also rings a bell once heard.

Higher Intellectual Functions

Higher intellectual functions test the cerebral cortex and its integrative capabilities. Because it is the most complicated and advanced area, it is the first influenced by general toxic events. Begin by *testing orientation to time, place, and person.* Immediate memory is tested by having him repeat a series of digits. Next have him count backwards from 100 by 7's. Abstract reasoning can be assessed by asking him to interpret proverbs, e.g. "a rolling stone gathers no moss."

The thought content comes out throughout the interview. Such things as morbid preoccupations, paranoia, and hallucinations may be alluded to. Do not be embarrassed to pursue these.

With all tests of higher intellectual function, bear in mind the patient's social and educational background. More specific tests now aim at more specific cortical areas. Agnosia, apraxia, and aphasia are the defects.

Agnosia is the inability to perceive objects through any of the special senses—mainly visual, tactile, and auditory. Each can be tested in turn. With *visual agnosia*, the patient will not recognize written words or pictures and will see but not comprehend what he sees. This points to the *occipital cortex. Auditory agnosia* prevents him from recognizing sound—bell ringing, clock ticking, etc. Suspect a *temporal lobe problem* but make sure that inability to act is not secondary to paresis, apraxia, or aphasia (vide infra). *Tactile agnosia* exists when touch senses cannot be interpreted by the cortex. With his eyes closed, place familiar objects in his hand and ask for identification. Dysfunction is located in the *parietal lobe* as is right-left discrimination.

Apraxia is the difficulty in executing purposeful, useful, or skilled acts in the absence of paralysis. In order to do this, of course, there must be no agnosia. Can the patient close a safety pin, set a wrist watch, put his left hand on his right ear?

Aphasia is deficiency in a language function—spoken, written, or gestures. Deficits are usually of all of these and as such do not localize well. When you encounter one deficit with the others intact, a specific localization is suggested. Again, before the act can be accomplished, the input must be received. Obviously, a patient with auditory agnosia will not be able to answer your questions. *Expressive writing and speaking* test the posterior frontal areas. If the patient cannot understand and act on a verbal command—auditory receptive aphasia—suspect the temporal lobe. If he cannot understand and act on the written command—visual receptive aphasia—suspect the parietooccipital region.

Cranial Nerves

Each cranial nerve (C.N.) is available for testing. Review the neuroanatomy of each nerve to understand more fully how specifically placed lesions might cause unique findings.

The *olfactory nerve* (C.N. I) is tested by the patient's ability to recognize the presence of an odor. He need not identify it, only recognize that an odor is present. Common mild substances such as coffee, tobacco, soap, or peppermint should be used. Test each nostril separately by occluding the other. A unilateral loss is more significant than bilateral reduction in smell. The latter accompanies aging alone. An occasional patient will deny the ability to smell anything. Try a noxious odor such as ammonia. The ability to recognize such substances is mostly a function of the mucous membranes of the nose rather than the olfactory nerve. Inability to smell noxious stimuli, as well as "soft" odors, suggests hysteria (Fig. 17.1).

The optic nerve (C.N. II) is tested both for acuity and for field of vision. The Snellen chart, with and without pinhole, and with and without the patient's corrective lenses, is first used as described in the section on the eye (see Chapter 7) (Fig. 17.2). Visual fields are tested by confrontation—matching the patient's field against your own, presupposing that your fields are normal. Sit opposite the patient with your head the same height as his and about 3 feet apart. The patient covers his right eye and you cover your left one. Look directly at each other. Now take a hat pin or cotton-tipped applicator and move it into your field of vision. Put the tip of the object into your blind spot. This will usually be found at about 9 o'clock, 6 inches away from

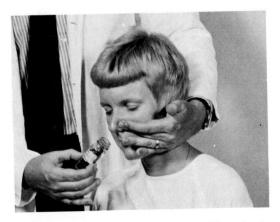

Fig. 17.1. Testing the olfactory nerves. The patient's eyes are closed, and the examiner occludes one nostril and tests the other with a substance with a faint odor.

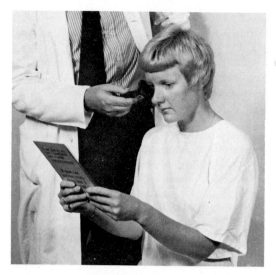

Fig. 17.2. Testing visual acuity. This simplified version calls for the patient to read a chart with one eye only.

The *oculomotor* (C.N. III), *trochlear* (C.N. IV), and *abducens* (C.N. VI) nerves are tested together since they jointly coordinate eye movements (Fig. 17.5). With his head straight, have the patient follow your finger with his eyes as it is moved in direction of gaze (Fig. 17.6). The oculomotor nerve controls the superior and inferior rectus muscles, the inferior oblique, and the medial rectus muscle, the ciliary body, and levator of the lids. Paralysis of the oculomotor nerve pro-

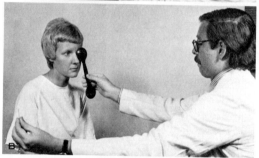

Fig. 17.3. *A*, Testing visual fields. The examiner first places the test object in the patient's blind spot and the examiner's blind spot. *B*, Testing visual field by confrontation. The patient and examiner are looking at each other, and a small cotton-tipped applicator is moved into various fields of vision. The patient's ability to see the object is compared with that of the examiner's.

the line connecting your eye and the patient's eye. Move the object back and forth until it is also in the patient's blind spot. Now you know that the object is equidistant from both of you. Now move the tip of the test object into the different quadrants of vision, finding the point at which it can no longer be seen by the patient (Fig. 17.3). Repeat the test covering the other eye. A field cut, when carefully mapped, will allow an accurate anatomic localization of the lesion since optic fibers follow a very precise course into the brain. Anterior to the chiasm, an optic nerve lesion will block out part of the field of that eye. A pituitary tumor gives a very characteristic field cut. As this tumor enlarges, it presses on the posterior aspect of the chiasm, interrupting the crossing fibers. The fibers that cross originate from the nasal part of both retina. Recall that the lens of the eye inverts an image so that the nasal retina sees the temporal field. The result of a pituitary tumor, then, pressing on the chiasm is a *bitemporal hemianopsia*. Behind the chiasm, the right optic tract carries the left field of vision and vice versa. A lesion in the right optic tract between the chiasm and the lateral geniculate will register as a left *homonymous hemianopsia*. The contralateral upper quadrants are vulnerable in the temporal lobes and the contralateral lower quadrants are vulnerable in the pareital lobes. Occipital blindness is distinguished by central sparing (Fig. 17.4).

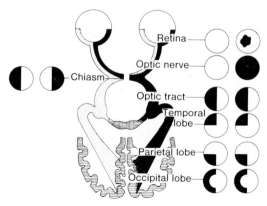

Fig. 17.4. Visual fields produced by lesions at various levels of the optic tract.

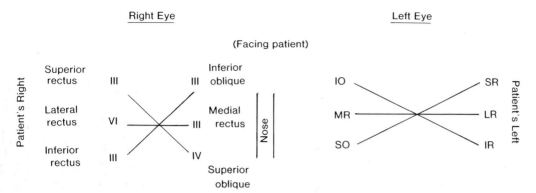

Fig. 17.5. Ocular muscles responsible for the motions of the globe. The Roman numerals refer to the appropriate cranial nerve.

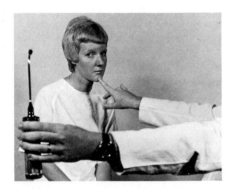

Fig. 17.6. Testing extraocular motion. The head is held in a central position and the patient asked to follow the beam of light. The examiner notes the conjugate eye movements.

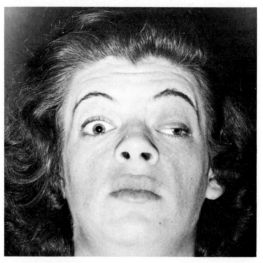

Fig. 17.7. Oculomotor palsy. The patient has a palsy of the third cranial nerve on the left. Ptosis is obvious. The lateral deviation of the left eye is due to the unopposed action of the lateral rectus.

hibits the patient from looking up, down, or medially. In the neutral position, the unopposed lateral rectus may drag the globe laterally; the lid will be ptotic and the pupil dilated. A trochlear lesion is characterized by inability to look down and medially. Abducens palsy results in lateral gaze restriction (Figs. 17.7 and 17.8).

Pupillary reactions involve both the optic and oculomotor nerves. Test in a semidarkened room by shining a bright light into each pupil excluding the other. First note the *direct response*—constriction of the illuminated pupil. Next, note the *consensual response*—constriction of the contralateral pupil. With an optic nerve lesion, the pupil will not constrict directly but will react consensually. With an oculomotor lesion, the affected side will not react either directly or consensually (Fig. 17.9).

Pupils also constrict with near vision *ac-commodation* (Fig. 17.10). Have the patient focus his vision first on a far object and then quickly focus on a near object. In so doing, the pupils will constrict. A peculiar abnormality is seen in tertiary syphilis. The pupil will not react to light but will react to accommodation—*Argyll Robertson pupils.*

The *trigeminal nerve* (C.N. V) contains both motor and sensory fibers. Sensory fibers arrive by way of three distinct trunks which fan out over the face and forehead. With the patient's eyes closed, test each division for light touch with a wisp of cotton and, for pain, with a pin point. Motor fibers of the trigeminal nerve control the temporal, masseter, and pterygoid muscles. Palpate the

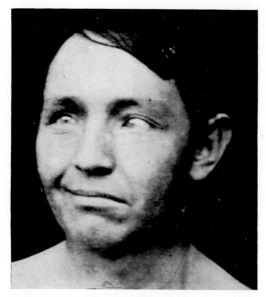

Fig. 17.8.   Combination of facial and abducens nerve paralysis (C.N. VI, VII). The left side of the face droops, and the nasolabial fold is flattened. The left eye deviates inward because of paralysis of the lateral rectus muscle. The combination of sixth and seventh nerve palsies points to a brain stem lesion. In this location, the seventh nerve passes very close to the nucleus of the sixth nerve. A small insult here results in paralysis of both nerves.

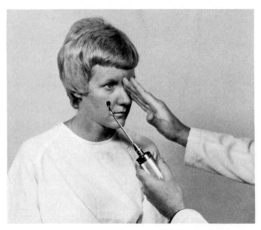

Fig. 17.9.   Testing the pupillary response. Note that the examiner has shielded the opposite eye from the light. He observes both the direct and consensual response.

masseters and temporal muscles as the patient clenches his jaws. Note whether the jaw deviates when he opens his mouth.

Two reflexes involve the trigeminal nerve. The *corneal* reflex initiates a blink when the cornea is touched (Fig. 17.11). Have the patient look to one side and slightly upward. A fine wisp of cotton is then touched to the cornea—coming at the patient from the opposite side so that he does not anticipate it. Be sure to touch only the cornea and not the sclera or eyelids. The *jaw jerk* is elicited by tapping lightly on the tip of the slightly opened lower jaw. The normal response is a brisk closure of the jaw.

The *facial nerve* (C.N. VII) controls facial musculature. Have the patient look at the ceiling, frown, smile, show his teeth, and raise his eyebrows (Fig. 17.12). With paresis, the affected side of the face appears flattened and expressionless. Spittle may drool from the corner of the mouth on the affected side. When paralysis is evident, pay particular attention to the forehead. The muscles of the forehead and eyelid receive contributions from both facial nerves since some crossover exists. Therefore, if facial paralysis is complete and involves the entire side of the face, it must be a lesion of the nerve trunk—*Bell's palsy*. A nuclear or supranuclear lesion, however, paralyzes only the lower part of the face. A pure nuclear lesion usually shows an abducens paresis as well since these two nuclei are so close together in the brain stem.

Sensory fibers originating lower in the medulla join the seventh nerve and supply the anterior third of the tongue. Have the patient extend his tongue and put a pinch of salt or sugar on the tip, first on one side then the other (Fig. 17.13). The patient must keep his tongue out until he identifies the taste. If he withdraws it, the substance will diffuse to other areas and be identified.

The *acoustic nerve* (C.N. VIII) has an auditory branch and a vestibular branch. Review the Weber and Rinne test for the audi-

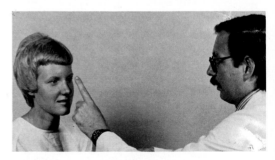

Fig. 17.10.   Testing the pupillary response to accommodation. The patient is first asked to focus on a distant object and then quickly focus on the examiner's finger. In so doing, a normal response includes pupillary constriction.

tory branch discribed in the section on the ear (see Chapter 7). The vestibular branch innervates the semicircular canals and provides great sensory input for the maintenance of balance. This branch is not routinely tested. A physical clue to the presence of disease here is *nystagmus*. Vestibular nystagmus is usually lateral, while brain stem nystagmus tends to be vertical. Have the patient look 45° to the left or right and carefully watch the eyes. If nystagmus is present, the eyes will slowly drift toward the center then snap back to the deviated position.

The *glossopharyngeal* (C.N. IX) and the *vagus* (C.N. X) nerves are conveniently tested together. Touch each side of the posterior pharynx with an applicator and confirm a gag reflex. Next, stroke each side of the soft palate close to the uvula—each side should rise when stimulated. The vagus can be said to be functioning normally if the patient can swallow and speak clearly.

The *spinal accessory nerve* (C.N. XI) provides motor supply to the trapezius, and sternocleidomastoid. Paralysis of shoulder girdle elevation and weakness in turning the head to the opposite side are found in unilateral lesions (Fig. 17.14).

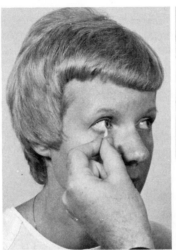
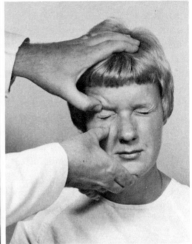

Fig. 17.11. (*Left*) Corneal reflex. A wisp of cotton is barely touched against the cornea. A positive normal response is an eye blink.

Fig. 17.12. (*Right*) Testing facial nerve strength. The examiner is trying forcefully to open the eyelid.

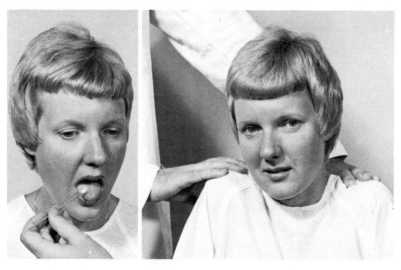

Fig. 17.13. (*Left*) Testing the lingual branch of the facial nerve. The cotton-tipped applicator has been dipped in sugar and touched to the tip of the patient's tongue. She is asked to identify the taste.

Fig. 17.14 (*Right*) Testing the spinal accessory nerve (C.N. XI). The patient is shrugging her shoulders, and the examiner exerts pressure against them.

The *hypoglossal nerve* (C.N. XII) controls the tongue muscles. Have the patient protrude his tongue. If there is hypoglossal paralysis, the tongue will deviate toward the side of the lesion (Fig. 17.15). Atrophy of the involved side occurs early. If in doubt, have the patient push his tongue against each cheek. Check the strength by pushing against the cheek.

Motor System

Emanating from the motor strip of the cerebral cortex, impulses traverse the internal capsule, decussate in the brain stem, run the cord in the pyramidal tracts, and synapse with the lower motor neurons which then terminate at the myoneural junction. Two other systems act as modifiers; the extrapyramidal system and the cerebellar system. From this cursory description, it is already obvious that motor function is an extremely complex circuit. In our review of the examination of the motor system, we will attempt to pick out features that point to disturbance in the *corticospinal tract, lower motor neuron, cerebellum,* or *extrapyramidal* systems.

Begin by testing muscle strength. Have the patient exert various muscle groups against your resistance (Fig. 17.16). A 5-point scale can be useful for repeated exams: 5 = full power of contraction, 4 = fair but not full

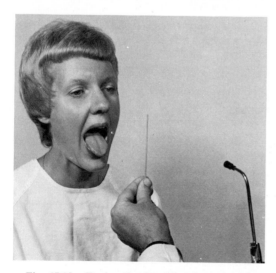

Fig. 17.15. Testing the hypoglossal cranial nerve (C.N. XII). The patient extends her tongue while the examiner lines up a vertical object to the midline of the face. In this fashion small deviations are readily apparent.

strength, 3 = just enough to overcome gravity, 2 = ability to move but not against gravity, 1 = very feeble flicker of contraction, and 0 = no contraction at all. If a muscle group is assessed to be weak, inspect it carefully. Atrophy will result from lower motor neuron death or from disuse if the upper motor neuron is damaged (Fig. 17.17). *Fasciculations,* small flickering involuntary contractions of groups of muscle fibers, are an important sign of *lower motor neuron disease.*

If a muscle or group of muscles is adjudged to be weak, try repetitive stimulation. With *myasthenia gravis* this will result in complete paralysis. A curious *myopathy associated with malignancy* demonstrates the reverse, i.e. increased strength with repeated use (Fig. 17.18).

Now, evaluate the postural tone in each limb by alternately flexing and extending the limb at each joint without active resistance by the patient. Normally a smooth tone will be felt. *Spasticity* accompanies corticospinal tract disease. True spasticity has a "clasp-knife" character. The resistance is at first great then quickly collapses. True spasticity is associated with hyperactive reflexes (Fig. 17.19).

Extrapyramidal disease, Parkinson's, for instance, causes *rigidity.* The resistance is more constant and is sometimes called plastic rigidity. The accompanying extrapyramidal tremor adds a jerking quality to the rigidity—"cogwheeling." Other features of extrapyramidal disease include the ability to maintain awkward positions, lack of associated body movements, blinking, arm swinging, and small facial movements—"mask-like facies" (Fig. 17.20).

Hypotonia is best tested by carefully watching the movements of the joints when a relaxed limb is suddenly shaken or displaced. The truly hypotonic flaccid limb has no breaking action. The hand dropped toward the face strikes it with full force. Such a state exists with a lower motor neuron lesion or early in a corticospinal lesion. The latter becomes spastic as time passes.

Involuntary movements must be noted. Some movements are apparent at rest while others appear with activity. The tremor of Parkinson's is a rest tremor and best seen in the hands as a rhythmical 7/second "pill-rolling" movement of the thumb and fingers. It lessens with motion but reappears in a new

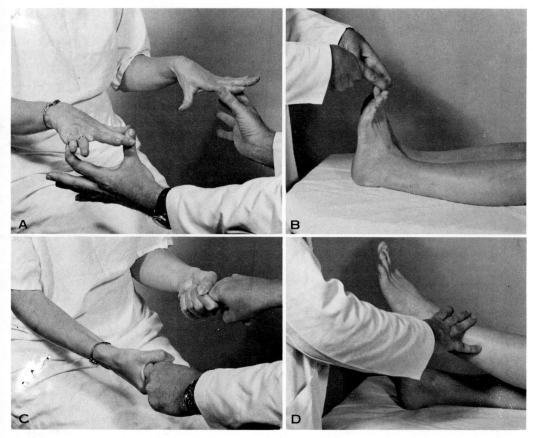

Fig. 17.16.   *A*, Testing the strength of the intrinsic muscles of the hand. The patient is attempting to maintain her fingers in a splayed position and the examiner is exerting pressure to close them. *B*, Testing the strength of the extensors of the great toes. By testing simultaneously, a difference in strength is more readily appreciated. *C*, Testing grip strength. The patient squeezes the index and middle finger of the examiner with both hands at the same time. Be careful to offer the patient only two fingers, since a strong individual can injure your hand if you give him three fingers. *D*, Testing the muscle strength of the hip extensors. The examiner is exerting pressure against the leg.

posture. It is, interestingly, absent in sleep. *Senile and familial tremors* appear most marked in the outstretched hands, disappear at rest, and are aggravated by tension or stress. *Cerebellar tremors* are intention tremors. An oscillating movement occurs. In an attempt to touch an object with his outstretched index finger, a tremor occurs back and forth at right angles from the line toward the object.

Myoclonus, athetosis, and chorea are less rhythmical. *Myoclonus* has many origins and is a sudden jerk of a muscle group. *Athetosis*, an instability of posture, usually indicates basal ganglion disease. With athetosis, the limb flies widely, usually from full extension to full flexion. *Chorea* also shows abnormal posturing, but it is more irregular and discon-

jugate, so that some fingers may come into extension while others flex.

Reflexes

In testing reflexes, we examine involuntary arcs consisting of afferent stretch receptors, spinal synapses, efferent motor fibers, and various spinal cord modifiers coming from above. Many reflexes can be tested, but the following are the most valuable. When testing reflexes, always compare one side with the other. The part being tested must be relaxed and the same intensity stimulus applied to both sides.

The *deep tendon reflexes* are elicited by tapping briskly on a tendon which suddenly stretches a muscle, resulting in its contraction. The *biceps reflex* runs through the fifth and

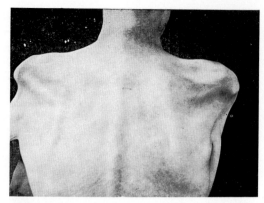

Fig. 17.17. Marked muscle wasting. This patient suffers from amyotrophic lateral sclerosis. This is a uniformly fatal disease of unknown etiology which affects both upper and lower motor neurons.

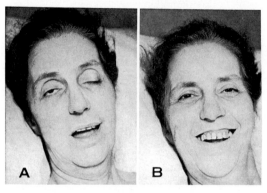

Fig. 17.18. Myasthenia gravis. *A*, Before treatment, this patient while trying to smile cannot elevate her eyelid nor the corners of her mouth. *B*, After treatment, the patient's facial expression while trying to smile is perfectly normal.

The *Achilles reflex* causes plantar flexion of the foot through the first and second sacral segments when the Achilles tendon is struck (Fig. 17.22). Bilateral loss of this reflex is common in old age. While testing the ankle

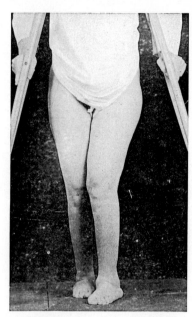

Fig. 17.19. Spastic paraplegia. The typical gait with spastic paraplegia is described as a scissors gait.

sixth cervical segments. To elicit it, strike the biceps tendon of the partially flexed elbow. Cradle the forearm on your own arm. Put your thumb on the tendon for accurate localization; then strike your thumb with the reflex hammer.

The sixth, seventh, and eighth cervical segments serve the *triceps reflex*. Strike the triceps tendon just above the olecranon. This is facilitated by having the patient put his hands on his iliac crests. This tends to relax the muscle groups and brings out the reflex.

The *patellar reflex* is brought out by striking the patellar tendon of the flexed knee (Fig. 17.21). This one circuits through the second, third, and fourth lumbar segments. You may need *reinforcement* to obtain the patellar reflex. Have the patient clasp his hands and pull while you tap.

Fig. 17.20. Parkinson's syndrome. This man is stooped and expressionless. When he walks his arms do not swing. His upper trunk tends to move faster than his legs, and he is propelled forward at an ever increasing rate. This is called festination.

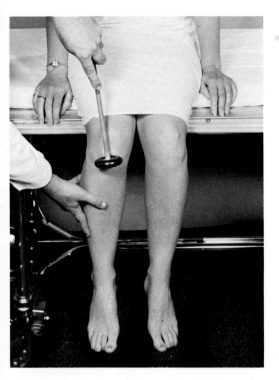

Fig. 17.21. Testing the knee jerk. The extremity must be totally relaxed.

reflex. It is seen *only* with corticospinal tract disease. To perform the test, firmly stroke the lateral sole of the foot from the heel up toward the little toe and then across the ball of the foot. Use a hard object such as a key. Take at least 3 seconds to perform the maneuver. The Babinski response is present if the big toe dorsiflexes and the other toes fan out. The normal response is plantar flexion of all toes. If the patient withdraws his entire foot, repeat the test using less pressure (Fig. 17.23).

The *Hoffmann reflex* is a less reliable indicator of corticospinal disease. To elicit this

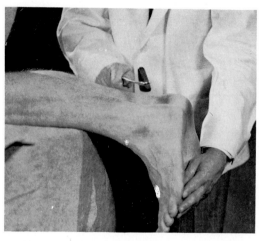

Fig. 17.22. Testing the Achilles reflex. The patient is kneeling on a convenient table or chair. This assures total relaxation of the Achilles tendon.

jerk, also test for *clonus.* Quickly dorsiflex the foot. With clonus, a sequence of rythmical beats occurs.

Hyperreflexia and *clonus* indicate loss of modifier control from the corticospinal tract. A unilaterally absent reflex means disease at the cord or below.

A few superficial reflexes aid in testing arcs at various cord levels. The *abdominal reflexes* are brought out by stroking the abdomen above and below the umbilicus. When so stroked, the umbilicus moves up and toward the area being stroked. The 7th, 8th, and 9th thoracic segments accomplish the reflex above the umbilicus and the 11th and 12th thoracic segments below the umbilicus.

The *cremasteric reflex* elevates the scrotum when the inner thigh is stroked. Loss of abdominal and cremasteric reflexes confirms a pyramidal tract lesion but their absence in otherwise normal individual has no significance. It is occasionally helpful to check for a perineal reflex if a lesion of the cauda equina is suspected. This test is performed by lightly pricking the perianal region with a pin. The normal response is a brief contraction of the anal verge called an *anal wink.*

The *Babinski sign* is the most important

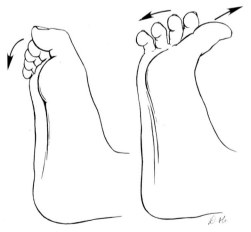

Fig. 17.23. Babinski response. (*Left*), The normal response to stroking the sole is plantar flexion of all the digits. (*Right*), The Babinski response includes dorsiflexion of the great toe and fanning and plantar flexion of all other toes.

reflex, hold the hand up by the middle finger, letting the other fingers hang freely. Flick the middle fingernail. A positive response is a brisk apposition of the thumb and forefinger.

Sensory Examination

The sensory tracts are even more complex than the motor system. In addition to the sensory afferent fibers, cord tracts, thalamus, and cerebral cortex, we are faced with the added complexity that fibers of different modalities traverse different routes. Therefore, different modalities must be tested both to check different tracts and to separate primary forms of sensation from those requiring cortical integration.

Pain and temperature fibers run together in the dorsolateral fasciculus for a short segment after entering the cord then they *cross* and run up the lateral spinothalamic tract. It is unnecessary to test routinely for temperature sense. Have the patient close his eyes, and touch various body parts with a pin. Use the same intensity and always compare one side with the other (Fig. 17.24).

Pain sensation is lost in *tabes dorsalis* in which the dorsal nerve routes are destroyed. This also interrupts the stretch reflex arcs so that patellar and Achilles reflexes are unobtainable.

Fibers for *simple touch* enter the cord and may travel a considerable distance upward before crossing to enter the anterior spinothalamic tract to the thalamus. Again, with the patient's eyes closed, touch the patient with a wisp of cotton and have him indicate when he was touched (Fig. 17.25). This sensation is

the most durable in the spinal cord. Both anterior spinothalamic tracts must be destroyed before light touch sense is lost, probably because of the many collaterals and tendency to run ipsilateral for some distance before crossing.

Vibration, proprioception, and *stereognosis* utilize the same afferent fibers (Fig. 17.26). These enter the cord and follow one of three routes. First, they may synapse directly with motor cells to complete a reflex arc. Second, they may run in the dorsal column to the cerebellum, where they provide valuable feedback on where various body parts are located. Third, they travel in the dorsal columns to the lower medulla where they cross, run to the thalamus, and thence to the somesthetic strip of cerebral cortex. The first path we have already tested with reflexes. We will note tests for cortical discrimination shortly. A tuning fork is the easiest way to test the simple sense. With dorsal column disease, the patient cannot tell a vibrating fork from a still one when it is pressed against a bony prominence. The most striking example of pure dorsal column disease is in *pernicious anemia.* When fully developed, this disease prevents the patient from recognizing a tuning fork or small manipulations of his toes or feet. The latter, *position sense,* is tested

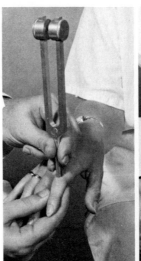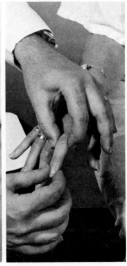

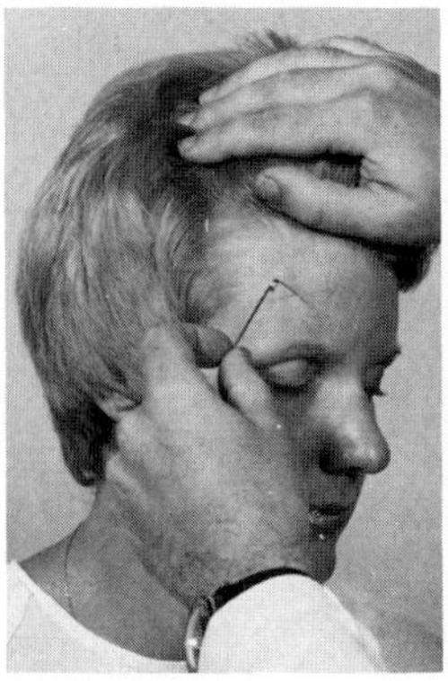

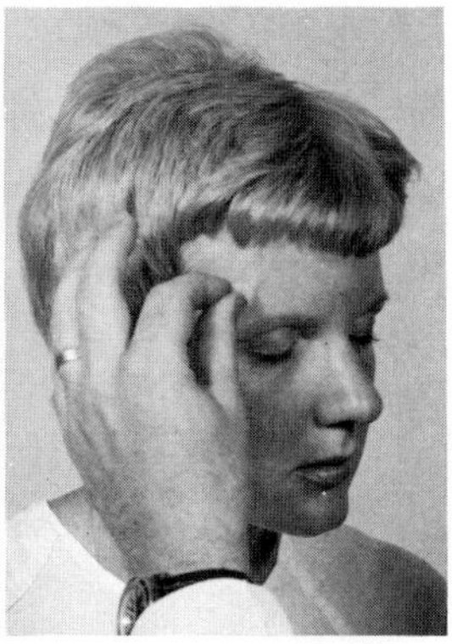

Fig. 17.26. (*Left*) Testing vibratory sensation in the upper extremity. The tip of the vibrating tuning fork is placed against a bony prominence.

Fig. 17.27. (*Right*) Testing position sense in the upper extremity. Without looking, the patient is asked to indicate in which direction her finger is being moved.

Fig. 17.24 (*Left*) Testing the sensation for pin prick.

Fig. 17.25 (*Right*)   Testing light touch with a wisp of cotton. The patient is asked to indicate when she has been touched.

by grasping a digit gently and slowly moving it, asking the patient to indicate the direction you are going (Fig. 17.27).

There are several tests of *cortical* or *discriminatory* sensation which ask, "if everything is intact below, is everything O.K. above?" These sensations require cortical sorting out or discrimination. *Two-point discrimination* is tested by touching simultaneously with two sharp objects (Fig. 17.28). The patient is asked if he is being touched by one or two pins. Calipers can be very useful for comparing two sides since the distance between the pins can be kept constant.

Point localization asks the patient to point to the spot where he was touched.

Stereognostic tests ask him to identify familiar objects placed in his hand—coins, keys, etc. Ask the patient to identify numbers or letters written on his palms with a blunt object. This is a test of *graphesthesia.* If a patient is touched simultaneously on opposite sides of the body in identical areas, he should recognize that he has been touched on two sides. Inability to so recognize is called *extinction.*

If the primary senses are intact, but the cortical modalities are abnormal, suspect *parietal lobe disease.*

Cerebellar System

The cerebellar system essentially provides balance and coordination, and tests of cerebellar function are designed to show abnormalities in these abilities.

Normal *posture* and *gait* depend heavily on cerebellar modifiers. First, observe the patient's normal gait. Does he sway or stagger? Especially note if he consistently falls to one side. Cerebellar *ataxia* requires the patient to maximize his stability. He does this by spreading his feet for a broad base.

Romberg's test helps sort out cerebellar disease from afferent neuron disease. Have the patient stand with his feet together and touching (Fig. 17.29). Be prepared to catch him if he falls. If he has afferent limb disease, he can compensate for the lost sensory input by using his eyes. When asked now to close his eyes, he will promptly reel from his position. Cerebellar lesions, on the other hand, show the same degree of difficulty if the eyes are opened or closed.

The patient with cerebellar disease demonstrates *dysmetria*—the inability to stop a movement at a desired point. He will either overshoot or stop short of the goal. Have the patient touch your finger with his finger then

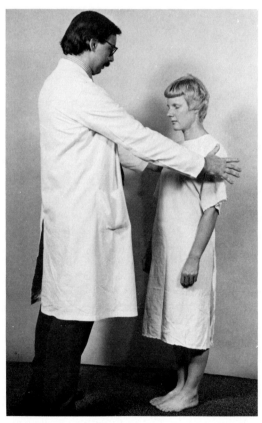

Fig. 17.29. Romberg test. The patient has her feet close together and her eyes closed. The examiner is cautious to avoid allowing the patient to fall. Inability to maintain this posture with the eyes open or closed indicates disease of the cerebellum. Difficulty with this posture with the eyes closed, but not with the eyes open, indicates disease of the afferent limb which can be compensated for by visual fixation.

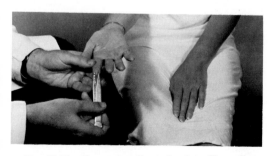

Fig. 17.28. Two-point discrimination. The calipers are used to find the closest distance at which two points can still be recognized as such. One side is compared against the other.

touch his nose and back again. Is he sloppy? Does he consistently *past point* to one side (Fig. 17.30)?

Dysdiadochokinesia is the inability to reverse gears rapidly. Test this with *rapid alternating movements* such as supination and pro-

Fig. 17.30. Finger-to-nose test. The patient is asked to touch the examiner's index finger, then her nose, then his finger once again in an increasingly rapid fashion. This is a test of cerebellar function.

nation of the hands or tapping quickly with the fingers.

Cerebellar tremor is an intention tremor. It will be noted during the finger to nose test. The heel to shin test examines the lower extremities. The patient is asked to place his heel on his knee and slide it slowly to the ankle along the shin. His movements will be coarse and wander off the line.

Meninges

There is no test for normal meninges. When inflamed, however, certain physical findings appear. The patient will resist attempts to stretch the meninges.

With the patient supine, lift his head so that his chin comes toward the sternum. With *meningitis*, there will be active resistance, and the hips and knees may flex—*Brudzinski's sign. Kernig's sign* is the inability to extend the lower leg when flexed at the hip.

The examination of the major cerebrovascular vessels and optic fundi are, of course, integral parts of the examination of the nervous system and have been covered previously.

PRACTICE SESSION

A thorough neurologic examination may take the experienced neurologist over an hour. If it takes you less than that, you have forgotten something. For purposes of this exercise, examine your partner by neurologic systems. Later, you will acquire the skill of integrating the neurologic examination into other parts of the examination. For instance, the cranial nerves involved with eye movement and acuity would be tested together with the funduscopic examination, and the lower limb reflexes could be combined with the orthopedic and vascular examinations of the legs. It is unwise, however, to begin that way because you will soon be doing things by rote and not in the context of the organ system being evaluated. Begin with the cortical areas and work down to the more primitive regions of the brain.

Test your partner's *higher intellectual functions*. Ask the questions to determine orientation to time place and person. Have him repeat a series of numbers or objects, subtract serial 7's from 100, and interpret a proverb or two. It may sound foolish to ask these questions when you know that these functions are intact, yet that feeling of foolishness must be dispelled or you will suppress this part of the examination sometime when it is really important.

Now, test the arcs of the cortex by assessing for the presence of *agnosia, apraxia,* and *aphasia.* Remember that if the receptor loop is abnormal the effector loop also will not function.

Now, go through the functions of each of the *cranial nerves.* Begin with C.N. I and test olfaction using non-noxious substances (vinegar for instance can really be tasted as well as smelled and is not useful). Cranial nerves II, III, IV, and VI can be tested as a group by checking visual acuity and oculomotor activity. Check the arc of C.N. II and C.N. III by testing the pupillary response both directly and consensually, and by testing the pupillary reaction to accomodation. Now, test the motor, sensory, and reflex actions of C.N. V by testing masseter strength, light touch to the three facial divisions, and the corneal reflex. Motor C.N. VII controls facial expressions. Note both the forehead and the facial muscles recalling that peripheral lesions of C.N. VII will obliterate the forehead action whereas central lesions do not because of crossover fibers. Taste on the anterior one-third of the tongue is also controlled by C.N. VII—use sugar or salt. Cranial nerve VIII you have already tested somewhat with the Weber and Rinne tests. Check for nystagmus. You can confirm the *doll's eye* phenomenon by having your partner shake his head "no" and watch the eyes which will remain stationary as the head rotates. C.N. IX and C.N. X are tested together by provoking a gag reflex, and by stroking the soft palate and noting elevation of that side of the palate. A strong shoulder shrug confirms an intact C.N. XI and midline tongue protrusion tells you that C.N. XII is functional.

The *motor examination* should focus on comparable groups; proximal, distal, large groups and small groups, such as the intrinsics and eye lids. Look carefully for evidence of wasting and for fasciculations, both of which are good markers of distal nerve disease. Now, assess postural tone by alternately flexing and extending around major joints. The normal response is a smooth passive resistance to your motion. Make careful note of any *tremors*. They will be most obvious in the head and in the hands. Are they present at rest, during motion, or both? Is the tremor fine or coarse, irregular or repetitive?

Practice eliciting reflex responses. Check brachial, triceps, patellar and achilles reflexes, abdominal reflexes, and Babinski and Hoffmann reflexes. Try reinforcement if you are unable to get one of the normal ones, and always compare one side with the other.

Sensory testing is next. Recall the fiber routes and the combinations. Pain and temperature run together so it is not usually necessary to check both unless you suspect malingering. Such a patient might deny pain but might feel temperature. A simple pin prick will suffice for most examinations. Deep pain is a little different in its tracking and is also difficult to produce. Simple touch is tested with a cotton wisp. When you detect a loss of light touch, be sure to check the midline closely. The loss should fade over the midline and not end abruptly since there is some crossover of fibers. Likewise, the level of loss should be higher closer to the spine. The sensory examination includes testing of vibration, proprioception, and stereognosis. Use a tuning fork, finger and toe positioning, and common articles in the hand. The latter, of course, brings us back to an assessment of cortical functioning. This is further evaluated by checking for two-point discrimination and extinction.

Cerebellar testing really also includes the extrapyramidal system and it is sometimes difficult to separate one from the other. First, assess posture and gait as the patient walks normally away from you, turns around, and walks back. Watch arm swing, foot placement, and trunk movement. Now, stress the gait by having him walk tandem, on the heels and on the toes in succession. Perform the Romberg test first with the eyes open then with the eyes closed. Be certain that the feet are close together. If the test is suspicious, bump the patient lightly on a shoulder to see if he looses balance.

Now have the patient alternately point to your finger tip and then to his nose, slowly and with eyes open. Watch for intention tremor and for past pointing. With the patient supine, have him place one heel on the other knee and slowly glide the heel down the shin. The motion should be smooth and without wobbles. Finally, test rapid alternating motions of the hands and of the tongue—la, la, la.

chapter 18

EXAMINATION OF PEDIATRIC PATIENT

Mark D. Widome, M.D., M.P.H.*

Continual growth and maturation distinguishes the pediatric patient from the adult. This dynamic process offers the physician both challenge and opportunity. In the assessment of normality in the pediatric patient, one must take into account both the child's age and individual pattern of maturation. In every anatomical region and in every organ system, there are physical findings which are acceptable at one age but cause for concern at another age. Likewise, given two children of the same age, there may be findings in one that are quite consistent with his past pattern of growth and family history yet in another represent a significant deviation and potential health problem. The above is true with regard to emotional and intellectual as well as physical growth. Recognition of deviation from optimal growth and maturation is the major objective of pediatric physical diagnosis.

In the patient who is actively undergoing physical, developmental, or emotional maturation, there are many opportunities for detecting problems at a reversible stage. Scoliosis recognized prior to the adolescent growth spurt may be treated to prevent cosmetic deformity, pulmonary restriction and, ultimately, a shortened life-span. Middle ear disease and its associated hearing loss in the infant may be detected and treated early enough to allow the normal acquisition of language skills. Simple emotional and behavioral problems may be recognized and managed before they snowball into irreparable social disabilities. Such opportunities are often fleeting and the child's physician must remain alert.

It is not the purpose of this chapter to comprehensively cover human growth and development or even all aspects of pediatric physical diagnosis. In a limited discussion, the reader will most benefit from emphasis on the aspects of pediatric diagnosis which most distinguish it from adult practice. Toward this end, the bulk of discussion will center on the infant and young child. Selected aspects of diagnosis in the older child will be mentioned. Beyond this introduction, standard pediatric texts should be consulted. Much of it is experience and one should not underestimate how much can be learned from the children themselves.

Approach to Patient

Your examination of the child is most informative and least traumatic if you are first able to win the young patient's trust and cooperation. Good rapport is not achieved through the use of simple tricks or deceptions; such strategies usually yield short-term gains that do not begin to offset the long-term losses. Rather, a positive doctor-patient relationship depends on the physician's understanding of and willingness to deal with the child's needs and fears. Time and effort expended in this direction is well spent. A familiarity with normal child development will help. Most important is a positive attitude toward children. Where there is a trusting relationship, the examination can usually proceed efficiently and atraumatically. It can be an educational experience for child, parent, and physician and serve as a foundation for good cooperation at future visits.

Walking into the Room

Child and parent begin forming impressions about you from the moment you enter the examining room. Mothers want someone

* Assistant Professor of Pediatrics, The Milton S. Hershey Medical Center of the Pennsylvania State University.

with whom they can easily talk and who will listen to their story. They want someone who is "good with children." Children, depending on their age, will search for nonverbal cues to answer some pressing questions. "Will he hurt me? Will he understand why I am unhappy or uncomfortable? Is he gentle?" Importantly, the child will want to know if mother and father like and trust the doctor because he often relies on their judgment in such matters. "If he is O.K. with Mom and Dad, then maybe I'll give him a chance."

Introduce yourself to *both* parent and child. Make sure that you know the name of each. Remember that, not infrequently, mothers and children will have *different* last names; insensitivity to this can create an unnecessary obstacle. Call the patient by the name or nickname that he prefers; call Mrs. Jones, *Mrs. Jones,* rather than "mom" or "mother." She is not your mother! Both parents and children are likely to be turned off by your referring to the patient as "sonny," "your kid" or any other obvious excuse for the forgotten name. Mothers and fathers will quickly detect the physician who looks upon their child as one would a household appliance that has been brought in for repair. Likewise they will appreciate the physician who understands that the child, any child, is someone who is special and precious and worthy of respect.

Handshakes are in order from the toddler years on up; many children appreciate the fact that you think to greet them in the same fashion that you greet their parents. However, be careful not to come on too fast and too strong to the shy or frightened child who needs extra time to become acquainted. Give him that time and let him observe you in a few moments of conversation with his parent.

A child who immediately becomes hysterical when the physician enters the room usually does so because of previous experiences. Earlier visits to the doctor may have been exceptionally traumatic—occasionally unavoidably. More frequently, the child senses the parents negative feelings (fear, anger, etc.) about the visit to the doctor. Regrettably, some parents use the visit to the doctor as a threat in their attempts to achieve discipline at home. "If you don't behave, the doctor will give you a shot."

An occasional child's temperament will be such that initial introductions are difficult, even in the absence of negative past experi-

ence. The parents will relate the child's lack of adaptability to a wide range of new situations often including simple changes in household routine. Parents occasionally need guidance in helping their child to master this difficulty.

Talking, Playing, and Working with the Child

Interaction between the physician and the somewhat shy or anxious child should proceed gradually, well before the beginning of the formal physical exam. It is all a matter of pacing yourself to the child's needs. After the initial introductions, begin some nonverbal interaction with the child as the parent relates the history. Sharing a small toy with the patient is a nonthreatening form of communication. You are telling the child that you are a real person who, like his parents, enjoys playing with him and sharing a happy moment (Fig. 18.1). The laying on of hands can proceed naturally. Each physician will develop a personal natural style for inviting the child to work *with* him.

The child approaching school age can also interact with you by active participation in the history taking. The informational value of the child's story will be discussed later, but recognize it here as one important means by which the child warms up for the physical exam.

History

In pediatrics, the history is a family affair. Depending on the child's age, the physician must rely on family members to provide part

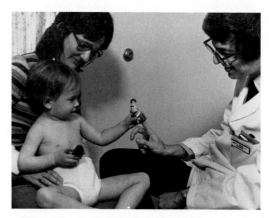

Fig. 18.1. Communication. A finger puppet is a familiar toy to which most toddlers can relate. Sharing a toy allows the patient to learn something about you and it thereby reduces his anxiety. The parent also learns something about how you relate to children.

or all of the story. With each patient encounter, this serves as a reminder that the child's health and welfare are inseparable from what is going on in his family. When, on occasion, the history of an adult patient must be obtained secondhand, we look upon this as a handicap; in pediatrics, such a history has its advantages. As you learn the story, you inevitably learn much about the storyteller. This storyteller is the most important single influence in a child's health. Parents largely control diet, activity, environmental exposures, and the emotional climate.

Example. Ten-month-old Margie has been brought in by her parents because of her inability to sleep at night. For the past month, she has awakened each night from a sound sleep and begun crying and coughing. At times she seems to be in pain. The problem began in the midst of a recent febrile respiratory illness which has since come and gone. Her daytime habits and diet are unchanged. Her parents rush to her aid for fear that she will choke or "hurt her lungs." A bottle of milk administered in the parents' bed usually relaxes her to the point where she is able to fall asleep. Her physical exam is normal.

You find upon further questioning that this "minor" respiratory illness was of major concern to the parents. It was their first experience with recognizable illness in their first child. Nocturnal cough received rapid attention for fear of choking. After the first night, the parents felt safer with Margie in bed with them. The problem started when the illness ended! Although Margie was considerably less distressed by her illness than were her parents, she did very much enjoy the midnight attention and was not about to give it up. Now when she occasionally wakes up in the middle of the night as she always has, rather than quietly dozing off, she finds that crying, especially if interspersed with an occasional cough, gets rapid results.

This is a family problem; it is the result of common behavior patterns of both child and parents. Even if Margie could tell the whole story by herself, she wouldn't have enough of the pieces to put the whole thing together. Reassurance, explanation, and recommendations to the *parents* resulted in rapid resolution of "Margie's" problem.

Problem Visit versus Health Supervision

The nature of the history will depend on the purpose of the visit. If you are being consulted about a specific acute or chronic problem that is of concern to parent or child, then it is best to proceed by allowing the parent to describe the problem in his or her own words much as if it were the parent who personally had the complaint. As you listen to the complaint, it is important to keep separate in your mind the symptoms' functional impact on the patient and their emotional impact on the parents. I have heard parents describe their child's cough and getting the impression that it was totally debilitating. The child playing nearby would proceed to do little more than clear his throat to which the parent will respond, "There, do you see what I mean?" In such a situation one must diagnose and manage both the symptom complex and the emotional and symbolic meaning that it may hold for the parents. Reasonable questions might include: Does Bobby's cough awaken him from sleep at night? Does it awaken *you* at night? For which of you is this symptom most disturbing? What sort of other things does this symptom (cough, rash, blood in the stools, etc) cause you to think or worry about? Not uncommonly the very symptom in question was the same as the presenting (or coincidental) symptom of serious illness in a friend or relative.

The child who comes for health supervision or a "regular checkup" is quite a different matter. There is presumably no pressing complaint but rather you are called upon to detect difficulties early. These interviews are best begun by giving the parents an opportunity to subjectively assess their child's general health since the last visit. This gives them an invitation to introduce any concerns, major or minor, that they may have. One can begin by asking, "Are there any health or growth or behavior concerns or questions about Sally that you would like to talk about?" The opportunity of a few moments of silence to think and respond will sometimes bring to the surface deviations from health that have developed insidiously. Be sure not to interpret a negative response to constitute a negative history. From there, one should proceed to determine that the child's psychosocial development is age appropriate and that the review of systems is unrevealing. At different ages, you will come to appreciate different sets of questions that satisfy these requirements with high efficiency. In the developmental area, the Denver Developmental Screening Test is the most widely used structured screening format.

Where there is an older child at home, take advantage of him for comparison purposes. Are there things that Robert was doing at Joshua's age that Joshua is not? In what ways do you see your two children differing? Likewise, parents keep tabs on neighbors' children and will offer comparisons which, if nothing else, reflect the degree of concern that most parents have that their child's development is normal.

Prenatal and Birth History

Much abnormal growth and development in the early years can be traced to prenatal and perinatal factors. One should inquire about illness during pregnancy and evidence of drug, alcohol, and tobacco use during gestation. Were there any infections or other illnesses during pregnancy? Was the child born on time or prematurely? What was the condition of the infant at birth? Birth weight? Were there problems with breathing, feeding, or jaundice in the newborn nursery? Did mother and baby leave the hospital at the usual time together?

Past Medical History

There are certain elements of the early history which deserve more attention in the pediatric than the adult interview. A review of early *feeding* can lend insight into aspects of general health, nutrition, allergy, development, and the nature of the parent-child relationship. Was breast-feeding attempted and was it successful? How did the bottle-fed infant tolerate his formula? When were solids introduced and how were they tolerated? Would you characterize your child as a "good eater?" Are mealtimes enjoyable or are they a battle?

An *immunization* history not only indicates probable immunity to specific childhood diseases but also whether the parents have conscientiously taken the child for regular health supervision visits. A history of *trauma* and *accidents* may indicate whether the child's social situation or environment put him at special risk for future injury.

Social History

Emphasis on the social history again reflects how closely the child's fate is tied to that of his family. Who lives in the house-hold? Who is the child's primary caretaker? What responsibilities does the father assume? Are there relatives living nearby who can help out? Do you find it difficult sometimes to make it from one day to the next? Why? General questions effectively serve to give the family permission to discuss social concerns relevant to the child's health.

Remember that a family's financial security does not immunize it against social disorders. Even in a relatively affluent community, you willl find an occasional household where there is not enough money put aside to buy the baby's formula. Abuse and neglect of children occurs in all social strata and are generally triggered by family disorganization and stress. Tension in any home can reflect itself in a wide range of somatic complaints in the child, and the symptoms of all forms of organic disease are exacerbated by the emotional dust that accumulates in some households.

Order of Examination

In the examination of infants, the rule is to take advantage of opportunities. Specifically, if the child is quiet and relaxed at the beginning of the exam, use that opportunity to listen to the heart and lungs and perhaps examine the abdomen. Begin with the least threatening and least uncomfortable procedures and save the potentially upsetting things for last.

Many physicians begin without instruments and simply inspect the hands and feet. Few children are unwilling to cooperate in this regard. The child is allowed to examine the stethoscope and familiarize himself with it. Some even wish to try it out (Fig. 18.2). You can demonstrate how it is used by auscultating the chest of a doll or stuffed animal. Some children may want to show you where their heart is and place the bell of the stethoscope on their chest for you.

Abdominal exam in the child requires some simple misdirection. Conversations about unrelated matters, a smile, and good eye contact allows the child to remain relaxed. Orthopedic and parts of the neurologic examination can follow, taking advantage of any degree of cooperation that is available. Head and neck examination, particularly examination of the throat and the tympanic membranes, should be saved for last. Again, the instruments used in these examinations

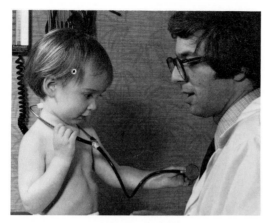

Fig. 18.2. The stethoscope. The tools of our trade are less threatening to the child who is given a few seconds to familiarize himself with them.

should be shown to the child so that he has a chance to become familiar with them.

Examination of Infant

In the first 2 years, the child undergoes a more dramatic physical and developmental change than in any subsequent equivalent time period. His length increases by 75% and his weight quadruples. From a nonverbal, nonmobile state of total dependency, evolves a toddler who can walk, talk, and both take an interest in, and have considerable control over, the world around him. One's technique of examination must be adapted to, and take advantage of, the particular stage of the infant's development.

Neonate

In the first few months of life, the infant will not mind being separated from his mother and may be examined while laying on the examining table. Too often, however, the physician does not remember and take advantage of the fact that even newborns are acutely sensitive to, and react to, their environment. They dislike cold hard surfaces, loud noises, and rough handling. Babies become irritable when they are surrounded by anxiety and tension.

A fretful infant will be soothed by gentle and confident handling. A calm voice and a smile are as important at this age as with older patients (Fig. 18.3). Mothers will often remark that their particular baby is either more comfortable on his stomach or his back; begin the examination in this position. Offer-

ing the baby the bottle, a pacifier, or a finger to suck on can greatly aid in eliciting cooperation for the examination of the chest and abdomen.

Six-Month-Old

When the infant reaches the second-half of the 1st year he undergoes rapid changes in his ability to perceive and evaluate his environment. Most important to the examiner, he is able to distinguish and therefore be anxious about strangers. Therefore, examinations beginning around this age becomes somewhat more challenging. It is best to let the parent do as much of the handling as possible and have her stand by you or across from you at the examining table to offer visual and vocal reassurance to the infant as the exam progresses. The 6-month exam is characterized by the infant nervously glancing back and forth between physician and mother, first on the verge of tears and then reassured. Take advantage of whatever cooperation that you have to do the parts of the exam that require quiet, e.g. auscultation of the heart. Again, at this age, gentleness and soft speech offer much reassurance to the patient.

One-Year-Old

Toward the end of the 1st year, stranger anxiety is complicated by mobility. Some children are quite outgoing at this age and willing to show off all their new skills. Others

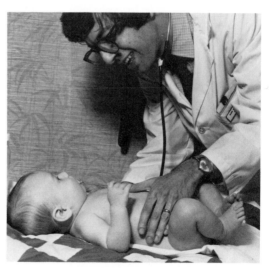

Fig. 18.3. The early infant exam. A warm room, a soft surface, eye contact, and a reassuring touch will calm a fretful baby.

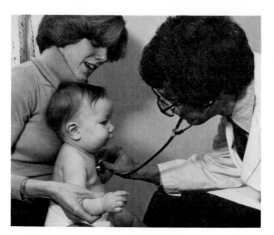

Fig. 18.4. Examination of the infant on his mother's lap.

remain anxious and cling to their parents. Much of this exam is best conducted on the parent's lap. There is no rule that the examining table must be used! The parent serves both to reassure and gently position and restrain the child during various parts of the exam (Fig. 18.4). Reserve for last those parts of the examination that require that the child lie flat on the table.

Eighteen to Twenty-Four-Month-Old

Parents often will relate this is a difficult period for dealing with their child. As children achieve greater independence in the 2nd year, they repeatedly test their parents authority in order to discover the limits of their autonomy. Children at this age are often described as "opinionated" and "contrary." Their favorite word is "no!" The physician should approach such a child in a friendly but firm manner. The physician who cajoles, begs, and bargains for cooperation is not as likely to achieve it as the one who can, in a friendly manner, firmly make clear to the child exactly what is expected of him. Recognizing self-assurance in the physician and having behavioral expectations clearly defined, the child is apt to be less anxious and more willing to cooperate.

General Observations

As in the adult, examination of the infant should begin with some general observations. Does the infant look healthy? One's subjective impression of health takes into account state of nutrition, color, activity, and degree of interaction with the environment (Fig. 18.5). Notice, also, the interest the parents

take in the child. Do mother and father *talk to* and *look at* their child? A child's state of general health is as much dependent on a healthy loving relationship with the parent as it is on good nutrition and the absence of organic disease.

Growth

Physical growth is quantitated by at least three parameters that should be measured at every infant health visit: length, weight, and head circumference. Length is measured with the infant laying flat. A specially designed rigid measure with a perpendicular slide is available. Alternately, marks may be made on the paper tablecover at the vertex and heels. The distance between the marks is then measured with a tape. These results should be plotted on normal growth curves such as

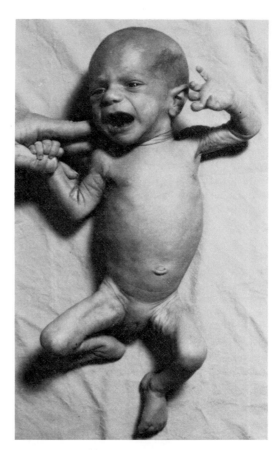

Fig. 18.5. This 6-week-old infant is irritable, poorly nourished and difficult to interest. A grossly inadequate home environment appeared responsible. In foster care, the child has subsequently gained weight normally and over the following 2 years demonstrated normal acquisition of social skills and developmental milestones.

those generated by the National Center for Health Statistics (Figs. 18.6–18.9).

It is not enough to simply demonstrate that the growth parameters are between the 5th and 95th percentiles. The growth measurements must be compared with previous measurements in order to demonstrate that the child is roughly paralleling percentile lines

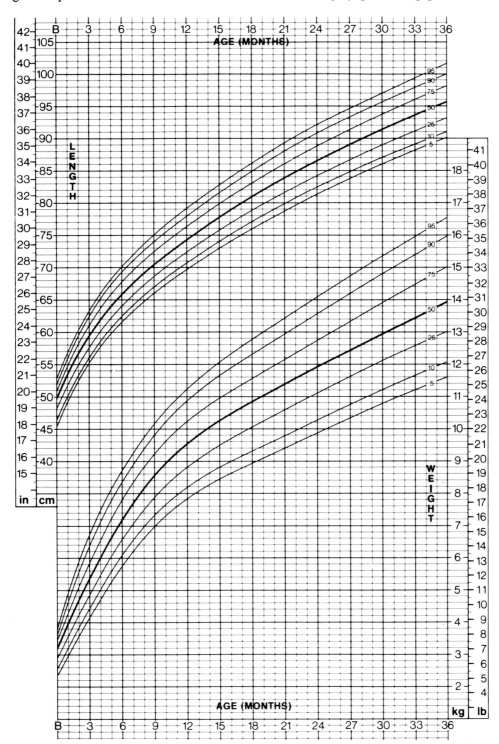

Fig. 18.6.   Girl's growth chart from birth to 36 months. (Adapted from National Center for Health Statistics; reproduced by permission from Ross Laboratories.)

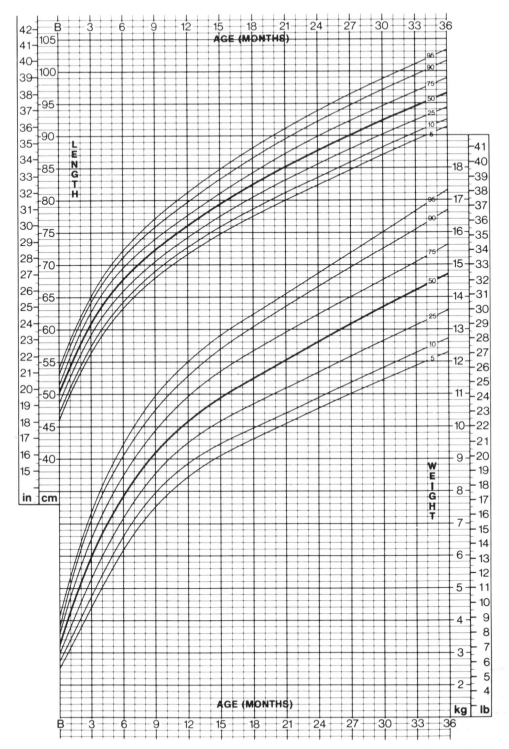

Fig. 18.7.   Boy's growth chart from birth to 36 months. This and the previous figure are based on data derived from body measurements of 867 United States children. Similar charts are available for children 2–18 years of age. (Adapted from National Center for Health Statistics; reproduced by permission from Ross Laboratories.)

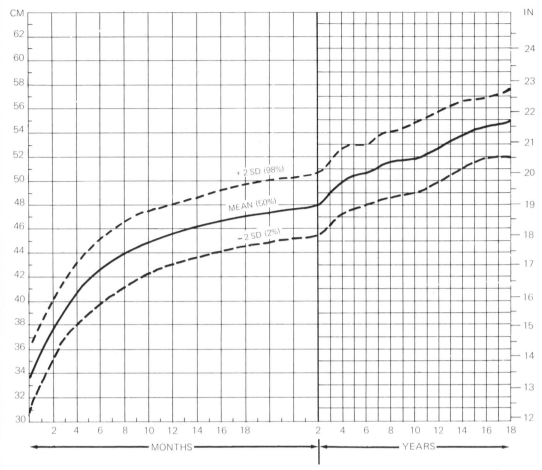

Fig. 18.8. Head circumference percentiles for girls. (Reproduced by permission from G. Nellhaus, Pediatrics, *41:* 106, 1968.)

consistent with his genetic endowment. A starving or chronically ill infant may have a weight at the 20th percentile for age. He may have moved to that position from the 80th percentile on his last visit! On the other hand, 10% of normal children track outside of the 5th and 95th percentile limit.

Body Proportions and Failure-to-Thrive

Failure-to-thrive is the term used to describe the infant or child who shows a pathologically suboptimal pattern of growth usually with one or more growth parameters falling below the fifth percentile. The etiology may be attributable to disease in virtually any organ system as well as being of genetic origin. In some series, the majority of cases of failure-to-thrive have a nonorganic etiology. It has been well demonstrated that children who have received inadequate parenting, even in

the presence of adequate food, will not grow well. Given an infant with poor growth, comparison of the three basic growth parameters can provide valuable clues to etiology (Table 18.1).

The *malnourished* infant will have a weight percentile disproportionately reduced when compared to length and head circumference. In this circumstance, one must consider inadequate caloric intake, malabsorption, chronic diarrhea, or increased metabolic needs (Fig. 18.10). Common conditions associated with increased metabolic needs include chronic infections of the urinary tract, tuberculosis, congenital heart disease, and malignancy.

The infant whose length and weight are proportionately small is suffering from a long-standing deficit. One must consider in this group a number of congenital syndromes associated with failure-to-thrive as well as

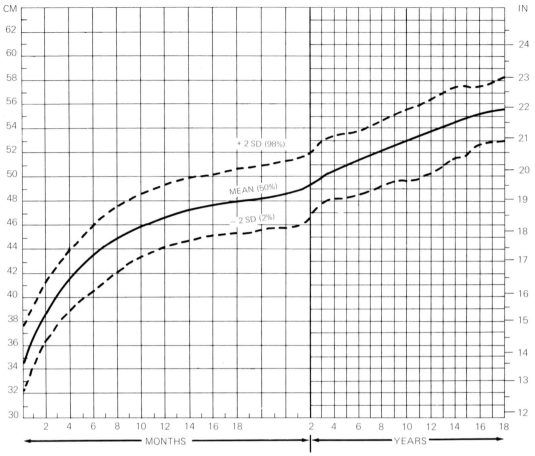

Fig. 18.9.   Head circumference percentiles for boys. (Reproduced by permission from G. Nellhaus, Pediatrics, *41:* 106, 1968.)

Table 18.1
Patterns of Failure-to-Thrive

Weight disproportionately reduced when compared to length. Normal head circumference
  Decreased caloric intake
  Malabsorption
  Increased metabolic needs
    Chronic infection
    Malignancy
    Hyperthyroidism
    Collagen disease

Weight and length proportionately reduced. Normal head circumference
  Deficiency in cellular metabolism
    Metabolic disease associated with acidosis
    Renal failure
    Chronic hypoxemia
  Syndromes (primordial growth failure)
  Constitutional failure-to-thrive

Subnormal weight, length, and head circumference
  Central nervous system (CNS) involvement
    Intrauterine infections
    Severe perinatal insult
    Structural CNS abnormalities

hypothyroidism. Identification of failure-to-thrive syndromes depends on recognition of distinguishing physical characteristics (Figs. 18.11 and 18.12).

The child with a central nervous system deficit will generally have a subnormal head circumference in addition to delayed growth in length and weight. Included in this group are infants with congenital viral infections and structural malformation of the central nervous system.

Hydration

Infants have a proportionately larger daily turnover of body water than do older children and adults. The typical 10-kg 1-year-old has about 6 liters of body water. He will take in and put out a liter of that every day. Because the majority of the output is obligate, vomiting or inability to drink can rapidly lead to dehyration. The dehydrated infant typically does not look well; he is lethargic and unin-

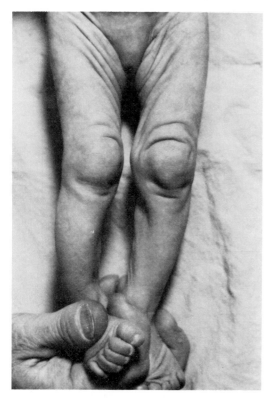

Fig. 18.10. The lower extremities of the malnourished child in Fig. 18.5. The characteristic loose skin and lack of subcutaneous tissue are evident in this infant whose weight is reduced out of proportion to her length and head circumference.

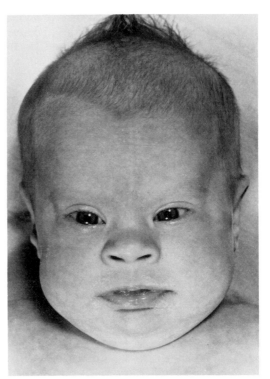

Fig. 18.11. Down's syndrome. Note the slanted palpebral fissures, prominent epicanthal folds, and flat nasal bridge. The protruding tongue is also characteristic. (Courtesy of Dr. Roger Ladda.)

terested in his surroundings. Tears and saliva are absent when the infant has lost fluid equivalent to 5% of his body weight. Moderate dehydration can be roughly defined as a loss of 10% of body weight in fluid. At this point, the eyes will have a sunken appearance and the fontanelle will be depressed. Skin turgor will also be decreased (Fig. 18.13). With more severe degrees of dehydration, pulse, blood pressure, and state of consciousness will be affected. When dehydration is hypernatremic, the fluid loss may exceed that suggested by physical examination as extracellular volume is preserved at the expense of intracellular fluid.

Neglect and Abuse

In the past two decades, physicians have recognized child abuse and neglect as a widespread problem. Its early detection demands constant alertness to the common presenting clues. The importance of the history in uncovering social disorganization and tension

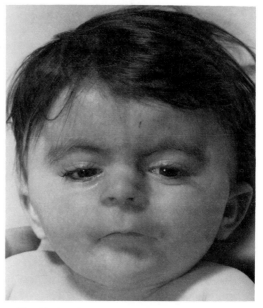

Fig. 18.12. Cornelia de Lange syndrome. This failure-to-thrive syndrome is characterized by hirsutism, continuous eyebrows, long curly eyelashes, a thin downturning upper lip, short stature, and mental retardation. (Courtesy of Dr. Roger Ladda.)

was previously discussed. Likewise, one must consider neglect in the differential diagnosis of failure-to-thrive. The infant or toddler who has been severely abused or neglected will often appear poorly cared for. One should assess general cleanliness as well as evidence of trauma on the skin surface (Fig. 18.14).

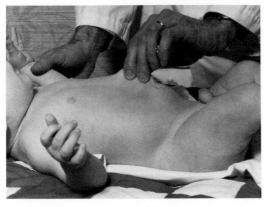

Fig. 18.13.   Evaluation of skin turgor. The skin over the abdomen is pinched and lifted. In the moderately dehydrated child, it will remain momentarily tented when the fingers are released. Tenting of the skin may also be observed in the chronically undernourished child.

Unusually severe diaper rashes may also be evidence of poor care. Often times, the parent will be inappropriately unconcerned about the child's physical condition or emotional flatness. I have seen children with obvious physical evidence of trauma whose parents would only express concern about the runny nose or teething problems.

The older infant or toddler may have inappropriate lack of *stranger anxiety.* The 9- or 12-month-old who passively allows you to approach and examine him is reason for concern. Such children will often have what is called "radar gaze." This is a penetrating emotionless stare that is unchanging even as you attempt to initiate some social interaction.

Vital Signs

Temperature, pulse, and respiratory rate are routinely measured. In the infant, the rectal temperature is most convenient as it does not require patient cooperation. Note that fevers are uncommon in the neonatal period even in the face of severe infection. Often, they are secondary to excessive environmental temperatures. Throughout the rest

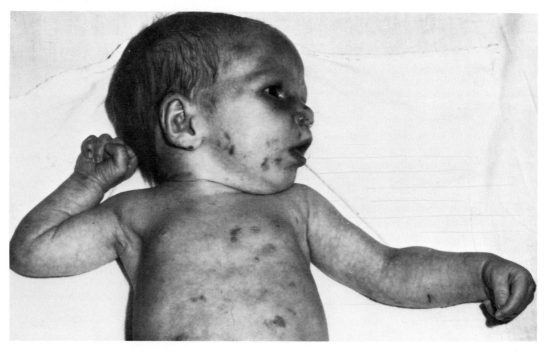

Fig. 18.14.   Child abuse. In this picture, there is clear evidence of physical trauma. More often, the signs are subtle. This infant was brought in for evaluation of a "rash." Scratches, scabs, and evidence of other physical trauma was found. Radiographs showed multiple fractures in various stages of healing. Note the swollen left forearm. (Courtesy of Dr. Glen S. Bartlett.)

of infancy and the preschool years, patients may have high temperatures (39°–40°C) with minor viral infections. Parents and physicians should pay attention to how sick the child looks rather than how high is the temperature.

Pulse rate is quite variable in infants and children and it is dependent on degree of activity. The newborn pulse may be 120–140/minute. Pulse rates in the 1-year-old vary between 80 and 140. Respiratory rate in the neonate is 30–50/minute and is normally quite irregular with periodic sighs. Respiratory rate varies between 20 and 40 throughout the rest of infancy and early childhood.

Blood pressure is routinely measured in children aged 3 and over, although it should be measured at any age if clinically indicated. Direct auscultation of systolic and diastolic pressures in infants is quite difficult but may be attempted. The most accurate method of determining systolic and diastolic pressures in infants is by the Doppler method. Special portable apparatus is required. In the absence of such an apparatus, mean arterial pressure may be approximated by the flush technique. In this technique, the appropriate size pressure cuff is applied loosely. The hand and forearm are wrapped with an elastic bandage or Penrose drain starting distally. The cuff is then inflated to a pressure above the estimated systolic, and the wrap is removed. The cuff is deflated and the pressure at which the arms turns from white to red (the flush pressure) is noted (Fig. 18.15).

Blood pressure varies directly with age.

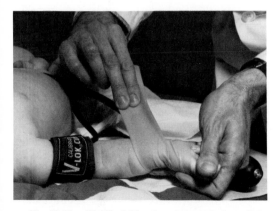

Fig. 18.15. The flush blood pressure technique. Prior to inflation of the cuff, the arm is wrapped with a latex surgical drain beginning distally. In pediatrics, a selection of suitably sized cuffs must be available for reliable blood pressure measurement.

Readings should be compared with established normal ranges for age and sex (Figs. 18.16 and 18.17).

Infant and Newborn Skin

While many adult dermatologic conditions can also be found in children, there are a number of additional findings peculiar to infants. Diaper dermatitis spares few if any infants (Fig. 18.18). Severe diaper irritation suggests chronic diarrhea or inadequate hygiene. In addition, there are a number of benign skin findings that are peculiar to infancy:

Salmon patches are macular birthmarks typically found about the head and neck. They blanch with pressure and become accentuated with crying. They usually, but not invariably, fade over the first few months of life. Occurring about the eyebrows and forehead, they are known as *angel kisses,* while those found at the nape of the neck have been attributed to *stork bites.*

Mongolian spots are bluish gray nonraised discolorations found in the lumbosacral area of the majority of black and oriental newborn babies. They are occasionally seen in white babies as well. Like salmon patches, these benign markings generally fade over the 1st year.

Milia are firm white papules, no greater than 1 mm in diameter, found across the bridge of the nose or elsewhere on the newborn infant's face. These small superficial epidermal inclusion cysts should be ignored as they clear spontaneously.

Cutis marmorata is a mottled vascular pattern seen in an occasional newborn and persisting for days to weeks. It is sometimes accentuated by cool environmental temperatures and is attributed to a normal variant vasomotor reactivity (Fig. 18.19).

Erythema toxicum is a common eruption in full-term newborns. It consists of yellow-white pustules on a broad erythematous background. The trunk is most commonly affected. A smear of the lesions reveals eosinophils but no organisms. The rash typically lasts 2–3 days.

Strawberry hemangiomas are not present at birth but appear over the first few months of life. They generally reach their maximal size by the first birthday and miraculously undergo regression and complete dissappearance by school age (Fig. 18.20).

Carotenemia is a generalized yellowing of

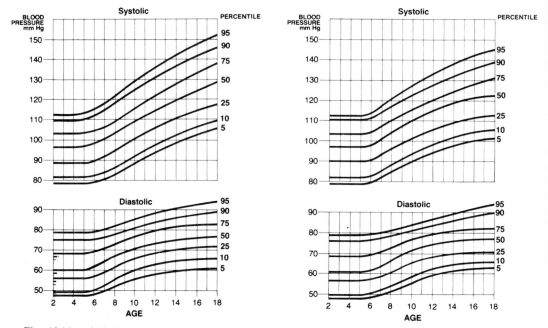

Figs. 18.16. and 18.17. Percentiles of blood pressure measurements for boys (*left*) and girls (*right*). (Reproduced by permission from the report of the National Heart, Lung and Blood Institutes Task Force on Blood Pressure Control in Children, Pediatrics *59:* 797, 1979.)

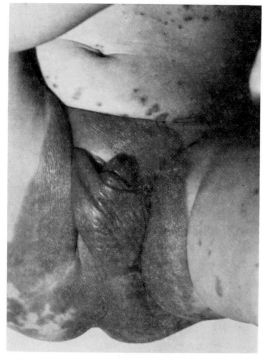

Fig. 18.18. Monilial diaper dermatitis. It is distinguished from the more common irritative diaper rash by the intense red color, well demarcated borders, and "satellite" lesions. Diagnosis may be confirmed by a KOH preparation.

the skin seen toward the end of the 1st year and attributed to excessive amounts of yellow vegetables in the diet. Though often misdiagnosed as jaundice, it may be differentiated by examining the sclera which are spared.

Head and Neck

An infant's head circumference increases about 9 cm in the 1st year of life and 3–4 cm over the next 2 years. This reflects considerable postnatal brain growth and is made possible by unfused skull sutures. A rapidly growing head may represent hydrocephalus while a head that is not growing is usually reflective of lack of brain growth or, less commonly, premature closure of sutures. Palpation of sutures and fontanelles can provide evidence for chronic increased intracranial pressure: the sutures may feel widely split and the fontanelle bulging.

The anterior fontanelle will close shortly after the first birthday. If it is not closed at birth, the posterior fontanelle will clinically close during the first 2 months of life. Large fontanelles and delayed closure is characteristic of *hypothyroidism* or other delay in skeletal maturation.

A head that is too large or growing too rapidly should be *transilluminated*. In a dark

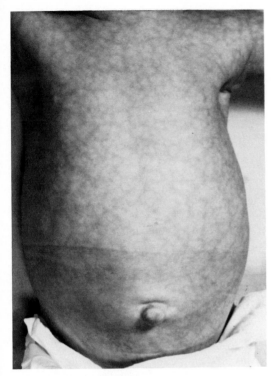

Fig. 18.19. Cutis marmorata. This asymptomatic bluish reticular pattern is seen on the trunk and extremities in some young infants. It is accentuated by cold.

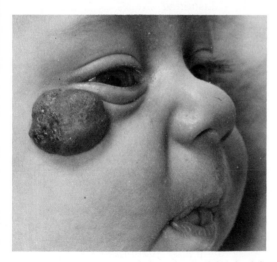

Fig. 18.20. Strawberry hemangioma of the cheek in a 4-month-old. This superficial capillary collection usually undergoes complete spontaneous involution by the 5th or 6th year of life.

room, a special halogen light source (e.g. "the Chun Gun") or a flashlight with rubber adapter is held to the infant's head at several different locations. Underlying fluid such as a subdural effusion or a cystic lesion will cause an abnormally large ring of transillumination around the light source.

Contrary to popular opinion, infants have some degree of visual discrimination from birth onward. Even in the 1st month, mothers should report that they are able to maintain eye contact with their infant during feeding and that the neonate seems interested in the mothers face. Most babies do not, however, have a consistent parallel gaze for the first several months. Until 6 months, it is not necessarily abnormal to have intermittently asymmetrical corneal light reflexes. On the other hand, a *fixed* internal or external deviation of either eye is abnormal at any age and may represent intraocular pathology interfering with vision or it may represent central nervous system disease. In infancy, examination of retinal detail is not routine but one should examine the eyes with an ophthalmoscope to assure that the *red retinal reflex* is not obstructed. A white rather than red reflex may represent *cataract* or an intraocular mass such as *retinoblastoma*. Most young infants will voluntarily open their eyes for examination if given a pacifier or finger to suck on.

Infants do not cry with *tears* before a month of age. Unilateral excessive tearing usually represents a blocked tear duct: *dacryostenosis*. The duct usually becomes patent by 6 months of age. Unilateral tearing associated with photophobia and an enlarged cornea represents *congenital glaucoma*, an autosomal recessive condition requiring the urgent attention of an ophthalmologist.

Examination of the ears in infants and children is often among the most traumatic parts of the physical examination though it need not be. Subsequent exams are much better tolerated if one proceeds with extreme gentleness. Infants are best examined lying down and gently restrained (Fig. 18.21). Gentle downward traction on the external ear is necessary in order to visualize the drum, as the infant's ear canal is directed slightly upward. The drum appears somewhat more opaque in infancy and crying or fever will cause it to turn red from hyperemia. For these reasons, lack of mobility (coupled with a supporting clinical history) is far better evidence of acute otitis than is the appearance of the drum. Use of a diagnostic otoscope and insufflation of air into the canal by a squeeze bulb or mouthpiece is therefore essential for

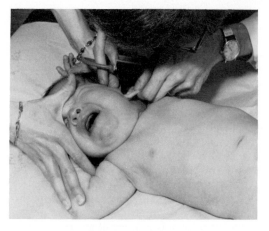

Fig. 18.21. Examination of the infant's ear. The parent or an assistant can restrain the infant's head and arms as shown.

the pediatric ear exam.

Vigor of the sucking reflex is best evaluated by placing a finger in the mouth of a young infant. When a mother reports that her baby is not nursing as well as usual, that is reason for concern as it is a nonspecific indication of an acute illness. Occasionally, a cleft of the soft palate which is difficult to visualize can be palpated by the examining finger. The eruption of primary dentition has been previously discussed; there is wide normal variation in timing.

Oral thrush has been previously discussed. In infants, milk can mimic monilial infection but may be differentiated because it is easily scraped from the tongue or bucal mucosa with a tongue depressor. Small white inclusion cysts may be seen on the gingiva or at the poster border of the hard palate where they are known as *Epstein's pearls*.

The *oropharynx* may be examined by eliciting a gag reflex. Tonsilar tissue is not seen in the infant. The quality of the *cry* should be noted. An unusually *high pitch* is characteristic of central nervous system disease. *Hoarseness* is typically due to acute viral inflammation (croup); it is also found as a manifestation of *hypocalcemia*.

The infant's neck is relatively short. When testing neck mobility, a *tonic neck reflex* may be elicited up until 3–5 months of age. When the head is rotated to the right or left, the arm to the side of rotation will extend and the contralateral arm will flex at least momentarily. Persistence of this posture is abnormal. The infant's ability to roll over at about 4 months of age is partially dependent upon dissappearance of this reflex.

In the young infant, meningitis or other meningeal irritation will *not* uniformly cause *nuchal rigidity*. *Opisthotonus* is posturing where the neck and back are rigidly extended; it represents severe central nervous system insult as in cerebral palsy, kernicterus, or encephalitis.

Chest

One should take advantage of opportunities as they present themselves to obtain a good examination of the infant's chest. Respiratory rate may be determined while the infant is feeding or quietly playing. Upper airway obstructive conditions such as *croup* will rarely cause respiratory rates above 40–50/minute. On the other hand, lower respiratory obstructive disease such as *bronchiolitis*, the most common lung infection in infancy, may cause respiratory rates as high as 80 or 90/minute.

Auscultation of the lungs is often best accomplished with the child in the parents arms. Even in the crying child, much can be learned by paying attention to inspiration. Breath sounds are considerably louder and more widely transmitted in the pediatric chest. *Bronchovesicular* sounds are typically heard throughout the lung fields. Well transmitted sounds from larger airways can be mistaken for evidence of parenchymal disease. The fine rales of pneumonia are usually localized and best heard at the end of inspiration. However, a consolidating pneumonia can occur in a small chest with no more evidence than fever, cough, and tachypnea. Rales and/or dullness may be undetectable, particularly in the first 6 months of life.

Excellent transmission of sound throughout the chest characterizes and complicates examination of the heart as it does the lungs. Heart murmurs are more difficult to localize although some clues to their origin are available by "mapping" the murmur and finding the area of maximal intensity. While the holosystolic murmur of a *ventricular septal defect* may be heard throughout the chest, it will become louder by moving the stethoscope down the left sternal border. *Peripheral pulmonary artery stenosis* produces a short diamond-shaped midsystolic murmur that is best heard as one moves his stethoscope *away* from the precordium toward the axillae. This murmur is a physiologic one in small infants which results from turbulence created by the right and left pulmonary arteries which come off of the main pulmonary trunk at acute angles.

Palpation of *femoral pulses* is an essential part of the cardiac exam in infants. Their absence or extreme weakness suggest *coarctation* of the aorta, even in the absence of a murmur.

Infants and young children manifest *congestive heart failure* differently than older children and adults. In infants, *poor feeding* may be the first sign. *Tachypnea, tachycardia*, and *hepatic enlargement* are the usual predominant signs. Enlargement of the cardiac silhouette on x-ray is also a sensitive early sign. Rales and edema in the extremities are very late findings of cardiac overload.

Cyanosis indicates mixing of the pulmonic and systemic circulations or a right-to-left shunt. Most likely etiologies are dependent on age at presentation. Cyanosis present at birth is most commonly attributable to *transposition of the great vessels*. Desaturation later in the 1st year may represent Tetralogy of Fallot. The appearance of cyanosis in a preschooler may represent right-to-left shunting across a ventricular septal defect secondary to increasing pulmonary hypertension (Eisenmenger's complex).

Abdomen

The infant and young child have protuberant abdomens. This is due to relative weakness of the abdominal musculature as well to the relatively large size of the intra-abdominal organs. In the toddler, a normal lordotic posture also contributes to the abdominal appearance. Infants and young children are abdominal breathers making relatively greater use of their diaphragms and abdominal muscles and less use of intercostal muscles.

A careful palpation of the abdomen is an essential part of the infant exam even in the absence of complaints. One may palpate a *spleen* tip during the first 2 years of life. This is in sharp contrast to the adult where a *spleen* is not palpable until it enlarges to several times its normal mass. Because it is easily palpable, enlargements may be recognized in minor infectious states as well as more serious conditions. Likewise, the *liver* is palpable 1– 2 cm below the right costal margin during infancy.

Kidneys are usually palpated by a bimanual technique similar to that described for adults. In newborns and young infants, some prefer to palpate by applying a gradually increasing pressure with the thumb. The fingers of the same hand support the flank posteriorly. As the thumb is moved caudad, the kidney can be felt to slip cephalad beneath (Fig. 18.22). Renal enlargement in infancy may indicate multicystic kidney, hydronephrosis, tumor, or renal vein thrombosis.

Wilms' tumor is a relatively common malignancy of the infant kidney occurring in about 1 in 10,000 births. It is a rapidly growing firm abdominal mass often detected on routine examination or even by the parent in the course of routine care at home.

Pyloric stenosis is a developmental anomaly presenting with projectile nonbilious vomiting, most commonly in males around 1 month of age. The pyloric mass or *olive* may be palpated in the right upper quadrant adjacent to the midline in the relaxed abdomen (Fig. 18.23). A right lower quadrant sausage-shaped mass associated with crampy abdominal pain and bloody-mucous stools are characteristic of *intussusception*, a frequent cause of intestinal obstruction in infancy.

Umbilical hernias are relatively common in infants and young children. Abdominal contents can be felt to protrude through the hernia ring especially when crying. It can be manually reduced. In black children, umbilical hernias are common throughout the preschool years. They are also a frequent finding in *hypothyroidism* and *Down's syndrome*.

Urogenital Tract

Inspection of the external genitalia is part of the routine exam in pediatrics. In newborn

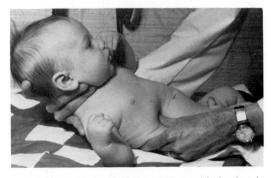

Fig. 18.22. Palpation of the kidney with the thumb. The opposite hand raises the infant to a semireclining position so that the abdominal muscles are relaxed.

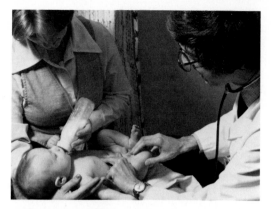

Fig. 18.23.   Abdominal palpation for a pyloric mass. The infant's abdomen is completely relaxed during feeding. The examiner's index and middle fingers apply steadily increasing pressure just to the right of the midline.

girls, the labia minora are relatively hypertrophied secondary to the maternal hormones; they then atrophy until puberty. Hormones are also responsible for a mucoid discharge, occasionally with some blood during the first weeks of life. Vaginal skin tags are common and quite normal. Adhesions between the labia minora are frequently seen and part either spontaneously or with gentle pressure.

The male genitalia are inspected for the presence of *hypospadius* or *epispadius* where the urethral orifice opens onto the ventral or dorsal surfaces of the penis, respectively. In either case, circumcision should be delayed pending surgical evaluation as reconstruction may require use of the additional skin.

One should palpate the scrotal contents. Undescended testes are more common in premature than full-term newborns. The majority will descend during the 1st year of life but spontaneous descent thereafter is uncommon. Oftentimes, the scrotum initially appears empty but the testis can be "milked" down from the external inquinal ring. In uncertain cases, the scrotum should be palpated with the infant or child in the knee-chest position. A small underdeveloped scrotum is good evidence of true *cryptorchism*. *Hydrocele* is common in infancy. These fluid collections are painless and not reducible; unlike hernias, they transilluminate.

Musculoskeletal System

Musculoskeletal development is responsible for the most obvious manifestations of physical growth. During periods of rapid

growth, detection of problems is most important if functional and cosmetic impairment is to be avoided. Likewise, the range of normal variation must be appreciated to avoid unnecessary parental anxiety or medical treatment.

Congenital dislocation of the hip is an important example of a developmental abnormality that responds well to early treatment but poorly if undetected until after early infancy. In the newborn period, *Barlow's maneuver* is used to test for instability. With knee fully flexed and hip flexed to 90°, pressure is applied by the examiner's thumb against the lesser trochanter in an attempt to posteriorly dislocate the femoral head from the acetabulum (Fig. 18.24). In the unstable hip, one can appreciate a sensation of the head sliding from the socket. Follow this procedure with the *Ortolani maneuver*. The hip is abducted while the fingers on the lateral thigh gently lift the femur. In the dislocated hip, a dull

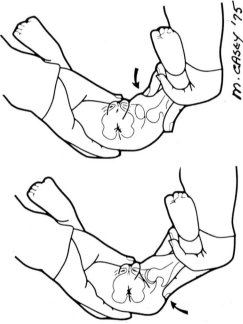

Fig. 18.24.   Examination of the newborn's hip. *Barlow's test* (*top*). The thigh is held with the middle finger on the greater trochanter and thumb opposite the lessor trochanter. Pressure with the thumb will dislocate the unstable femoral head across the posterior lip of the acetabulum. *Ortoloni's test* (*bottom*). One can feel the "thud" of the reduction of the femoral head as the hip is abducted. (Poster, reproduced by permission from the Orthopaedic Division, The Hospital for Sick Children and Ontario Crippled Children's Center, Ontario, Canada.)

thud will be felt as the hip relocates. This test is no longer useful after the first few weeks of life. After 1 month of age, tightness of the adductor muscles is the most reliable sign of hip dislocation. The hips are abducted while flexed at 90°. Normally, they will abduct to about 75°; lesser abduction is suggestive of hip disease.

Early detection of fixed deformities of the feet is extremely important. *Club foot* consists of metatarsus adductus, inversion of the heel, and plantar flexion. It is a deformity that cannot be passively corrected. Metatarsus adductus can occur independently although this is most often a flexible deformity which can be passively corrected.

Infants normally have a slightly bowed appearance to their legs secondary to intra-uterine position. This slowly improves particularly in the 2nd year as the child assumes weight-bearing and walking. Bowing which is severe or progressive suggests specific bone disease such as *rickets* or an asymmetry of proximal tibial epiphyseal growth (*Blount's disease*).

Flat feet are the rule rather than the exception in infants and young children. This is due to normal fat pads on the soles and flexibility of the arch. The longitudinal arch begins significant development around the age of three. *Intoeing* (pigeon toe) is commonly caused by some degree of internal tibial torsion. This may be detected by having the infant or child lie face down with knees bent to 90°. A line down the length of the thigh should line up with a similar line down to the long axis of the foot. Inward deviation of the foot reflects "twisting" of the tibia. Intoeing may also be caused by metatarsus adductus (vida supra) or anteversion of the femoral head.

Neurological Examination

Observations of cranial nerve and motor and sensory intactness are made as in the adult although often in a less orderly fashion, as one must take advantage of observations and opportunities as they present themselves. In addition, a major purpose of the neurological examination in infants and young children is to document the normal maturation of the nervous system. This is done by assessment of specific reflexes and documentation of the acquisition of specific skills or behaviors. A few representative examples will be given:

The *Moro response* is a *primitive reflex* seen in full-term newborns. It disappears around 3–4 months of age. With the infant supine, raise him by the arms until his back is just lifted off the table. Then, release his arms. A positive response consists of symmetrical abduction of the arms and extension of forearms followed by adduction at the shoulders (Fig. 18.25).

The *placing response* is elicited with the child held in the standing position. The dorsum of the foot is made to brush against the edge of the table. The infant will step onto the table with the tested extremity. While holding the infant in the standing position with his feet on the table top, a *stepping reflex* may be elicited (Fig. 18.26). The place and step reflexes normally disappear at 6–8 weeks. Persistence of these or other primitive reflexes suggest neurologic abnormality. It should be noted that the *Babinski response* in normal infants is variable and has no particular neurological significance. Symmetrical absence of normal reflexes may be due to central nervous system depression or to generalized weakness in the infant. Generalized hypotonia is referred to as the *floppy infant syndrome*. When supported by chest and abdomen in the prone position, these infant's bodies will form an inverted "u" shape. The extremities can be put through a range of motion with little resistance and are often

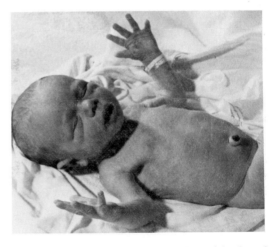

Fig. 18.25.   The startle (Moro) reflex. Abduction of the arms and extension of the forearms is triggered by sudden withdrawal of head or body support or by a loud noise. An asymmetrical response is seen in brachial plexus injury or fracture of the clavicle.

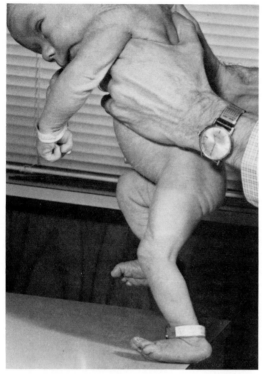

Fig. 18.26. The stepping reflex. The walking motions that are elicited by placing the soles on a flat surface and inclining the infant forward are mediated at the brainstem and spinal levels.

hyperextensible. The cause of the hypotonia may be cerebral such as in hypoxic perinatal brain insult (atonic diplegia) or progressive spinal cord disease (spinal muscular atrophy). Myopathies, congenital syndromes, and metabolic disease may also have hypotonia as a prominent finding.

Acquisition of specific skills may be documented by report of the parent or observation of the child. The skills may be grouped to identify progress and problems in four general areas: gross motor skills, fine motor skills, language, and social behavior (Table 18.2). One must keep in mind that delays in any of these areas may be the result of an unstimulating environment as well as organic disease.

Aspects of Examination of Older Child

Physical development from the 2nd year to maturity is not simply a matter of getting bigger. If it were, physical diagnosis would be a relatively simple matter. In fact, different organ systems mature at differing times and rates, and body proportions undergo substan-

tial changes (Fig. 18.27). Areas of the exam to be emphasized at health supervision visits at various ages depend on a knowledge of the common development-related problems likely to be encountered at those ages. Several examples will be discussed.

Head and Neck

Development of the *paranasal sinuses* occurs largely after birth. Maxillary and ethmoid sinuses are present but small at birth. Frontal sinuses first appear around 3 years of age and the sphenoids develop around age 5. These facts must be kept in mind in the interpretation of signs and symptoms. For

Table 18.2
Average Timing for the Acquisition of Certain Developmental Milestones in Infants[a]

Milestones	Age Range
Gross motor characteristics	
Holds head off of bed for a few moments while lying prone	Birth–4 wks
Rolls over from front to back, or from back to front	2–5 mos
Sits without support	5–8 mos
Stands holding on	5–10 mos
Stands alone well	10–14 mos
Walks holding onto furniture	7½–13 mos
Walks alone	11–15 mos
Fine motor characteristics	
Brings his hands together to midline	6 wks–3½ mos
Grasps a rattle	2½–4½ mos
Transfers a toy from one hand to the other	5–7½ mos
Brings together two toys held in his hands	7–12 mos
Scribbles with a pencil or crayon	12–24 mos
Language	
Laughs	6 wks–3½ mos
Turns toward voice	4–8 mos
Says "Dada" or "Mama"	6–10 mos
Uses Dada or Mama to mean one specific person	10–14 mos
Social behavior	
Smiles responsively	Birth–2 mos
Smiles spontaneously	6 wks–5 mos
Tries to get a toy that is out of reach	5–9 mos
Feeds himself crackers	5–8 mos
Drinks from a cup by himself	10–16 mos
Uses a spoon, spills little	13–24 mos
Plays pat-a-cake	7–13 mos
Plays with a ball on the floor	10–16 mos

[a] Modified from *Infant Care*, Children's Bureau, United States Department of Health, Education, and Welfare, Washington, D.C., 1973.

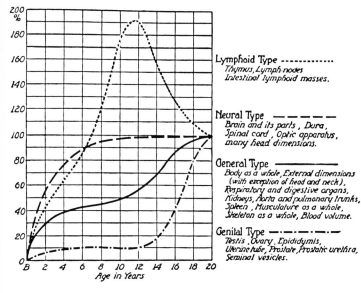

Lymphoid Type ------------
Thymus, Lymph-nodes
Intestinal lymphoid masses.

Neural Type — — — —
Brain and its parts, Dura,
Spinal cord, Optic apparatus,
many head dimensions.

General Type ——————
Body as a whole, External dimensions
(with exception of head and neck),
Respiratory and digestive organs,
Kidneys, Aorta and pulmonary trunks,
Spleen, Musculature as a whole,
Skeleton as a whole, Blood volume.

Genital Type —·—·—·—·—
Testis, Ovary, Epididymis,
Uterine tube, Prostate, Prostatic urethra,
Seminal vesicles.

Fig. 18.27. Relative postnatal growth rates for various organ systems. Tonsils are an example of lymphoid tissue that are "normally" hypertrophied in the child when compared to adolescent or adult. (Reproduced by permission from J. A. Harris, *The Measurement of Man*, University of Minnesota Press, Minneapolis, 1930.)

example, transillumination of the frontal sinuses often cannot be accomplished until after the 7th year.

Some estimation of *visual acuity* can be made at any age based on observations of behavior. Children 3–5 years of age can usually be tested with the "Illiterate E" Snellen eye chart. Formal testing of acuity is an important part of a preschool examination as good academic performance is so dependent on correction of refractive error. As *myopia* generally develops, in those genetically disposed, around the age of 9 or 10, rescreening at that time is most useful.

The examination of the ears of the older child who still needs some reassurance and steadying can generally best be accomplished on the parents lap (Fig. 18.28).

It is unusual for extensive *caries* to develop in the central and lateral incisors. When found in the deciduous teeth of preschool children, it suggests *baby bottle syndrome*. These children typically fall asleep each night sucking on a bottle of milk. The long periods during which the teeth are incubated in milk promote the overgrowth of decay causing bacteria.

Tonsils and lymphoid tissue in general reach maximal proportional size in the early school years. By adulthood, considerable atrophy has occurred. It is perhaps for this reason that those unaccustomed to the care of children are apt to overdiagnose "tonsilar hypertrophy" in the young child. Size of tonsils is best estimated on a 1+ to 4+ scale, 4+ indicating that they touch in the midline.

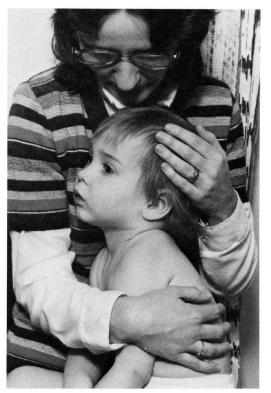

Fig. 18.28. Ear examination of the toddler. Parent may gently restrain the child and steady the head by wrapping one hand around the head and the other around arms and body.

Chest and Heart

Innocent heart murmurs are extremely common in infants and children. They are unrelated to heart disease and represent normal turbulence well transmitted through a thin chest wall. They are short, soft and systolic. Characteristically, they change over time and may sound markedly different with the patient in different positions. They usually are well localized at the lower left sternal border, are low pitched, and have a distinct vibratory or musical quality. Diagnosis depends on absence of any other signs or symptoms of cardiovascular disease.

Sexual Maturation

Precocious or delayed onset of puberty should be recognized as each can represent significant underlying conditions. Breast development in girls normally beings between ages 8 and 13 (average 11) and is completed between ages 13 and 18 (average 15). Appearance of pubic hair generally follows onset of breast development within a year. Onset of menses typically follows about 2 years later.

Testicular enlargement signals the onset of puberty in boys between 9½ and 13½ years of age (average 11½). Appearance of pubic hair and penile enlargement follow testicular enlargement within a year. Within the next 2 years, axillary and facial hair appears.

INDEX

Page references in *italic* type refer to illustrations